Alexanderson

Johns Hopkins Studies in the History of Technology

Merritt Roe Smith, Series Editor

Alexanderson

Pioneer in American Electrical Engineering

James E. Brittain

The Johns Hopkins University Press

Baltimore and London

This book has been brought to publication with the
generous assistance of the Georgia Tech Foundation, Inc.

The Johns Hopkins University Press
701 West 40th Street
Baltimore, Maryland 21211-2190
The Johns Hopkins Press Ltd., London

Library of Congress Cataloging-in-Publication Data
Brittain, James E., 1931–
Alexanderson : pioneer in American electrical engineering / James E. Brittain.
p.   cm. — (Johns Hopkins studies in the history of technology ; new ser., no. 12)
Includes bibliographical references and index.
ISBN 0-8018-4228-X
1. Alexanderson, Ernst Fredrik Werner, 1878–1975.   2. Electric engineers—United
States—Biography.   3. Inventors—United States—Biography.   I. Title.   II. Series.
TK140.A42B75   1992   91-36731

For Jo Ann

Contents

Preface   ix

Acknowledgments   xiii

1. From Uppsala to Schenectady: The Heritage of an Engineer-Inventor   1

2. Dialectical Engineer: Power and Wireless Engineering   20

3. High-Frequency Alternators and Wireless Politics   55

4. Corporate Consultant: The Magnetic Amplifier, Applied Electronics, and the Phase Converter   73

5. Stentorian Alternator: The Creation of the Alexanderson Radio System   104

6. "Alexanderson the Great": The Origins of the Radio Corporation of America   137

7. Television, Thyratron Applications, and Engineering Philosophy   180

8. Electronic Engineering: The Invention of the Amplidyne   213

9. "Inventor for Victory"   256

10. "Inventors Never Stop!"   286

Epilogue   307

Appendix   311

Bibliographic Note   323

Notes   325

Index   371

Preface

This is a biographical study of one of the premier electrical engineers and most prolific inventors of the twentieth century, Ernst Fredrik Werner Alexanderson. Born in Sweden in 1878 and educated in Sweden and Germany as an engineer, Alexanderson in 1901 came to the United States, where he enjoyed a professional career of extraordinary duration and achievement. Except for occasional brief visits to Sweden, he remained in America until he died in 1975, only two years after having received the last of his more than 340 U.S. patents. His inventions helped bring about many of the remarkable changes in electrotechnology that occurred during his lifetime.

Alexanderson's inventive capacity and engineering interests embraced radio, television, electric power transmission, radar, electric railways, ship propulsion, electronic power conversion, and computers. Certain of his inventions in the fields of radio and television captured public attention, especially between 1919, when he became the first chief engineer of the newly organized Radio Corporation of America, and around 1930, when his highly publicized television experiments ended.

The diversity of Alexanderson's technical interests and the ever-changing population of engineers, scientists, managers, and patrons with whom he interacted make this a biography that also serves as a case study in the history and sociology of twentieth-century technology. Several themes emerge from the tracing of a career that began in the drafting room, led to a high-profile position as chief engineer of a major corporation, and then returned Alexanderson to relative obscurity as an in-house inventor and a mentor of young engineers at General Electric.

A recurrent theme concerns the issue of creativity in a corporate setting, and I have attempted to discover the sources of Alexanderson's inspiration for some specific inventions. I have also tried to trace the complex process through which clusters of inventions became elements of technological systems, such as the Alexanderson system of transoceanic radio communication. Alexanderson himself became quite interested in such matters and attempted to influence corporate policies that affected the environment of invention and creative engineering at General Electric. He expressed opposition to excessive compartmentalization, believing that a broad perspective and continual exposure to problems in several areas would tend to stimulate the inventive imagination. His own experience seemed to support this view, for a kind of dialectical exchange between the concepts of power engineering and those of communication engineering served to inspire many of his inventions, notably the amplidyne.

Intellectual exchange between the power engineering and radio engineering cultures went both ways. Alexanderson utilized power engineering concepts and principles in designing his radio alternator system. He characterized the alternator transmitting stations as being like electric power plants in which kilowatts were converted to transmitted words. He employed concepts drawn from radio engineering in several inventions in the power engineering field, notably a phase converter used in railroad electrification. Other Alexanderson inventions, such as the multiple-tuned antenna and the barrage receiver, provide further evidence of the creative exchange between radio and power engineering.

Another theme concerns the distinctions between science and engineering and the interaction between scientists and engineers in a corporate setting. Alexanderson in 1910 was a charter member of the Consulting Engineering Department founded and directed by Charles P. Steinmetz. A comparison of the Consulting Engineering Department and the General Electric Research Laboratory helps to illuminate exchanges between scientists and engineers. Alexanderson came to share Steinmetz's philosophy of the importance of having in-house consulting engineers to provide leadership in the creation of new products and systems. Alexanderson consistently opposed any attempt to change the consulting engineering tradition at General Electric by placing departmental restrictions on the most creative engineers.

The researches that Alexanderson orchestrated in the 1920s on shortwave propagation, radio facsimile, and mechanical-scan television also featured the dynamic interaction of fundamental scientific research, invention, and engineering development. He viewed radio facsimile as a necessary stepping-stone on the way toward commercial television. This was consistent with his belief that the introduction of a complex technological system could best be done in a series of stages, of gradually increasing complexity. Later he followed the same pattern in his work on radar-controlled guns.

The history of the Alexanderson radio alternator is a good example of creative interaction between science and engineering. The machine and the system of which it was part had a somewhat mixed ancestry in physics, radio engineering, and power engineering. The alternator proved its value as a scientific tool, as it was used to conduct high-frequency research on the properties of materials. From Alexanderson's perspective, this research was directed primarily toward filling the needs of the designer of radio-frequency circuits and machines. The

Alexanderson alternator played an unanticipated role in the initiation of research and development regarding electronic tubes at General Electric in 1913. Somewhat ironically, the vacuum tube would supersede the alternator as a generator of radio-frequency energy during the 1920s.

In the political context of World War I and its aftermath, the Alexanderson radio alternator had both strategic and symbolic value. It provided the technological basis for a global communication network intended to be anchored by a complex at Radio Central on Long Island, New York, and operated by the Radio Corporation of America (RCA), which had been formed in 1919. Alexanderson served the new American communication company as its chief engineer while the alternator network, which included stations in several countries, was being installed. Alexanderson and his radio alternator became well known—at least in part owing to RCA's public relations activities as the new corporation sought to define its identity and legitimize its mission. As a result, the Alexanderson alternator became for a time something of a symbol of American patriotism and a vehicle of cultural nationalism.

Alexanderson's international connections and his active participation in professional engineering societies made him an effective gatekeeper for knowledge transfer both at General Electric and at RCA. He made several trips to Europe, and each time he brought back information on engineering developments in Sweden, Germany, and elsewhere. While he was chief engineer at RCA, he served as president of the Institute of Radio Engineers and also was active in the American Institute of Electrical Engineers. He attended professional meetings regularly and published numerous technical papers throughout his career. These papers counter the view held by some that, unlike scientists, engineers in corporations tend not to publish.

Tracing Alexanderson's career over five decades also involves an examination of corporate style and culture. For example, Alexanderson was a participant in a battle of electric traction systems, a competition between the use of direct current and the use of alternating current for the electrification of railroads. It was in the context of this competition that he invented a series-repulsion motor to overcome a problem that had plagued the motors used in traction. His modification of a repulsion motor represented a continuity of a tradition at General Electric which reveals the persistence of corporate style in engineering design. In this instance as in others, General Electric and its foremost competitor, Westinghouse, exhibited distinctive styles and traditions. Alexanderson's use of a rotary scanning disk in television also showed a stylistic parallel with the rotary disks used in his earlier radio alternator designs.

The role of military enterprise in technological change is a theme that has received considerable attention among historians of technology in recent years. Much of Alexanderson's inventive and engineering activity may be viewed from this perspective, although he was always alert to the possible industrial applications of systems being developed for military patrons. The foundations for a military-industrial alliance were clearly evident in projects that Alexanderson worked on at General Electric well before the outbreak of World War II. The torque amplifiers he and his colleagues developed for naval gun control were also found suitable for industrial uses in the automatic control of machine tools. While engaged in military-supported work on gun-control systems, he learned about radar and how it could be used in conjunction with analog computers to control antiaircraft guns. He soon recognized the potential for computers in nonmilitary applications such as fuel control in diesel locomotives.

Alexanderson was much like Thomas Edison in having a persistent and competitive urge to invent and to obtain patents. Clearly, this characteristic is not restricted to independent inventors who expect to profit economically from patents. Alexanderson's rewards seemingly were more psychic than monetary, and his inventive addiction survived his excursions into management and even his reluctant retirement from General Electric. He was 90 years old when in 1968 he filed his final patent application, for a motor-control system that used a combination of semiconductors and magnetic amplifiers. This invention was a virtual paradigm of his career as an engineer-inventor, since it was a synthesis of one of his first inventions, a variable-speed electric motor, and the latest solid-state electronics. He ended his career as he had begun it, inventing at the technological frontier and at the interface between power and communications.

Acknowledgments

I initially decided to undertake this biography in 1971 after contacting the late Philip L. Alger to ask him about source materials on Charles P. Steinmetz. Alger had retired from an engineering career at General Electric and had become interested in electrical engineering history. In response to my inquiry, he suggested that I consider instead a biography of Ernst F. W. Alexanderson. Alger informed me of the large collection of Alexanderson's papers at the Schaffer Library at Union College, in Schenectady, New York, and said that no one as yet had shown much interest in them. He subsequently told Alexanderson of my interest and obtained a letter from the inventor expressing his approval of the project. I then applied for a postdoctoral fellowship in the Smithsonian Institution's Program in the History of Science and Technology and began research on the biography.

The Smithsonian fellowship enabled me to devote the period from July 1972 through July 1973 to the biography. This proved a most rewarding year, and I am especially indebted to Bernard S. Finn, Elliot Sivowitch, and Otto Mayr for their advice and encouragement. Many other members of the Smithsonian staff contributed to this valuable experience by facilitating my access to both documents and artifacts in the Smithsonian collections. While in Washington, I also spent considerable time at the Library of Congress, where I worked in the Langmuir Papers and several other collections.

During my tenure as a Smithsonian Fellow, I also made extended trips to Schenectady, New York, where I visited Alexanderson on two occasions, gaining factual information as well as insight into his remarkable personality and inventive drive. I spent many enjoyable days consulting the Alexanderson Papers at Union College and am especially indebted to Frances Miller, who willingly shared her knowledge of the Alexanderson Papers and made my time there so rewarding. Edwin K. Tolan and other members of the Schaffer Library staff at Union College also facilitated and encouraged my work.

Additional support, which enabled me to continue my research and writing, was provided by the National Science Foundation (Award no. SOC 78-00104). This made possible further research at Union College, where I found that the Alexanderson Papers had been increased by an additional body of manuscripts discovered after Alexanderson's death in 1975. I also examined documents in the General Electric corporate archives. George Wise of General Electric provided me with numerous photographs relating to Alexanderson and his inventions, transcripts of interviews of several of Alexanderson's colleagues, and copies of equipment manuals. Other information came from interviews with Thyra Alexanderson and with Alexanderson's children, Verner Alexanderson, Amelia Alexanderson Wallace, Edith Alexanderson Nordlander, and Gertrude Alexanderson Young. I also visited and interviewed William W. Brown, who had once worked with Alexanderson. Mr. Brown provided additional photographs and followed my suggestion that he donate a collection of his notebooks and other records to the Smithsonian. Another engineer, Edward D. Sabin, who had helped install Alexanderson radio alternators in the 1920s, kindly donated a photograph album and discussed his reminiscences.

Information gathered during the research supported by the National Science Foundation was used in writing a major revision of the first several chapters of the biography, including a lengthier discussion of Alexanderson's life and education in Sweden. Dagmar von Perner of Sweden's Royal Institute of Technology provided useful information about Alexanderson's time there as a student. Thorn L. Mayes and Kaye Weeden freely shared their expert knowledge of Alexanderson alternators. I have corresponded with Hugh G. J. Aitken about this project at least since 1973, and he was kind enough to read and comment on several draft chapters. Thomas P. Hughes read an earlier draft of the entire manuscript, and I have followed many of his suggestions in later revisions. Ronald Kline has also read and commented on the manuscript and has kindly permitted me to read portions of his manuscript study of Steinmetz.

Much of the research for the biography has been done at the Price Gilbert Memorial Library at the Georgia Institute of Technology, where I located most of Alexanderson's published papers and studied his patents. I am greatly indebted to the library staff and to Georgia Tech for its support and encouragement throughout the project. Although my colleagues in the School of History, Technology, and Society may have wondered if this book would ever be completed, they have been indulgent, and I owe them much for the intellectual environment that has been stimulating in this and other research endeavors.

I also wish to acknowledge the contributions of the editorial staff of the Johns Hopkins University Press, and of Miriam Kleiger, who copy edited the manuscript. I especially appreciate the continued encouragement and helpful suggestions of Henry Y. K. Tom and Robert J. Brugger.

I typed several drafts of this biography on a 1950s-vintage Smith-Corona portable typewriter. However, the final draft was word-processed by Jo Ann Layne Brittain, who became my best friend and wife during my year as a Smithsonian Fellow. She met Alexanderson and was my companion on nearly every research trip. Without her help and encouragement, this book would still be unfinished.

Alexanderson

Chapter One

From Uppsala to Schenectady:
The Heritage of an Engineer-Inventor

Ernst Fredrik Werner Alexanderson was born on January 25, 1878, in Uppsala, Sweden, where his father taught at the University of Uppsala. In retrospect, 1878 was a propitious year for the birth of a future inventor of electrical devices and systems. The Edison Electric Light Company—the corporate ancestor of the General Electric Company, where Alexanderson would spend most of his professional career—was organized later the same year. Thus he was born near the dawn of the electrical age and grew to maturity along with the new industry. The conjunction had significant consequences for his educational opportunities, his ultimate choice of profession, and his decision to leave Sweden to practice his profession. By the time he completed his secondary schooling, college programs in electrical engineering were available and the United States had become a virtual mecca for Swedish engineering students. Alexanderson recalled long afterward that the glamour of America had been almost constantly present in the minds of his fellow engineering students. His college song had described the wonders of California, he recollected, but Schenectady had been the attraction for electrical engineering students, even though they hadn't known how to pronounce it.[1]

The roots of the Alexanderson family have been traced to Aron Alexanderson (1724–74), a Swedish sea captain who was the great-great-grandfather of the inventor and the first to adopt the surname. The great-grandfather, also named Aron (1763–1822), was an ordained minister, and the paternal grandfather, Fredrik (1802–79), served as a judge in Stockholm. Aron Martin Alexanderson (1841–1930), the inventor's father, was born in Stockholm and earned a Ph.D. degree in classical languages at the University of Stockholm. In 1873 Aron Martin married Amelie von Heidenstam (1846–1924). He was a professor in Uppsala when their only child was born five years later. The von Heidenstam family was of the Swedish nobility and had a tradition of military service. Karl Verner von Heidenstam (1811–95), the inventor's maternal grandfather and an important early influence on him, served as head of the Swedish marines and became a railroad executive after retiring from military service.[2]

1

In 1880 the Alexandersons moved from Uppsala to Lund, where Aron Martin Alexanderson continued his academic career as a professor at the University of Lund, the university for southern Sweden. His son spent the next 17 years in a culturally advantaged environment as the only child of a university professor and an indulgent mother who was fluent in several languages, including English. The stimulating environment proved fertile for the growth of a creative instinct and a reflective mind. Alexanderson's early experiences caused him to suggest that he had discovered himself and become conscious of a strong urge to do creative work as a small boy.[3]

Under his father's expert tutelage, Alexanderson acquired the easy familiarity with shop tools and processes that was to prove so important in his subsequent career as an engineer-inventor. A well-equipped home shop provided his father with a pleasant diversion from his duties at the university. Alexanderson later speculated that his initial interest in becoming an engineer probably had been stimulated by his enjoyment of the time spent in this shop. He remembered having made a toy steamboat there, and recalled that as a boy he was unable to look at any mechanism without thinking about how it was made and how it worked.[4]

Sailing in small boats was another favorite activity the young Alexanderson shared with his father. Alexanderson later recalled being told that he first had been taken sailing at the age of two. When he was nine he was given his own sailboat.[5] Sailing remained a lifelong avocation for Alexanderson, who introduced competitive yachting at Lake George, New York, and taught his own children to sail.[6] The father-son relationship and the nurturing of a questioning mind is well illustrated by an episode that Alexanderson recalled having taken place when he was about six. He and his father had been becalmed while sailing and had to row to shore. The child had asked why his grandfather, a railroad executive, couldn't send a locomotive to tow them in. His father explained that locomotives could not float because they were made of iron. The boy persisted, and won the admission that steamboats, which were also made of iron, could float.[7]

Karl von Heidenstam, the grandfather mentioned in the anecdote, provided a second important role model for the young Alexanderson. Alexanderson stated in an interview that his grandfather's personality was one of the strongest impressions he retained from his childhood. Conversations with his grandfather about trains and ships probably served to reinforce his interest in becoming an engineer. His grandfather tutored him in algebra during summer vacations spent with his grandparents.[8]

Alexanderson received his first structured education in an informal "little school" organized by his father in their home. A private tutor taught the young Alexanderson and five other boys for about three years, until they were old enough to enroll in the Cathedral School in Lund. From 1890 to 1896 Alexanderson attended the Cathedral School in a class of about 30 students. All students at the school were required to take Latin, but they were given the option of Greek or English as a second language. Since his father was a specialist in Greek, Alexanderson was encouraged to study that language despite his own feeling that English would be more appropriate for an aspirant engineer. Before the end of his first year at Cathedral School, he managed, with the support of his mother, to persuade his father to permit him to change from Greek to English. During the Christmas vacation, his mother tutored him in English so that he could make the change successfully at mid-year.[9]

At the Cathedral School, Alexanderson found geometry to be his favorite subject. His early love for geometry, followed by years of experience in engineering, led him to the conviction that an ability to solve geometrical problems was a good indicator of creativity in engineering. He came to believe that descriptive geometry required the use of that part of the creative imagination which frequently was expressed in invention, and that those with geometric aptitude were most apt to possess creative engineering ability.[10] More recently, similar views have been expressed by Ferguson (in a stimulating essay dealing with the importance of nonverbal thought in the creative process in technology)[11] and by the authors of recent studies of the creativity of Morse and Edison.[12] Alexanderson's enthusiasm for geometry was reinforced by a teacher at Cathedral School whom he remembered having been so inspiring that the future inventor never came to class without having solved all the problems assigned.[13]

Although he had already decided to study engineering at the Royal Institute of Technology in Stockholm, Alexanderson decided, with his parents' encouragement, to attend the University of Lund for a year. This would permit him to strengthen his background in mathematics and chemistry before enrolling in the difficult and highly competitive engineering curriculum in Stockholm. Also, his parents felt that he would profit from a year in a university environment, with its humanistic ambience and traditions.[14]

In the fall of 1897, Alexanderson began a three-year program in engineering at the Royal Institute of Technology. It contrasted sharply with his year at the university in Lund: less than one-third of the applicants were admitted, the curriculum was highly structured, and class attendance was compulsory.[15] The curriculum provided a thorough grounding in engineering mechanics, drawing, and machine design, as well as in physics, chemistry, and mathematics. Although topics in electrical engineering were treated in a few courses, the institute did not yet offer a separate program in that field. Gustaf R. Dahlander, the nominal professor of electrotechnics, also was serving as administrative head of the institute and was nearing retirement. Karl Walling, adjunct professor of electrotechnics, was active professionally as an editor of and contributor to *Teknisk Tidskrift,* the leading Swedish engineering journal. Alexanderson fondly remembered Anders Lindstedt, who taught mechanics at the Institute as "an outstanding teacher."[16] Alexanderson's senior thesis was titled "Method for Calculating Water Turbines."[17]

During the summers of 1899 and 1900, Alexanderson gained valuable practical experience working at the factory of Sweden's largest electrical equipment manufacturing firm, the Allmanna Svenska Elektriska Aktiebolaget (ASEA), in Westeras. He worked as an apprentice in the armature-winding shop during July and August 1899 and spent the summer months of 1900 testing alternating current (AC) and direct current (DC) machinery.[18] He kept and later brought to America several test data sheets and diagrams of apparatus.[19] During the summer of 1900 he received a letter from his father warning him to be very careful in the "handling of those very dangerous machines."[20] Alexanderson considered whether it might be beneficial to remain at the Westeras factory for a few months to gain more experience before continuing his engineering education in Germany. By late August 1900, however, he had decided to go to Berlin in October. His father urged him to take speech lessons in the interim, since hearing "real German" would be "something quite different from old Gustav Petersfont."[21]

At the time of Alexanderson's graduation from the Royal Institute of Technology in June 1900, a career in electrical power engineering seemed attractive, especially in Sweden. The Edison DC system was already being superseded by the AC system of power transmission, the technical advantages of which had been demonstrated in both Europe and America during the 1890s.[22] The AC revolution had important implications for Sweden because it opened the prospect of exploiting the country's abundant water power resources. Sweden was endowed with large, undeveloped water power sites in the north and was deficient in fossil fuels. Swedish electrical engineers such as Jonas Wenstrom and Ernst Danielson were among the pioneers in advocating and designing polyphase AC power systems. Wenstrom was regarded in Sweden as an independent inventor of the polyphase principle along with Nikola Tesla, and Wenstrom's polyphase patents were dominant in Sweden and Norway.[23]

Ernst Danielson had spent two years in the United States in the early

1890s before returning to Sweden to become a designer of large electric power plants. It was Danielson who recommended that Charles P. Steinmetz be hired by the General Electric Company.[24] Danielson was an early proponent of the Wenstrom AC system and had a consulting engineering practice in Stockholm during 1895–1900. In 1900 Danielson was appointed chief design engineer of ASEA, where Alexanderson worked for several months after his graduation the same year.[25] It seems likely that Alexanderson became acquainted with Danielson at that time, since he later sought Danielson's advice on several occasions.

By the turn of the century, Danielson and other informed Swedish engineers were well aware of the enormous quantity of relatively inexpensive water power which could be harnessed electrically and transmitted from remote sites. Sweden began to follow the example of Switzerland in its hydroelectric projects.[26] The impact of the coming of the electrical age must have been very evident to a young engineering student in Stockholm, a city whose night illumination was described as spectacular in 1899.[27]

A comparison of the industrial exhibitions held in Paris in 1878, the year of Alexanderson's birth, and in 1900 showed the magnitude of "that leap in electrical practice which has since given rise to the dominant industry of the world."[28] It was the Paris Exposition of 1900 which stimulated the American historian Henry Adams to write his strikingly evocative essay on "The Dynamo and the Virgin." As Adams reflected on the symbolism of the 40-foot-diameter dynamos displayed in Paris, Alexanderson was going to Germany to learn how they were designed.[29]

Alexanderson's Year in Germany

It was not unusual for Swedish engineering students with adequate means to complete their education in Germany. The Koenigliche Technische Hochschule in Charlottenburg, Germany, where Alexanderson studied, was regarded as one of the finest engineering schools in the world at the time. Foreign students were admitted under special regulations and were permitted to take only the fourth year of the program if qualified by previous preparation. The electrical engineering curriculum included considerable practical work in the design and construction of electric dynamos and transformers. It also included such subjects as electric lighting, telegraphy, electric railways, electrochemistry, and potential theory. Students at Charlottenburg enjoyed somewhat greater freedom of choice in their courses than was common in engineering programs in America and elsewhere at the time.[30]

Adolf Slaby (1849–1913), sometimes described as the "German Marconi," was a professor of electrotechnics at Charlottenburg, where he had occupied a chair since 1882. He had become interested in wireless communication during the 1890s and had observed some of Marconi's experiments in England in 1897. Slaby subsequently acquired several German patents on wireless apparatus and joined Georg von Arco (1869–1940) in the development of the Slaby-Arco communication system, which provided the technical basis for the organization of Gesellschaft für drahtlose Telegraphie (commonly known as the Telefunken Company) in 1903.[31]

Alexanderson attended Slaby's lectures, and one of his student notebooks reveals that the topics discussed included the Branly coherer, Popov's wireless apparatus, and the Braun cathode-ray tube. Slaby also discussed Maxwell's electromagnetic theory, and Alexanderson included sketches of gears in his notes on the Maxwellian displacement current. Another of Alexanderson's student notebooks mentioned the Edison effect and the potential theory, and contained simple problems illustrated by sketches.[32] In view of Alexanderson's later achievements in wireless engineering, it would seem natural to assume that Slaby was an important stimulus. However, Alexanderson recalled that wireless had been of only peripheral interest to him at the time and that he had concentrated mainly on electrical power engineering.[33] Despite this caveat, his entry into the field of wireless communication must have been considerably facilitated by this early exposure to its principles and techniques.

Gisbert Kapp (1852–1922), an eminent electrical power engineer, was among those who taught Alexanderson during his year in Germany. In his lectures, Kapp could draw on a rich background of industrial and consulting engineering experience. He was born in Austria, the son of a German father and a Scottish mother, and graduated in mechanical engineering from the Polytechnic Institute in Zurich. He worked for an engineering firm in London before entering the electrical industry as manager of the Crompton Company in 1882. Kapp left Crompton in 1886 to establish a consulting engineering practice. He was among the first to apply the theory of magnetic circuits to dynamo design and was a pioneer in AC engineering. In 1894 he moved to Germany, where he served as secretary of the Verband Deutscher Electrotechniker and as editor of the journal *Elektrotechnische Zeitschrift*. He returned to England in 1904 and became a professor of electrical engineering in Birmingham.[34] Alexanderson had studied some of Kapp's publications on electrical machine design even before leaving Sweden.[35]

A student notebook reveals that Alexanderson heard Kapp lecture on the design of AC motors and transformers during the winter semester at Charlottenburg.[36] In a design text published a few months later, Kapp stated that the trend toward larger electrical machines would soon limit their manufacture to those companies that could afford large and expensive facilities. He wrote that the designers of such machines no longer could rely on "haphazard or rule-of-thumb methods" but instead "must thoroughly understand the scientific principles involved in dynamo-electric machines."[37] Alexanderson soon would demonstrate a mastery of those principles as an engineer-inventor.

Gustav Roessler taught a general course in electrical engineering and was Alexanderson's favorite professor at Charlottenburg.[38] He invited advanced students to informal sessions in his home and arranged field trips to allow them to observe current practice in industry. It was Roessler who encouraged Alexanderson's interest in learning the "new mathematics" contained in Charles P. Steinmetz's book on the theory of AC circuits. Alexanderson had bought the book and discussed it with Roessler although it was not a required text in his course.[39] The novelty of Steinmetz's method was his heavy reliance on complex-number algebra in the analysis of AC circuits and machines.[40] The alternative method of AC analysis generally used by Kapp and most other authors before 1900 stressed graphical techniques. Roessler seems already to have been somewhat familiar with Steinmetz's work, since he had cited Steinmetz's theory of magnetic hysteresis in a paper published in 1895.[41]

The excitement Alexanderson felt at his first encounter with Steinmetz's analytical method is documented in his notes for Roessler's course in 1901. The notebook begins with differential equations, phasor diagrams, and wiring diagrams. Many of the diagrams were carefully and esthetically rendered in color. More than 100 pages of the notebook were devoted to laborious manipulation of differential equations and graphs. Then, at the top of a page, Alexanderson recorded a note on the treatment of AC motors by "*Steinmetz*!!" (italics and emphasis his). He wrote the expressions for electrical impedance and admittance using Steinmetz's complex number notation and began to use them tentatively. A page in another of Alexanderson's student notebooks dates his discovery of the Steinmetz method to May 13, 1901, and includes Ohm's law for AC circuits expressed in complex numbers—followed by an exclamation point.[42] Alexanderson also wrote a thesis comparing the use of the circle diagram and of Steinmetz's symbolic method in the analysis of the induction motor.[43]

An American Immigrant

Completing his studies in Charlottenburg in August 1901, Alexanderson returned briefly to Sweden. He now confronted the problem of where to seek employment. Although the prospect of an engineering career in Sweden offered the attraction of living in a familiar cultural setting near his family, he was drawn to America, as were many of his peers. He later estimated that approximately 25 former classmates from the Royal Institute of Technology were employed in Schenectady when he arrived there. He recalled that the young Swedish engineers had "naturally, wanted to see how things were done in a big way in America."[44] On another occasion, Alexanderson mentioned that America had been an attraction because of its reputation for resourcefulness when confronting new situations and its drive to put new ideas to practical use."[45] He also felt that the greater importance of class and social rank in Sweden had contributed to his decision. He acknowledged that this had resulted in an unusually stable society but recalled that, being "too much of an experimenter and individualist by temperament," he had found it somewhat too stable.[46] Perhaps most importantly, he reminisced, he had been motivated by a "spirit of adventure."[47]

In deciding to go to America, Alexanderson became a participant in a mass demographic phenomenon of the early twentieth century. The net Swedish emigration during the years 1901–5 was over 120,000, with most emigrants going to the United States.[48] An investigation of the reasons for this mass migration produced a multivolume report published in 1908–13. The principal reasons given by the emigrants were greater economic opportunity and greater democracy.[49] Although the typical Swedish emigrant was of lower socioeconomic status than Alexanderson, the drain of trained engineers was substantial and reflected an industrial economy that could not absorb the supply of recent graduates of the Royal Institute of Technology.

Alexanderson made his initial journey to America in the company of a college acquaintance who had been working in the United States and was returning from a visit to Sweden. They went by way of England and crossed the Atlantic on the *Campania* from Liverpool, arriving in New York on September 1, 1901.[50] Alexanderson spent his first few days looking for work. His first purchase was a small, circular slide rule that he described more than 40 years later as his most prized possession.[51] During his job search, he traveled by rail to Garwood, New Jersey, on September 3, and to Orange the following day.[52] Presumably it was on this trip to Orange that he met the legendary inventor Thomas A. Edison, who was in the early stages of a long and frustrating effort to develop an efficient electric storage battery. Alexanderson later recalled that he had been offered a job by Edison but had declined it since it was not the sort of work he wanted.[53] A copy of an undated letter addressed to the Edison Company by Alexanderson stated that since the "position of assistant wattmeter tester is not of the kind I wanted, I withdraw my application."[54]

The encounter between the young Swedish engineer and Edison is suggestive. They shared certain qualities that later became more evident. Both men were obsessive inventors who were attuned to the potential of new technological systems and who systematically invented components needed for the successful introduction of those systems.[55] They differed in cultural heritage, education, and style. Unlike Edison, Alexanderson showed no interest in becoming an entrepreneur-inventor, involved in raising capital and founding companies to exploit his inventions. By the time of their meeting, Edison was becoming increasingly autocratic in his relations with his assistants. Alexanderson perhaps perceived that the excitement and informality that had characterized Edison's laboratory at Menlo Park no longer existed in the West Orange facility. The experience of Bertil Hauffman, a gifted young Swedish engineer who later came to work under Edison, is indicative. Edison and Hauffman proved incompatible, and only the intervention of Edison's son Theodore kept Hauffman from being summarily dismissed.[56] Alexanderson's withdrawal of his application spared him such an unpleasant experience and was an action he never had cause to regret.

A Draftsman with the C&C Electric Company

The day after his trip to Orange, Alexanderson accepted a position as a draftsman with the C&C Electric Company in Garwood, New Jersey. He reported for work on September 9, 1901. C&C had been founded by Charles G. Curtis and Francis B. Crocker, both graduates of the Columbia School of Mines. Curtis received his degree in civil engineering in 1881 and a law degree from the New York Law School in 1883. After practicing patent law, he joined Crocker in organizing the C&C Electric Company in 1886 to manufacture electric motors. In 1896, Curtis patented a steam turbine; he sold the rights to General Electric Company in 1901. Crocker graduated from Columbia in engineering in 1882. In addition to being a cofounder of C&C, he later became a partner of Schuyler S. Wheeler, organizing the Crocker-Wheeler Company in 1888. Crocker helped establish an electrical engineering program at the Columbia School of Mines of Columbia College in 1889 and served as the program's director for approximately 20 years.[57] The founders no longer were involved actively with C&C when Alexanderson worked there in the fall of 1901.

Alexanderson arranged to board in the home of the Beebee family in Westfield, New Jersey. He later recalled fondly that the Beebees had virtually treated him as a member of their family; their kindness had given him a perspective on America unlike that of most immigrants and had influenced his decision to remain in the United States.[58] Soon after his arrival in Westfield he purchased a bicycle, and wrote in his address book that a young Swede who had spent a year in Berlin wanted to exchange German for English conversation.[59]

Alexanderson's work in the drafting room at C&C proved unexciting, as he was assigned such tasks as tabulating the results of commutator heat tests. He did find time to undertake a long analysis of the causes of commutator heating, including brush friction, ohmic losses, and sparking. During his spare time he also continued to study Steinmetz's book on AC theory.[60] By December 1901, Alexanderson had concluded that his future prospects at C&C were too limited, and he began to look elsewhere. The renewed job search led to a weekend trip to Schenectady, New York, home of the General Electric Company (GE) and of Charles Steinmetz. He called on Steinmetz and also met Ernst J. Berg (1871–1941), Steinmetz's Swedish-born assistant. Steinmetz and Berg provided him with a letter of introduction to Albert L. Rohrer (1856–1951), the principal recruiter of young engineers for GE's Test Department, and Alexanderson submitted an application for employment to the Test Department.[61]

Charles P. Steinmetz (1865–1923), himself an immigrant, was a leading theoretician in the GE organization; in addition, he received 195 United States patents during his career. Since coming to the United States in 1889, he had risen to the top of the electrical engineering profession, serving as president of the American Institute of Electrical Engineers (AIEE) during 1901. He came to enjoy a special relationship with Alexanderson—as mentor, advocate, and role model of the creative engineer in a large corporation. Steinmetz provided important guidance and encouragement that helped to facilitate Alexanderson's career development at GE during the early years.[62] However, the close relationship did not develop for several months, as Alexanderson's application to GE's Test Department met with a long and frustrating delay.

During the early months of 1902, Alexanderson continued his efforts to find a new job with higher pay. He wrote to Albert Rohrer at GE to ask about the status of his application and was informed that the application was so recent that it had not yet been considered.[63] Early in February, Alexanderson applied for a job with the Westinghouse Electric and Manufacturing Company, explaining that he had worked as a draftsman at $15 per week since coming to the United States and would like a better chance for advancement. He added that he intended to become a naturalized citizen as soon as possible. The chief draftsman at Westinghouse sent him an application form and asked whether he spoke and understood English. When he returned the form, Alexanderson stated that he was conversant with English, German, Swedish, and French. He mentioned also that he hoped for a monthly salary of $75 but would accept $60.[64] At about the same time, he wrote to the chief draftsman of GE stating a willingness to begin at $15 per week, his present salary. This application was accepted promptly, and he was asked to report for work immediately. He decided to accept the offer from GE and promised to be in Schenectady by February 21.[65] Soon afterward, he received word of an offer from Westinghouse at $70 per month, but he replied from an address in Schenectady that he had accepted another job and was withdrawing his application.[66]

A Draftsman-Inventor with General Electric

Alexanderson reported for work in GE's Drafting Department on February 23, 1902.[67] Despite some initial discontent, he would remain a GE employee for almost 50 years. At first he found his duties at GE to be as uninspiring as his previous work at C&C. He had been at GE for only a few days when he applied for an advertised position as test engineer with the DeLaval Steam Turbine Company in Trenton, New Jersey. He summarized his work experience in Sweden, Germany, and the United States, and gave his expected salary as $22.50 per week.[68] Apparently nothing came of this, since a few weeks later he responded to an advertisement published by Westinghouse in *Electrical World,* mentioning an expected salary of $100 per month. He received a curt reply saying that since he had declined an earlier offer, they had no need for him. He then drafted a response explaining the circumstances that had led to his rejection of the previous job offer, mentioning that the chief engineer of C&C had told him that he had made a mistake in asking for too low a salary at GE, and claiming that someone in Sweden had advised him to work for a small firm for a time before going to GE or Westinghouse.[69]

During his tenure as a draftsman at GE, Alexanderson experienced some of the disillusionment that caused some of his former classmates to return to Sweden, for at GE even well-educated engineers had limited opportunity for advancement unless they could somehow gain admission to the Test Department, which had become the accepted gateway into responsible engineering at the company. One of Alexanderson's first assignments at GE was to produce detail drawings of control units for railway and industrial motors.[70] He mentioned his work on motor controllers in a job application letter written to the Stanley Electric Manufacturing Company in September 1902. He stated that dissatisfaction with his salary was the principal reason for the application and that he would expect a salary of at least $20 per week. He received a favorable response from the Stanley Company which indicated that he could start work immediately on controllers and related apparatus and that he would be paid what he had asked if he were worth it: if he were not, they would not keep him. He declined the offer, explaining that since he had applied, conditions at GE had improved so that he now had no reason to leave.[71] The improved conditions evidently were related to his first invention, conceived at about that time, rather than to an increase in salary. He was about to invent his way out of the drafting room and, in the process, attract the attention of Steinmetz.

Alexanderson's first successful patent application grew out of his work on a relay in a protective circuit. He had conceived the invention by October 1902, when he discussed it with Edward M. Hewlett, a GE engineer who was a leading designer of switching and control circuits.[72] With Hewlett's encouragement, Alexanderson wrote a description of the invention, which was signed by two witnesses and dated November 6, 1902. In the description, he discussed the use of reverse-current relays for protection against short-circuit faults and explained why the relays did not always work satisfactorily. As an alternative, he proposed a circuit arrangement that would respond to very small fault currents and would open a circuit breaker only in the line where the fault occurred. He noted that an advantage of the proposed protective circuit was that it would avoid the need for an expensive instrument line between stations. His patent application eventually was filed on July 16, 1903; it was issued on May 1, 1906, as U.S. Patent no. 819,627. It was only the first of some 344 inventions for which Alexanderson received patents.[73]

Alexanderson's second invention followed a more erratic course and provided the leverage for his transfer from the Drafting Department. The exact date of the conception of this invention, a method to change the speed of AC motors by varying the number of active poles, is uncertain, but he may have discussed it with Steinmetz during a conversation on December 6, 1902.[74] Two days later, Alexanderson filed a description of the invention with GE's Patent Department, and he was asked to discuss his idea with a member of the AC Engineering Department.[75] This engineer in turn asked Steinmetz to assess the invention, and the latter identified a secondary magnetic effect that seemed to limit the method's practical value. After Steinmetz's pessimistic evaluation, Alexanderson received word that the Patent Department would be advised to terminate the application process. He subsequently discovered a way to overcome the difficulty but decided not to disclose the modification to the GE engineers. Instead, he concluded that the invention was apt to be valuable and decided to try to obtain the patent himself and possibly sell it to another company. He contacted a patent lawyer in New York City, reviewing the history of the invention and asking for the attorney's assistance in getting the patent.[76]

Alexanderson approached the Stanley Company about its buying the rights to his motor speed-control invention. He reviewed the case history and stated that the great interest the invention had stimulated at GE had convinced him that it was too valuable to give away. To avoid a possible misunderstanding, he explained that he worked in GE's Drafting Department, did not have a contract with GE, and had not got the idea from GE. Cummings C. Chesney, the chief engineer of the Stanley Company, responded to Alexanderson's inquiry indicating a willingness to negotiate.[77]

Meanwhile, Alexanderson's continued dissatisfaction with his position and salary at GE again became manifest. A patent solicitor in Rochester, New York, offered him a job as an expert assistant and assured him that his technical abilities would be put to greater use. He was offered a salary of $15 per week for an eight-hour day with the promise of an increase to $18 per week after six months.[78] He also drafted a letter to the pioneer wireless experimenter Reginald A. Fessenden, expressing interest in a position advertised in *Electrical World*. Alexanderson claimed a thorough familiarity with AC phenomena and mentioned his two recent inventions.[79] It is uncertain whether he actually sent this letter or received a response, but within a few months the paths of Fessenden and Alexanderson would intersect, leading to a long and stimulating interaction that had global consequences.

Alexanderson also explored the prospects for getting a transfer within the GE organization. He asked Hewlett for advice on whether he should again apply for admittance to the Test Department. Alexanderson stated that he personally would prefer to transfer directly into an engineering department, since he already had considerable test experience.[80] He also approached the director of the GE Research Laboratory, Willis R. Whitney, about a position in the laboratory. Whitney offered him the opportunity to participate, at a salary of $15 per week, in the construction of an electric furnace needed for research on lightning arrestors. Whitney recorded in his notebook that Alexanderson had made a good impression but could not be released immediately from his present job. He notified Alexanderson that he would not be permitted to come to the laboratory for at least two weeks but that a temporary arrangement might be possible until he could be released. Alexanderson, however, replied that in view of the difficulty it would be better not to make the special arrangement; moreover, since he had no experience as a shop foreman, he probably was not well suited for the furnace project. He asked that Whitney keep his name on file until there was some electrical work of a more scientific nature for which he would be better suited.[81]

Having declined the Research Laboratory assignment, Alexanderson asked that his salary as a draftsman be raised to $18 per week, mentioning that he had turned down an offer of $20 per week from another firm in the expectation of receiving a promotion at GE.[82] When there was no immediate response, he resumed his effort to obtain a patent for the variable-speed motor invention outside GE channels. He then learned that GE's Patent Department had not terminated its efforts even though he had asked that the application be dropped. He wrote to his outside patent attorney expressing concern that he might be discharged if GE became aware of his independent effort to obtain the patent. Mentioning that he planned to be on a trip from June through September, he asked that the patent lawyer delay further action until October.[83]

In February 1903, Alexanderson received an offer of a formal contract with GE which would make him a designer and draftsman at a salary of $19 per week. He responded that he was willing to sign a contract if GE met certain conditions, which included allowing him to spend the coming summer in Sweden, and giving him the assurance that he would be admitted to the Test Department when he returned. He regarded signing the contract as something of a sacrifice, he stated, because it would require that all his future inventions be assigned to GE. He mentioned that he had recently done a theoretical analysis of high-speed turbine wheels which might be of value, and that he had discovered that the method generally used to calculate the strength of rings to hold commutator bars in place was "entirely wrong." His study, he continued, had revealed that turbine wheels might be designed for much higher speeds than those at which they usually operated, and that the same analysis could be applied to the design of high-speed alternators.[84]

Alexanderson's theoretical investigation of the strength of high-speed rotary disks sheds new light on the genesis of the Alexanderson radio alternator, one of his major innovations. The problem, as he formulated it, was to determine the shape and dimensions of a disk such that it would be equally strong in all its parts. He found that it was theoretically possible to design wheels to run at several times the peripheral speed of 400 feet per second which had been standard practice. Also he considered the effect of slots cut in the rim of a disk and found that the slots did not weaken the disk very much.[85] He was to have an opportunity to use the theory almost two years later when he designed high-speed disk rotors as part of a radio alternator.

Having signed a contract with GE, Alexanderson abandoned the attempt to patent the variable-speed motor invention without going through GE's Patent Department. He explained to the GE engineers how he had overcome the field distortion caused by armature reaction, the difficulty Steinmetz had identified. Alexanderson submitted some further sketches and explanation to the Patent Department before the application, which included 42 claims, was submitted in July 1903. The patent was finally issued more than three years later.[86] Six months after the application was filed with the U.S. Patent Office, Alexanderson reported to a GE executive that it had come to his attention that the variable-speed AC motor ought to be marketed. He recounted a conversation with an engineer from the Brown and Sharp Company who had told him that such a motor was needed in machine shops.[87]

Just before leaving to spend the summer of 1903 in Sweden, Alexanderson submitted a new application to Albert Rohrer of GE's Test Department. Rohrer promised Alexanderson that he could enter the test program when he returned.[88] Despite this assurance, there is some evidence that he considered remaining in Sweden. A two-day visit at the home of Ernst Danielson, chief design engineer of ASEA, apparently helped Alexanderson come to his ultimate decision. He later recalled that Danielson had predicted that he would never come back to work in Sweden if he stayed in the United States any longer. Despite his parents' effort to persuade him to stay, he decided to return to Schenectady.[89] Danielson's prediction proved accurate, since Alexanderson never returned to Sweden except for brief visits.

The trip to Europe did give Alexanderson the opportunity to publish his first technical paper. His article appeared in the *Electrotechnische Zeitschrift,* the electrical engineering journal still edited by Gisbert Kapp, his former teacher at Charlottenburg. Kapp is known to have solicited articles from former students who had gone to the United States.[90] Alexanderson's paper gave a brief account of American practice in the design of rotary converters, motor-generators that were used to convert AC to DC or vice versa. Rotary converters facilitated the exchange of power between the newer AC power systems and the older Edison DC systems and also were used in railroad electrification. In his paper, Alexanderson discussed such features as starting torque and included a diagram showing a typical system connection. As references, he cited Steinmetz's book and recent papers by Steinmetz and Ernest Berg.[91]

The GE Test Department

The Test Department, which Alexanderson finally entered in October 1903, was in effect the final hurdle in the education and selection process of the ambitious young corporate engineer at GE. The test engineering program served as a test of the aspirant engineer as well as of the machines that were tested. A discussion of activities in the test program, published in 1904, called it the "best graduate course in electrical engineering in the world."[92] Steinmetz later wrote that the engineering colleges did not graduate engineers but only individuals who were capable of becoming engineers by completing their education in corporate programs such as the GE test course.[93]

The Test Department furnished a continuous supply of young engineers who were expected to fill responsible positions either at GE or elsewhere in the electrical industry. The participants were encouraged to become active members of a professional engineering society such as the American Institute of Electrical Engineers, which had its largest section in Schenectady, and they were given opportunities for social contacts with GE executives. The test program served as an initiation "into the privileged world of the professional and industrial elite," and stressed the business side of corporate engineering as well as the principles of product design and manufacture.[94] The experience tended to help "Americanize" the graduates of European colleges. A British engineer observed in 1904 that GE's test program quickly stimulated engineers trained in England to "drop into step and hustle."[95]

The engineers assigned to the Test Department were responsible for testing all the machines and apparatus manufactured by GE. The total test force of several hundred was organized into sections. Small squads were formed to carry out specific tests, with each squad being led by a more experienced engineer. Whenever a new product was to be tested, the designer was expected to participate in the initial testing. The data collected might lead to modifications in the design or to an entirely new design. The test instruments used were calibrated frequently in the GE Standardizing Laboratory. Each test engineer was given gradually increasing responsibility. The engineer's steadiness, judgment, and ability to manage a team while working with expensive machines and delicate instruments were observed continually. Resourcefulness was sometimes needed in the development of appropriate methods for testing new products. Each test engineer was encouraged to take the critical point of view of an ultimate user of the product. By 1905 the Test Department at the Schenectady works had grown to a total force of almost 600 people, who were responsible for testing equipment each month with a capacity equivalent to 60,000 kilowatts (kw).[96]

When Alexanderson entered the Test Department, he signed an agreement stipulating that he was to receive 17.5 cents per hour for a period of six months, after which he was to receive an additional 2.5 cents per hour at six-month intervals until he left the department.

Figure 1.1. Alexanderson as a test engineer at General Electric, c. 1903, youthful and well armed with the tools of his new trade. Courtesy of the Smithsonian Institution, Washington, D.C.

While it might seem incredible that he now was willing to accept what amounted to a drastic pay reduction, his future prospects at GE were far better than they had been when he was in the Drafting Department. In fact, only five months elapsed before he left the Test Department to enter an engineering department at a salary of $19 per week.[97]

William B. Potter was one of the more experienced engineers Alexanderson encountered during his tenure in the Test Department. Potter once had worked for the Thomson-Houston Company and continued with GE after the merger of Thomson-Houston and Edison GE in 1892. He served as an engineer and later as manager of GE's Railway Engineering Department, and received more than 130 patents during his career.[98] Alexanderson reminisced about the "grand old days" as a test engineer in a letter to Potter when the latter retired in 1931. Alexanderson recalled an exciting episode when they had run for their lives when a large series-wound motor had broken its coupling to the load and accelerated rapidly. According to Alexanderson, the motor had been saved by a "self-sacrificing test man just before it was going through the roof." He recounted another occasion when he and several other test engineers had driven an old car known as "Queen Lil" at 60 miles per hour to get to a ball game on time. Alexanderson ended his letter by stating that it had been a great privilege to be among those who had "played the game of engineering" under Potter's guidance.[99]

A novel test method devised by Alexanderson resulted in his first publication of a paper in an American journal. The paper appeared in *Electrical World and Engineer* in August 1904. Before it was published, it was read by a GE executive, who approved it with minor changes, such as deleting any mention of GE.[100] In the paper, Alexanderson explained a method that could be used to measure the power delivered by an induction motor under normal operating conditions. His method was based on a stroboscopic effect and required only that an observer count the apparent number of rotations per minute of a disk with alternating black and white sectors. The disk was mounted on the rotor shaft of the motor. The output power could then be calculated easily by means of a simple equation that Alexanderson derived in the paper. He reported that he had used the method during two months of tests on motors and had verified its accuracy by comparing his data with that obtained using conventional instrumentation. He noted that a major advantage of his method was that the motor output could readily be determined at any time under normal shop conditions without instrumentation.[101]

Alexanderson's unusually brief internship in the Test Department ended early in 1904, when he signed a new contract and was assigned to work in the AC Engineering Department. He was well prepared to embark on a productive career as an engineer-inventor. His assets, already considerable, included a network of friendships with fellow alumni of the Test Department, the attention of Steinmetz and other senior engineers, and a growing confidence in his own inventive ability. By the end of 1904, he had applied for nine patents that eventually were issued. He would maintain an annual average of about seven patented inventions for the next 40 years. His exceptional ability to incorporate elements of power, communication, and electronics technologies into an integrated system would ultimately involve him in international power politics and make him a celebrity. But in 1904 these developments still were years in the future, and he could look forward to a few years of comparatively "normal" engineering and invention.[102]

Chapter Two

Dialectical Engineer: Power and Wireless Engineering

A fundamental tenet of Alexanderson's mature philosophy of technology was that the most fertile soil for creative engineering and invention was near the boundaries between technological systems. He acquired empirical evidence for his conviction early in his career at General Electric, when he was given the opportunity to work simultaneously on problems in such seemingly unrelated fields as railroad electrification and wireless communication. The dialectical exchange of ideas between electrical power and wireless systems which Alexanderson experienced beginning in 1904 perhaps resembled the dialectic between communication and power systems which marked the early inventive career of Thomas Edison. The creative tension that was manifest in Alexanderson's 66 patent applications during 1904–10 existed at other levels. There was the inevitable tension in the inventor's mind between a perceived problem and an ideational solution that might be resolved by one or more patentable inventions.[1] Alexanderson also experienced an additional degree of tension due to the internal corporate patronage of his work on AC power machines and circuits, and the external patronage of his early work on the high-frequency alternator for wireless applications.

Alexanderson's mastery of the principles of AC machines and circuits was his greatest single asset as an engineer-inventor. Most of his early patents were in this area. Their specifications and related documents reveal a remarkable grasp of intricate combinations of armature and field windings, rectifying commutators, and external circuitry. He was able to visualize and analyze spatial and temporal relationships of rotating magnetic fields as a function of multiple variables of design and load conditions. His fundamental understanding of the principles of AC power remained a touchstone throughout his career, even when he was dealing with radio and electronic systems.

Current theories of invention and innovation have tended to reflect the findings of case studies of heroic independent inventors or inventor-entrepreneurs such as Thomas Edison or Elmer Sperry.[2] Alexanderson is more representative of the twentieth-century inventor employed by a large corporation. His inventive activity will be presented in considerable detail to permit comparison with other inventors both within and without the corporate environment. Insofar as possible, the sources of problems and ideas for solving them will be identified and the role of the corporate patent department and managerial policy will be assessed. Alexanderson's participation in the commercialization and diffusion of his inventions also will be examined.

The advent of the revolutionary high-speed turboelectric generator for power plants and the possible adoption of AC motors in interurban railroad service were providing challenging problems for engineer-inventors at GE in 1904, as Alexanderson completed his tenure in the Test Department. GE engineers had in fact installed the first large Curtis turbogenerator in a central station in Chicago the previous year.[3] Railroad electrification offered the prospect of an enormous market for GE if the company could develop the appropriate technology and persuade the railroad industry to convert from steam locomotives to electric. By comparison, the field of wireless communication seemed to hold little attraction for GE or its engineers. Yet Alexanderson unexpectedly became involved in a wireless-engineering project because of the activities of an outside patron, Reginald Fessenden.

A comparison of Alexanderson's inventive activities in railroad electrification and in the high-frequency alternator project is quite revealing, since one system failed to live up to expectations, while the other became a brilliant success. In retrospect, he was fortunate in having the wireless project almost thrust upon him. The project brought opportunities to work with engineers and scientists both in other departments at GE and outside the company. The Alexanderson radio alternator that had its origins in the project sponsored by Fessenden ultimately became the focal point of an intense interaction of technological, political, and business systems. One result was that Alexanderson had opportunities as a system engineer and a leader in the engineering profession which probably would have escaped him if he had restricted his activities to power engineering. By avoiding excessive specialization, he was in a favorable position to focus his attention and creativity on the more successful system as the winner gradually became apparent.[4]

AC Power Engineering: The Invention of a Self-Exciting Alternator

Alexanderson joined the AC Engineering Department at GE in February 1904. The department was responsible for designing and developing motors and generators and was headed by Henry G. Reist (1862–1942). Reist had graduated in mechanical engineering from Lehigh in 1886 and had worked for the Thomson-Houston Company be-

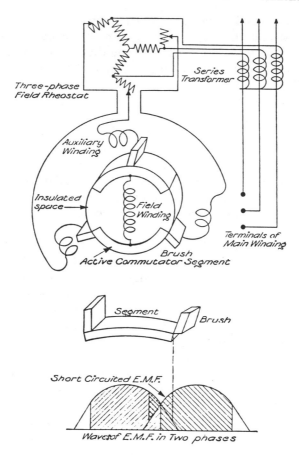

Figure 2.1. Circuit diagram of the self-exciting alternator, one of Alexanderson's first patented inventions. A rectifying commutator converted a portion of an alternator's output into direct current that energized the alternator's field winding. From Ernst F. W. Alexanderson, "A Self-exciting Alternator," *Transactions of the American Institute of Electrical Engineers,* 25 (1906): 63.

fore the merger with Edison GE in 1892. An engineer-inventor as well as a manager, he received approximately 75 patents during his career and remained with GE until his retirement in 1930.[5]

Soon after Alexanderson came to the AC Engineering Department, he invented a method for achieving self-excitation of alternators. It was the first of his inventions to attract considerable attention outside GE. It resulted in the publication of his first paper by the American Institute of Electrical Engineers and was the basis of several patent applications that he filed during a three-year period beginning in August 1904.[6] In the invention, he used a rectifying commutator to convert the alternating current produced by an alternator into direct current that could be used to energize the alternator's own field winding and also supply external DC circuits if desired. This technique had been tried before, but its success had been limited due to problems with commutator sparking and alternator voltage regulation. Thus an important determinant of the invention's commercial success was the inventor's success in explaining how these problems could be eliminated, or at least minimized.[7]

In late April 1904, Alexanderson requested authorization to spend $100 for experiments on a compensated self-exciting generator.[8] This request was made approximately four months before he filed a patent application covering the invention. His experiments proved encouraging enough to cause him to initiate the process of seeking a patent, and he sent a brief written description of the invention, along with a drawing, to Albert G. Davis, manager of GE's Patent Department, in late June 1904. As his experiments continued, Alexanderson kept Davis informed of each improvement and its significance. In one memorandum, dated July 1, 1904, Alexanderson referred to "my system of rectifying commutators" which would supply DC to mercury-arc lamps as well as to an alternator, for self-excitation. He went on to point out that his proposed method would make the alternator independent of a DC supply that might not be readily available in the AC power plant. By this time there was already discussion at GE of a commercial machine that would embody the method of self-excitation.[9] During the course of his experiments, Alexanderson borrowed some test apparatus from the GE Research Laboratory, beginning a cooperation that would persist throughout his career.[10]

The documents prepared by the employee-inventor for the corporate patent department often are quite revealing of the social and intellectual history of an invention; from its conception in the mind of the inventor to the filing of an application and the eventual issuance of a patent—or in some cases, the rejection or abandonment of the application. The contribution of company patent specialists such as Albert Davis at GE, whose role during the gestation period of inventions was analogous to that of an attending physician, perhaps has not been appreciated adequately in case studies of invention. Davis exercised a significant influence not only on engineer-inventors but also on corporate research policy and corporate receptivity toward departures from traditional product areas.

As an engineer-lawyer, Davis was in a favorable position to assess the potential of young inventors and to educate them in the patent process.

He had graduated in engineering from the Massachusetts Institute of Technology in 1892 and worked briefly as an engineer before becoming an assistant patent examiner in the U.S. Patent Office. While in Washington, Davis studied law, and in 1897 he joined GE as manager of its Patent Department. Along with Steinmetz and Elihu Thomson (1853–1936), a prominent GE engineer-entrepreneur, Davis was instrumental in the establishment of the GE Research Laboratory in Schenectady in 1900. Later he became a GE vice-president and he remained with the company until his retirement in 1933.[11] Alexanderson held Davis in high esteem and attributed to him the doctrine that executives in the company should follow the lead of engineer-inventors and scientists in deciding what innovations to develop or what research to support.[12]

Davis discussed his philosophy on inventors, patents, and the corporation in a talk prepared for a meeting of the AIEE in 1907. He explained the role of the patent attorney in serving as a mediator between the inventor and the patent examiner, and remarked that a good patent lawyer needed to be "a man who is a great deal of a lawyer, a great deal of an inventor, and a prophet." He believed it rare for an invention to "spring completely from one mind," he continued; more typically, the invention process was a long one, involving considerable cost and cooperative effort. Davis felt that substantial credit for the patent that finally issued might properly belong to people other than the designated patentee. He estimated that the average patent awarded to GE inventors cost the company at least $1,000. He defended the policy of assigning patents to the corporation as "eminently fair and proper."[13]

The nature of GE's patent strategy during this period is indicated by a close reading of the claims in patents issued to Alexanderson on the self-excitation of alternators and other inventions. An individual patent might contain 30 or more separate claims, many of which differed only in minor detail. The company strategy seems to have been to blanket any innovation of commercial potential with numerous patents and dozens of claims. In contrast, an innovation that, like the high-frequency radio alternator, initially appeared to have limited market potential received much less patent coverage until perceptions of its relevance to corporate goals were altered.

In January 1905, as the development of the self-exciting alternator continued, Alexanderson contacted Edwin W. Rice, Jr., to request an additional appropriation of $500 to support further experimental tests.[14] Rice, a protégé of Elihu Thomson, was the top engineer-administrator at the Schenectady works. He had studied under Thomson at Central High School in Philadelphia and then served as Thomson's principal technical assistant at the American Electric Company and its successor, the Thomson-Houston Company. Shortly after the formation of GE, Rice was named its chief engineer; in 1896 he became a vice-president in charge of the engineering and manufacturing departments.[15] He was credited with having instituted the "development jobs" system of budgeting at GE. In this system, each technical innovation received support in the form of a series of appropriations. The Rice budgeting system led to automatic reviews of each project at the time of each request for a new appropriation. This enabled management to consider periodically whether further support of a specific project was justified and also made possible the convenient calculation of the cumulative developmental costs of each product or innovation. Rice's principle of "integrity of costs" thus permitted the total developmental cost to be reflected accurately in the market price of a machine or other product.[16]

Rice's managerial style became legendary among the engineer-inventors at GE. He was said to have fostered "teamwork of a spontaneous and joyous nature" and a sense of "loyalty of the kind that men have for a cause."[17] Willis Whitney, who headed the GE Research Laboratory, wrote that Rice had formed a "great, coherent group of electrical specialists" which "included highly individualistic engineers, strong-minded egoists, weak-minded altruists, sanguine inventors, phlegmatic pluggers, systematic workers, optimistic spenders, and pure researchers. . . . It was a larger group of more varied personalities, more temperamental artists than any impresario ever tried to direct."[18] Rice followed closely the work of his engineers and often came to see their latest inventions, in keeping with his expressed conviction "One look is better than ten talks."[19] In later years, Alexanderson recalled that he had soon discovered that he "had [Rice's] good will and backing."[20]

As mentioned previously, earlier attempts to achieve self-excitation of alternators by means of the rectifying commutator had been frustrated by excessive sparking at the brushes and poor voltage regulation. Alexanderson attacked these problems by a combination of theoretical and experimental analysis. He employed a method Walter Vincenti has aptly labeled "parameter variation."[21] After identifying seven factors that influenced commutation, Alexanderson did an empirical investigation that consisted of an incremental variation of a single factor while the remaining ones were kept constant. The resulting data eventually enabled him to arrive at a design such that the optimum brush position was the same for full load and no load on the alternator. For this brush setting, he found that relatively sparkless commutation and adequate voltage regulation were obtained over the full range of load variation, from no load to full load. He next tested the method of self-excitation, first on a 100-kw alternator and then on a 500-kw turboelectric generator.[22] On at least one occasion, he notified Steinmetz that a commercial machine using self-excitation was undergoing final tests and suggested that Steinmetz might wish to observe them.[23]

The diffusion of information about Alexanderson's method of alternator self-excitation began in 1905. William L. R. Emmet (1859–1941) was among the senior GE engineers who observed tests of the method. He became its enthusiastic proponent, and discussed the innovation with British engineers during a trip abroad. In March 1905, one of the engineers who had heard about the method from Emmet contacted Alexanderson to ask for details on how to apply the technique. A few months later, Alexanderson advised another British engineer on the modification of an alternator for self-excitation.[24]

During Alexanderson's early years at GE, Emmet provided a role model that was perhaps second only to that of Steinmetz. After graduating from the U.S. Naval Academy in 1881 and serving for two years in the navy, Emmet had worked as an engineer for the Sprague, Westinghouse, and Edison GE companies. Following the formation of GE, he played a major role in the successful introduction of the Curtis turboelectric generator.[25] His interest in the Alexanderson self-excitation technique may have been influenced by the possibility of its use in an electric propulsion system for naval ships which he hoped to introduce. Alexanderson discussed the ship propulsion concept with Emmet during 1905 and later contributed to the development of the system.[26] Alexanderson credited Emmet with having shown him how the entire works at Schenectady could serve as a resource for the engineer-inventor. Emmet's doctrine was to have only a small staff of permanent assistants while drawing on all the talent and other resources of the company as needed. Emmet and Alexanderson occupied adjacent offices and often collaborated or exchanged ideas as each pursued the problem solving needed in the development of his systems or inventions.[27]

The diffusion of information was also aided by Alexanderson's first AIEE paper, which was presented at a meeting held in New York City in January 1906. In a letter proposing the paper, Alexanderson mentioned that it would concern a new line of self-exciting alternators that GE was prepared to market in sizes up to 500 kw.[28] In the paper, he discussed some potential advantages of adopting his method. It would, for instance, simplify power plants by eliminating the necessity for a separate DC exciter unit, with its steam piping and other apparatus. Pointing out that existing alternators could easily be modified for self-excitation by adding an auxiliary winding and a commutator, he suggested that the technique should result in substantially lower costs and labor requirements in both large and small installations.[29]

The discussants of Alexanderson's paper were favorably inclined toward his self-excitation method, although some reservations were expressed by a Westinghouse engineer. Arthur E. Kennelly of Harvard commented on the apparent simplicity of the innovation and concluded that if it worked as well as Alexanderson had indicated, both he and the AIEE were deserving of congratulations. Charles F. Scott of Westinghouse agreed that the method was ingenious and seemed superior to other proposed methods for self-excitation. However, he observed that changes in speed or load would tend to cause a greater fluctuation in the output voltage than would be seen if separate excitation were used. Consequently, he cautioned, the method probably would not prove to be a "complete and universal solution of the problem of excitation of alternators." Alexanderson's GE colleague William Emmet reported that he personally had witnessed tests of the method conducted under very severe conditions and that the "performance was astonishingly good." He added that he saw "almost no limit to the application of the idea." While agreeing that speed variation was a limitation for any method of self-excitation, Emmet emphasized that an alternator using the Alexanderson method tended to compensate automatically for changes in speed and load.[30]

Alexanderson's self-excitation technique enjoyed some success, especially in isolated plants, and in 1907 Emmet recommended that GE prepare a special bulletin on the method because it was achieving some prominence.[31] Self-excited alternators using the rectifying commutator and auxiliary field winding became commonly known as composite, or compound, alternators. The method proved to be not particularly suitable for alternators rated higher than about 500 kw and for large central power plants where relatively constant and precise control of output voltage and frequency were necessary. The authors of a textbook on alternating current published in 1910 advised that "it is therefore not usual to compound large alternators." They explained that the voltage output from large alternators generally was controlled manually by a switchboard attendant, although an automatic voltage regulator, the Tirrell regulator, was being introduced at some installations.[32]

Wireless Engineering: The Radio Alternator Project

While still engaged in developing the self-exciting alternator, Alexanderson received an unusual assignment that had far-reaching consequences for his career. The assignment was to design an alternator capable of producing power at a frequency of 100,000 cycles per second (100 kilohertz [kHz]), a frequency well above anything attempted previously at GE. The substantial increase in rotor speed that would be required to achieve such a frequency from an alternator presented difficult mechanical—as well as electrical—problems for the designer. Since Alexanderson lacked experience in wireless engineering, he tended to approach the project by extending established power-engineering principles to higher frequencies. This power-engineering perspective led to novel solutions and ultimately to the creation of a remarkable global communication system based on high-power radio alternators.

An examination of the genesis and development of the Alexanderson radio alternator sheds light on stylistic and cultural characteristics of creative engineering. Thomas P. Hughes has defined technological style as "the technical characteristics that give a machine, process, device, or system a distinctive quality." He added, "Out of local conditions comes a technology influenced by time and place, a technology with a distinctive style."[33] The Alexanderson radio alternator and the communication system in which it functioned seem to exemplify well the concept of technological style. A superb machine is a testimony to the personal style of the designer, just as a great painting reflects the skill and style of the artist. Reginald Fessenden, who initially was the outside patron of the alternator project, recognized the stylistic quality in engineering design. In 1900 he compared the engineer-designer to the artist and claimed that an experienced designer could look at a machine and recognize immediately that the "machine was designed by so and so" or else was "modelled after his style by a man who had worked in such a place."[34] In its mature form, the Alexanderson radio alternator was a machine that could be recognized as his design.

Power and wireless engineering tended to develop as distinctive cultures within electrical engineering during the early twentieth century. Alexanderson became one of the rare individuals who was willing and able to cross the cultural barriers, which included distinctive vocabularies and modes of thought. The cultural separation was, in effect, institutionalized by the development of separate organizations, with the power-engineering culture represented by the AIEE and the wireless engineering culture by the Institute of Radio Engineers (IRE). Alexanderson became active in both of these organizations. A substantial degree of cultural separation also existed at GE, as the specialized departments tended to produce and maintain barriers. With its mixed ancestry in wireless and power engineering, however, the radio alternator is a good example of the creative interaction between cultures. The contrasting views of the cultures were especially apparent in a dispute between Fessenden and Alexanderson over whether iron should be used as the armature core in the alternator, a dispute that eventually was resolved by using the alternator itself as an instrument.

Reginald Fessenden, a leading pioneer in wireless engineering, was chiefly responsible for stimulating GE to embark on the development of the radio alternator even though in 1900 the company was a center of the power engineering culture. Born in Canada in 1866, Fessenden received a liberal education at Bishop's College in Quebec and undertook considerable self-study in science and mathematics. He taught school in Bermuda and then worked for about two years in Thomas Edison's laboratory in New Jersey. Fessenden also worked for Westinghouse and for the Stanley Electric Company before teaching electrical engineering for a year at Purdue University during 1892–93. From 1893 to 1899 he taught at the Western University in Pittsburgh, where he and some advanced students began to experiment with wireless telegraphy.[35]

During the 1890s, Fessenden often engaged in theoretical and speculative physics and developed an "electrostatic doublet" theory of atoms in solids. He used the theory in an effort to link data on the size and spacing of atoms to material properties such as tensile strength and elasticity.[36] In 1900 he published in the *Physical Review* a long paper dealing with fundamental theories of matter, electricity, magnetism, and the ether. In the paper he revealed some familiarity with the work of Maxwell, Hertz, Helmholtz, and other leading nineteenth-century physicists.[37] While at Western University, Fessenden also did research on incandescent lamps for the Westinghouse Company, which paid a portion of his salary.[38]

By 1899, Fessenden's poor experience with the Marconi system of wireless telegraphy, which relied on an intermittent-spark-discharge transmitter and a rather erratic coherer device as a receiver, convinced him that the system had fundamental flaws. He conceived and became a determined advocate of a radically different system, which required a reliable source of continuous-wave power at frequencies suitable for wireless transmission.[39] Almost 20 years were to elapse before his prophetic insight on the superiority of the continuous-wave system over the damped-wave spark system was demonstrated conclusively in practice. In the short run, it was his need for a powerful source of continuous-wave energy that brought him to GE as a customer in 1900.

Flushed with enthusiasm over the potential of his continuous-wave wireless system, Fessenden resigned his teaching position with Western University to accept an offer from the U.S. Weather Bureau, which agreed to support him in developing his system for possible use in the collection of weather data. He established experimental facilities on the coast of North Carolina and devoted about two years to the effort. It was during this period that he invented the so-called barretter wireless detector, which proved far superior to the infamous coherer detector and also was suitable for reception of continuous waves. He also conceived and patented the heterodyne receiver, which required two sources of continuous-wave energy to function, and eventually proved to be of fundamental importance in radio systems.[40]

In June 1900, Fessenden contacted both the Westinghouse Company and GE with regard to their possible interest in designing and constructing a high-frequency alternator for use in his wireless experiments. A Westinghouse engineer, Charles Scott, responded that his company was too busy with other work to undertake such a project.[41] The response from GE was more favorable, and it came from Charles Steinmetz. In his initial letter to Steinmetz, Fessenden reported that he already had obtained results superior to those of Marconi. Fessenden enclosed some general specifications for a high-frequency alternator and expressed the hope that Steinmetz would persuade GE to submit a bid. Fessenden went on to say that he might eventually need 40 or 50 of the alternators and suggested that Steinmetz might like to undertake the design as an experiment. Steinmetz inquired about the possible use of an induction coil as an alternative to the alternator, and Fessenden responded with an explanation of why he believed an alternator would perform better than either the coil or a rotary interrupter that he had also tried. He also asked whether Steinmetz could provide him with formulas for the capacitance and inductance of a vertical wire antenna.[42]

In January 1901, Steinmetz wrote to Fessenden that he now agreed that an alternator was a more satisfactory source of wireless power than induction coils or "other such unreliable apparatus." Steinmetz mentioned that he and the GE engineers had experienced difficulty trying to make an interrupter work. Later the same month, the manager of the Schenectady works received word that construction of a 10-kHz alternator under the direction of Steinmetz had been approved, at an estimated cost of $250.[43]

Early in June 1901, Henry Reist, head of the AC Engineering Department, notified Steinmetz that the high-frequency alternator was now ready to be tested. Steinmetz wrote on the bottom of Reist's memorandum that the machine should be assembled and the saturation curve and synchronous impedance obtained.[44] Meanwhile, Fessenden in May 1901 had filed a patent application covering the use of the alternator as a generator of wireless waves.[45] The experimental Steinmetz alternator employed a rotating armature with a conventional winding. The armature was driven at a nominal speed of 3,750 revolutions per minute (rpm) by an induction motor. The alternator was to produce a designed power output of 2,000 watts, although the actual measured power was around 1,200 watts.[46]

In July 1901, from his wireless station at Roanoke Island, North Carolina, Fessenden wrote to Steinmetz to inquire about progress on the high-frequency alternator. Fessenden commented that he anticipated that the alternator would be used a great deal and that he would want to place an order. He mentioned that he had been experimenting with wireless telephony but so far had achieved only short distances. About two weeks later, Steinmetz informed Fessenden that the 10-kHz machine had been operated and seemed to work, although he added that GE's instruments were "absolutely useless at this frequency."[47]

Elihu Thomson, who already was almost legendary at GE in 1901, learned of Fessenden's wireless telephony experiments and wrote to Fessenden in July 1901 that he found the prospect of wireless transmission of speech "deeply interesting." Thomson was so interested that he had GE construct a second high-frequency alternator, apparently a duplicate of the one designed for Fessenden, for his own experiments.[48]

Despite his professed enthusiasm for the high-frequency alternator, Fessenden apparently did not see the 10-kHz machine until the summer of 1902. Meanwhile, his relationship with his sponsors at the U.S. Weather Bureau deteriorated. It ended as of September 1, 1902. In November 1902, Fessenden formed a private corporation known as the National Electric Signaling Company (NESCO) to continue the development of and to market his continuous-wave wireless system. The new company was backed financially by two Pittsburgh entrepreneurs, Thomas H. Given and Hay Walker, Jr. Over the next several years, their investment grew from an initial $30,000 to approximately $2 million.[49] With the infusion of capital, Fessenden soon arranged to purchase the 10-kHz alternator from GE. The machine, along with associated apparatus, cost $575. It was shipped to the NESCO station at Old Point Comfort, Virginia, in March 1903.[50]

Fessenden's experiments with the Steinmetz alternator and other methods confirmed his belief that high-speed rotating machinery could be used as a source of continuous waves for wireless communication but that alternators capable of still higher frequency and power would be required. Early in December 1904, he again contacted GE to ask whether the company would undertake to construct an alternator capable of generating 25 kw at a frequency of 150 kHz.[51]

Fessenden's efforts to have GE design and develop high-frequency alternators are well documented, but the exact circumstances that led to Alexanderson being assigned the task of designing a 100-kHz alternator remain uncertain. Alexanderson later recalled that the initial assignment had come through routine channels with no specific suggestions as to how the design should be carried out. He had completed a preliminary design on December 3, 1904, and had been in touch with GE's Patent Department before he learned of Fessenden's interest in such a machine.[52] This version seems to be supported by the exchange of letters that followed Fessenden's December 8 proposal that GE undertake to build a 150-kHz alternator. A GE sales representative responded to Fessenden's inquiry that the AC Engineering Department had already begun design of a 100-kHz machine and that the engineers would prefer to complete it before trying for a still higher frequency. Fessenden seemed puzzled by the mention of a 100-kHz alternator and asked whether its design costs would be charged to NESCO.[53] The GE representative replied that it was questionable whether even the 100-kHz alternator would be developed unless Fessenden were willing to pay the costs.[54]

If Alexanderson's assignment to design the 100-kHz alternator did not originate with the Fessenden proposal, who was responsible, and for what purpose? The evidence suggests that Steinmetz may have been involved and that his principal assistant, Ernst Berg, served as intermediary in delegating the design to Alexanderson. Perhaps the machine that became known as the Alexanderson alternator would not have been built if the timely inquiry from Fessenden had not revealed his willingness to assume the cost of development. It was Berg rather than Steinmetz or Alexanderson who corresponded with Fessenden about the alternator project during its first few months.

Berg's background was quite similar to that of his younger Swedish compatriot. He had graduated in mechanical engineering from the Royal Institute of Technology in Stockholm before coming to the United States in 1892. He then worked as a draftsman at the Thomson-Houston Electric Company until he was moved to Schenectady, where he became Steinmetz's assistant. He mastered the Steinmetz method of mathematical analysis of AC circuits and helped in the writing of the book Alexanderson had acquired while a student in Charlottenburg. His brother, Eskil Berg, also worked at GE, and the brothers, along with Alexanderson, enjoyed weekend recreational activities at Steinmetz's rustic camp on the Mohawk River.[55] Ernst Berg left GE in 1909 to become an electrical engineering professor at the University of Illinois.[56]

In his initial design of the 100-kHz alternator, Alexanderson chose to utilize an "inductor alternator," a type of generator in which both the armature and the field windings were stationary. He anticipated that this type of machine would avoid difficulties due to centrifugal forces acting on windings rotating at very high speeds. At the time, inductor alternators for use at normal power frequencies were manufactured by the Stanley, Westinghouse, and Warren companies. The advantages claimed for this type of generator included better insulation of windings, the elimination of the need for moving wire, and the absence of collecting brushes. In the typical commercial generator, the rotor had around its periphery pairs of iron projections called inductors, which were magnetized by current in the stationary field winding. As the rotor turned, the inductors passed the poles of the armature winding, which projected from the frame of the generator. The inductors formed part of a magnetic circuit in which the reluctance reached a minimum when an inductor was directly opposite an armature core and a maximum when the inductor was at an intermediate location. Rotation caused the magnetic flux linking the armature winding to change in a cyclical manner from a maximum to a minimum, although it did not change in sign. The frequency of the voltage induced in the armature winding was in direct proportion to the rotor speed and the number of poles.[57]

Alexanderson's proposed design called for a stationary armature winding on a laminated iron core situated between a pair of rotating steel disks which served as the inductor element. He intended for the disks to be shaped for maximum mechanical strength and to be made of nickel steel so as to withstand the stresses produced by rotational speeds of the order of 20,000 rpm. The inductors took the form of projecting poles or teeth around the rim of the disks, and a very small pole spacing, of the order of an eighth of an inch, was required to produce the design frequency, 100 kHz. Noting that such high speeds had only been reached in DeLaval steam turbines, Alexanderson proposed that the alternator be driven by a DeLaval turbine. He recommended that the alternator be enclosed in steel armor to provide greater safety while under test.[58]

The proposed design differed radically from that of the Steinmetz 10-kHz alternator, but it resembled in several respects the design of a 10-kHz alternator described in an AIEE paper presented in May 1904 by Benjamin G. Lamme.[59] Alexanderson had joined the AIEE in February 1904 but probably was not at the meeting at which the paper was presented. However, he may have had access to the published paper by the time he undertook the design of the 100-kHz alternator later in the year. Interestingly, Lamme's alternator was designed and constructed at the Westinghouse Company, which two years earlier had declined to undertake a 10-kHz alternator for Fessenden.[60] As in the case of the Steinmetz alternator, an outside patron had provided the stimulus for the Lamme high-frequency alternator. Maurice Leblanc, a French engineer, persuaded Lamme to undertake the experimental alternator, which Leblanc planned to use in research on a high-frequency system of telephony by wire. Leblanc spent some time at the Westinghouse works in Pittsburgh in 1902, perhaps on matters related to some of his patents which were licensed to Westinghouse.[61]

Lamme, who in 1896 had designed an inductor alternator for use at power frequencies, decided to adopt the inductor-type machine for the high-frequency output. He later explained that his choice had been motivated by mechanical rather than by electromagnetic considerations.[62] His design used a single rotating steel disk with 200 poles cut into the outer circumference and an armature core of finely laminated sheet steel. The disk inductor was 25 inches in diameter and it was driven at speeds up to 3,000 rpm. Lamme's AIEE paper included test data and dimensions. He concluded that the machine should be regarded as "a piece of laboratory apparatus" that had "no commercial value."[63] The Lamme alternator later was shipped to Leblanc in France. In 1921, Lamme wrote that he had since heard nothing of the alternator but that it still might exist.[64]

Although both the Lamme and Alexanderson designs employed a rotary inductor, there were differences in detail. Alexanderson proposed the use of two disks with the armature between them, while Lamme's machine had a single steel disk located between two concentric armature windings. Both designs stressed the need for very thin laminations in the armature core to minimize iron losses at high frequencies. Alexanderson planned to achieve the order-of-magnitude increase in frequency from 10 kHz to 100 kHz by using a slightly larger number of poles and an increased rotor speed. Surprisingly, the patent examiner's file on the high-frequency alternator patent issued to Alexanderson contains no mention of the Lamme alternator which Lamme apparently had not patented.[65]

Ernst Berg forwarded a copy of Alexanderson's proposed design of the dual-disk alternator to Fessenden for his evaluation in late December 1904. In his cover letter, Berg stated that GE was not willing to make any guarantees on such a machine but might undertake to build it if Fessenden approved the design and agreed to pay the costs. Fessenden replied that he could see no reason the machine should not work. He commented that he had tried earlier to persuade Steinmetz to build an inductor alternator but that Steinmetz did not like the type. However, Fessenden raised the question of whether an iron core should be used for the armature winding and suggested that he and the GE engineers discuss this. After receiving Alexanderson's comments, Berg wrote to Fessenden that the GE engineers still felt that construction of a 100-kHz alternator was questionable: they preferred to undertake a 50-kHz machine and to use iron as the armature core. Fessenden then urged them to try for an even higher frequency—150 kHz—using inductor poles shaped like sawteeth and eliminating iron from the armature. Alexanderson worked up a preliminary design of an armature without the iron core but again emphasized his personal preference for a laminated iron core. His patent application, filed a few days later, also stressed the advantage of using iron as the armature core to increase the induced voltage.[66]

The disagreement between Fessenden and Alexanderson over the use of iron in the high-frequency alternator deserves attention since it illustrates cultural differences that may exist within an engineering discipline as well as between science and engineering. Together with copper, iron was the *sine qua non* of electrical power engineering because it provided a minimum reluctance in the magnetic circuits of electric motors, generators, and transformers. Because impurities strongly affected the magnetic properties of iron and because of the metal's nonlinear behavior, machine designers had found it expedient to rely on empirical constants and graphs for the analysis and design of magnetic circuits that included iron. The designers of AC machinery frequently were skeptical about the value of analytical methods that did not use graphical techniques. Fessenden himself alluded to this skepticism in a paper on the use of magnetic formulas in design when he mentioned the "gibes" that design engineers had directed at those who employed "ironless mathematics."[67]

In light of Fessenden's background in physics and his practical experience in radio engineering, his conviction that iron should be avoided where possible at radio frequencies was quite reasonable given the state of knowledge at the time. Steinmetz and others had established in the early 1890s that the hysteresis and eddy-current losses of iron increased rapidly with an increase in the applied frequency. These losses could be tolerated and somewhat alleviated at power frequencies through the use of insulated laminations and through the selection of iron alloys that reduced magnetic hysteresis. Fessenden was well informed on the subject of magnetic science and evidently thought that his electrostatic doublet theory of atoms in metals could serve as a basis for a deeper scientific understanding. He had become convinced that an equation relating magnetic flux to the magnetic field intensity in ferric materials was the "expression of a physical law" and not "merely an empirical statement."[68] He undertook a series of careful experiments in an effort to provide convincing proof that the equation, formulated by his former colleague Arthur E. Kennelly, did express a "physical fact" and that it was "the touchstone" of modern magnetic circuit theory.[69]

Alexanderson and other members of the power engineering culture were not overly concerned about whether magnetic circuit equations were merely empirical or whether they were based on atomic or molecular theories so long as they served the purposes of engineering design. Alexanderson was familiar with the Steinmetz iron loss equation (for hysteresis and eddy current loss) but remained an agnostic regarding its implications for the use of iron in the high-frequency alternator. In the absence of empirical data, Alexanderson relied on his experience and instincts for the belief that the use of iron as the armature core would at least be superior to the use of wood, proposed by Fessenden. The lengthy dispute over what material should be used as the armature core was not so much a matter of personal taste as it was a reflection of the cultural divide that existed between power and wireless engineering in the early twentieth century.

By late February 1905, the GE engineers decided to attempt the construction of a 100-kHz alternator for Fessenden using Alexanderson's design as modified on the basis of Fessenden's critique. Alexanderson provided the descriptive data needed to arrive at a cost estimate of $1,200 for construction of the experimental alternator, not including the cost of a steam turbine to drive it. He stressed the need for very accurate machining of the two disks and a special tool to cut the polar projections around the circumference of the disks.[70]

The developmental process of the high-frequency alternator did not run smoothly during 1905, in part because of Fessenden's active role. He tended to favor more radical departures from conservative design principles than did the GE engineers. Frequently, he vacillated in his goals, and he consistently underestimated the time needed to convert new ideas into practical apparatus. This may in part have been due to his background, which was stronger in electrical science than in mechanical design, but it also reflected a difference in style. A man of many ideas and inventions, he seemed always to be impatient with a deliberate process of development. Since he was not overly concerned with costs, Alexanderson and GE attempted to satisfy his requests even when they seemed ill advised.

Just as developmental work began on the 100-kHz alternator, Fessenden submitted an order to GE for the construction of three or four 10-kHz alternators to be used in wireless telephony, and suggested that they be built before the construction of the higher-frequency machine was undertaken. He proposed the use of a 3-inch-diameter rotating armature mounted on a high-speed spindle and urged that the construction be completed quickly since NESCO hoped to have the alternators on the market by April 1905, or about two months after his order was placed. Alexanderson prepared a preliminary design of a 3-inch-armature alternator, and Berg forwarded it to Fessenden. The latter made no immediate response to the design proposal but instead asked whether GE could supply the armatures and fields for six 150-kHz alternators if NESCO supplied the driving apparatus. GE's response was that it was willing to attempt such a project but would make no performance guarantees and wanted the option of canceling the order after the first unit was completed. Early in April 1905, Berg sent Fessenden a copy of a design worked out by Alexanderson for a 150-kHz alternator. Fessenden replied that he found the design "very ingenious" but not practical. He suggested an alternative design, which he felt could be constructed with "very little difficulty," although he conceded that a "few minor changes in the mechanical arrangement" might be necessary.[71]

The developmental process became even more complicated when Fessenden learned that extremely high-speed rotation of machine tools had been achieved by Edward Rivett, a Boston toolmaker. Fessenden decided to subcontract the construction of the shaft and bearing system of the high-frequency alternators to Rivett's firm, the Faneuil Watch Tool Company, while obtaining the electrical elements from GE. Fessenden assumed that it would be a simple matter to mount a small armature on one of Rivett's high-speed spindles and generate substantial power. After observing a demonstration at Rivett's shop of a 3-inch brass disk driven at 60,000 rpm, Fessenden immediately decided to develop a 200-kHz alternator using mechanical components from Rivett's company and electrical components from GE. He placed a new order with GE for the electrical apparatus needed for six 10-kHz alternators and four 200-kHz alternators, stating that he planned to use the 10-kHz alternators in wireless telephony and the 200-kHz units in wireless telegraphy. He recommended that rotating armatures be used for all the units and foresaw "absolutely no difficulty whatever" in generating 1 kilowatt at 10 kHz with a 3-inch armature. He anticipated that Rivett would be able to achieve a speed of 120,000 rpm with a 3-inch phosphor-bronze disk in the 200-kHz alternators. Fessenden requested delivery by August 1, 1905, and recommended that GE send an engineer to visit Rivett's shop to determine what conditions would need to be satisfied. He received a cautious reply stating that GE could not promise a definite delivery date or price, and would prefer to build only one unit for each frequency before proceeding with the rest of the order.[72]

Alexanderson visited Rivett's shop in July 1905 and again in early August while continuing to work on the design of the various alternators ordered by Fessenden. When Berg began a trip of several weeks to Europe, Alexanderson began for the first time to correspond directly with Fessenden. On August 10, 1905, he notified Fessenden that a 10-kHz alternator had been completed but that he expected its output to be nearer to 100 watts than to the kilowatt predicted by Fessenden. Work was progressing on a 200-kHz alternator of the revolving armature type, Alexanderson reported, and he would obtain data on its no-load voltage and short-circuit current before completing the machine.[73] Fessenden continued to express optimism about the 3-inch rotor design but suggested that to reduce flux leakage the poles in the 10-kHz alternator be made twice as long as in Alexanderson's initial design. Alexanderson provided a numerical analysis showing why the shorter poles were superior and again voiced his belief that a great increase in rotor diameter would be required to increase the power appreciably.[74]

In mid-October 1905, Alexanderson reported the results of preliminary tests of the 200-kHz alternator. The tests had confirmed his doubts about the small-diameter rotating armature: the alternator had delivered only a few watts at low speeds and could not be run at the rated speed owing to problems with the armature windings. The tests had indicated that a maximum power of only about 15 watts would be produced even if the necessary speed could be reached. He recommended abandoning the 3-inch rotor and instead developing a machine with a larger rotor that did not use revolving wires. Fessenden declined to accept the recommendation, asserting that he felt "quite confident that success will be obtained if we continue along our present line." In mid-December 1905, he again wrote to inquire about progress on the small armature alternators. He commented that he believed that the GE engineers were "working too much around the shop" instead of following "the lines I laid out some six months ago."[75]

Having finally decided that the customer was not always right, Alexanderson was no longer willing to continue along the lines laid down by Fessenden. The December 1905 letter from Fessenden elicited a confident rejoinder from Alexanderson stating that he had concluded that the small, revolving armature was "entirely unfeasible" from both a mechanical and an electrical point of view. He listed the numerous faults of the design and stated that he did not "consider it worth while to spend any more time on this machine." Moreover, he already was working on a new design that would not employ moving conductors and would be of sufficient dimensions to produce the power level needed in Fessenden's work.[76]

Alexanderson now resumed development of a dual-disk inductor alternator similar to that proposed in his original design a year earlier. He adopted a rotor diameter of 12 inches and a nominal frequency of 100 kHz. He made one concession to Fessenden by adopting a wooden core instead of a laminated iron one for the armature winding. Development of the new alternator proceeded at an accelerated pace once the small, rotating armature was abandoned. In mid-January 1906, Alexanderson informed Fessenden that the alternator was nearing completion and that he expected to have preliminary test data within two weeks.[77]

Conscious of the potential dangers of testing a new high-speed machine, Alexanderson arranged to have it placed in a pit surrounded by sandbags in case the rotor should fly apart. He advised that during the tests, no one be permitted to stand along the line where broken pieces might be thrown. Some years later, a newspaper story mentioned that participants in the early tests remembered how they had clenched their teeth and glanced at the floor as the alternator rotor turned faster and faster. An engineer who operated the alternator after it was acquired by Fessenden recalled that anyone who adjusted the length of the air gap was apt to have to extract splinters from the armature core from his face; he and his coworkers had expected to "be cut lengthwise" if the steel rotor were thrown out of the alternator.[78]

While developmental work continued, Alexanderson undertook a theoretical study of the mechanical properties of rotating disks of various shapes. His results were incorporated in an unpublished report entitled "Strength of Revolving Discs,"[79] in which he applied the general theory of stresses and analyzed a variety of forms, including a solid free-running disk, a disk with a hole in the center, a disk with a hub, and a disk with a continuous rim. He considered the problem of how a disk should be tapered for maximum strength, and derived an expression for the optimum thickness as a function of the radius, mass, peripheral velocity, and material strength. One conclusion Alexanderson drew from the investigation was that prestressing a disk caused a considerable increase in the speed it could withstand without further permanent deformation. His explanation for this was that the process of prestressing caused a molecular rearrangement such that the stresses became more evenly distributed instead of being concentrated at a hole. He proposed that the prestress method be used in the manufacture of turbine disks, but he was diverted by a trip to Europe in the summer of 1906 and evidently did not pursue the matter after his return.[80]

Another interesting problem that Alexanderson encountered with the alternator rotor was that of mechanical resonance at certain "critical speeds," which could cause severe and potentially destructive vibrations. By March 1906, he found a practical solution to this problem. He mounted the rotor on a flexible and tapered shaft with end bearings and an auxiliary set of "middle bearings" located near the steel disks. The middle bearings were made slightly larger than the shaft, so that they functioned only when vibrations occurred in the vicinity of critical speed.[81] This bearing arrangement proved to have the additional and unanticipated virtue of tending to correct automatically the gap between the rotor and the armature. The automatic correction resulted from a differential thermal expansion of the shaft. Alexanderson's analysis of the problem of critical speeds showed that a first resonance occurred when the shaft vibrated like a loaded string, while a second resonance was due to vibration of the rotor around an axis perpendicular to the shaft. The two critical speeds could be fixed at predetermined values during the design.[82]

Another issue that surfaced during the developmental process was what method would be used to drive the alternator rotor at speeds of up to 20,000 rpm. Fessenden decided that he preferred not to use direct drive by a steam turbine, since this would require operating a steam plant at each transmitter site. As an alternative, Alexanderson proposed that an electric motor be used in combination with a DeLaval speed-reduction gear that was operated in reverse so that the alternator rotor would turn at a higher speed than the motor.[83] However, he used a belt-drive system during preliminary tests until the gear system could be obtained. Despite some problems with belt slippage and heating, he achieved a frequency of about 50 kHz by late March 1906. Fessenden was excited by this news and urged completion of the developmental work on the new alternator as quickly as possible. He predicted that the alternator would overcome wireless communication's existing difficulties and enable it to become "entirely commercial."[84] Alexanderson attributed delays in the project to crowded conditions in the shop and to the fact that modifications often required the use of special tools and were held up until the tools became available.[85]

In June 1906, Alexanderson notified Fessenden that 50 kHz seemed to be the best that could be achieved with the belt drive and asked whether he wanted the alternator in its present condition pending arrival of the gearing. Alexanderson suggested that GE could build a second alternator for the gear tests and might undertake to build a still larger machine with a 3-foot rotor if Fessenden were interested. Fessenden immediately placed an order for two additional alternators with 12-inch rotors, two alternators with 3-foot rotors, and several spare armatures. Alexanderson discussed the new order and outlined the general features of the proposed 3-foot-rotor machine in a memorandum to Henry Reist, head of the AC Engineering Department, stating that the larger alternators were to be of the same type as the dual-disk alternator already constructed but would be run at a lower speed, about 6,700 rpm. He noted that a different manufacturing method might be required but recommended that any necessary changes be made in the shop to avoid delays due to "waiting for further engineering advice."

Since some changes might be made as the work proceeded, he suggested that final drawings not be made until construction was completed.[86] The important but frequently anonymous contributions by shop personnel during the development process seem manifest in this instance.

Alexanderson departed for an extended trip to Europe in July 1906, leaving Henry Reist in charge of the high-frequency alternator project. In late August, GE shipped the original dual-disk alternator to Fessenden's wireless station at Brant Rock, Massachusetts. Alex Dempster, who had worked with Alexanderson during construction of the alternator, was assigned to assist in installing the machine and in instructing the NESCO engineers how to run it. By mid-September 1906, Fessenden notified GE that a woven cotton belt had been used to operate the alternator up to a frequency of 76 kHz without trouble. The power delivered was lower than had been anticipated, however; he suggested that this might be caused by the armature wire being too large. The alternator had produced less than 50 watts at or above 30 kHz when it should have delivered about 250 watts at 50 kHz. This anomaly did not seem to worry Fessenden, for he concluded that the problem of generating wireless waves mechanically had been solved. He wrote that he had been surprised at how simple it had been to operate and adjust the machine, and that all that remained was to standardize the parts. He congratulated the GE engineers on having brought the project to "such a successful issue."[87]

Fessenden's continuing experiments with the alternator at his Brant Rock station reinforced his conviction that a continuous-wave system would prove far superior to the intermittent-spark system. He concentrated his attention on wireless telephony rather than wireless telegraphy since the spark system was least suited to the former area. However, in the fall of 1906 his enthusiastic expectations of immediate commercial success with the alternator system were overly sanguine. In September, he wrote to a representative in Europe that NESCO planned to abandon the spark transmitters in favor of alternators. Fessenden asserted that wireless telephony was now a complete success and would soon supersede current methods on ships and supplant long-distance telephone wires. He predicted that there would be no difficulty in telephoning to Europe by wireless using the larger alternators that were being developed.[88] At the same time, he notified GE that he anticipated many orders for alternators and asked for a delivery date for 50 alternators of the 12-inch-rotor design. He mentioned that his experiments at Brant Rock had achieved wireless telephony at distances of more than 3 miles using a 60-foot mast and an alternator power of 30 watts.[89]

Both the NESCO and the GE engineers continued to be perplexed by their inability to obtain greater output power from the alternator at the higher frequencies. In early October 1906, Reist notified Fessenden that experiments were planned using the second alternator being constructed at GE in the hope of isolating the problem. Fessenden suggested three possible causes of the sagging of the voltage curve at high speed: a loose key ring, which might allow the inductor teeth on one side to be opposite the slots of the other disk, eddy currents in the armature windings, and a slight springing of the inductor teeth due to centrifugal force. He expressed confidence that any of these causes could be corrected easily. The problem proved more difficult than he anticipated, however, and in mid-November 1906 he sent Reist a voltage-speed curve that showed a pronounced droop above 50 kHz. Fessenden admitted that he and his assistants had been unable to determine the cause but mentioned that an assistant thought it might be due to the two disks being separated further at higher speeds.[90] This problem with the dual-disk alternator never was overcome satisfactorily and led Alexanderson to change to a single-disk alternator the following year.

Despite the unresolved problem with the dual-disk machine, Fessenden tried to discourage further experiments or modifications by the GE engineers. He wrote to Reist that he had heard that design changes had been proposed but that he thought none of them should be made. Fessenden argued that the present design was entirely satisfactory except that the key should be replaced by pins to keep the two disks in alignment. Two new armatures received from GE could not be used because of unnecessary changes, he complained. After a puzzled response from Reist, who assured him that GE was building identical armatures, Fessenden adopted a more conciliatory stance. He speculated that the changes probably had been made in the shop without consultation with the AC Engineering Department. In early December 1906, he informed Reist that the alternator-generated signal had spanned a distance of 50 miles at 50 kHz.[91]

The encouraging performance of the high-frequency alternator as a source of wireless waves led Fessenden to schedule a public demonstration of his continuous-wave system. He invited the press and the American Telephone and Telegraph Company (AT&T), viewed as a potential customer, to send representatives to witness wireless telephone tests between Brant Rock and Plymouth, a distance of about 10 miles. In the invitation, he stated that transmission of both speech and music would be demonstrated. He also mentioned that telephone messages brought by wire to Brant Rock would be sent by wireless to Plymouth and retransmitted from there by wire line. He anticipated that this part of the demonstration would be of particular interest to the telephone company.[92] Greenleaf W. Pickard represented AT&T at the NESCO demonstration held December 21, 1906; Elihu Thomson of GE and two reporters from the Associated Press also were present.

In his report on the wireless telephony demonstration, Pickard termed the dual-disk alternator a "most remarkable piece of apparatus" which had disproved the conventional belief that a substantial power output from a high-frequency alternator was impossible. He assessed the quality of speech received by wireless at Plymouth as "distinctly commercial," and he suggested that a range of several hundred miles might be obtained using a more powerful alternator.[93] Pickard's evaluation of the Fessenden system was interesting in light of a series of experiments Pickard had conducted for AT&T four years earlier. At that time he had submitted a rather pessimistic report that pointed out that many elements needed for a practical system still were lacking.[94]

Pickard's favorable report stimulated a renewed interest in high-frequency alternators within AT&T, and it was this interest that brought Alexanderson back into the development of high-frequency alternators in 1907. After reading Pickard's report, Hammond V. Hayes, chief engineer of AT&T, contacted Alexanderson in March 1907 to ask if he would design a special high-frequency alternator for the telephone company.[95] If this inquiry had not rekindled Alexanderson's interest in high-frequency technology, it is questionable whether his contribution to wireless engineering would have been long remembered. Even before Fessenden received the dual-disk alternator from GE, much of Alexanderson's time was taken up by problems related to railroad electrification.

Railway Traction Engineering: The Battle of the Systems

Alexanderson's invention of and developmental work on the self-exciting alternator and a variable-speed AC motor revealed his abilities as a designer of power machinery. As a result, in April 1906 he was transferred from GE's AC Engineering Department to the Railway Engineering Department. He thus became an active participant in an effort mounted by GE to overcome the Westinghouse Company's lead in the application of AC traction to interurban railroads.

The movement to replace steam locomotives with electric locomotives on American interurban railroads reached a high-water mark during the first decade of the twentieth century. The failure of the movement to become the multibillion-dollar enterprise forecast by some of the proponents of electric traction has been attributed to several factors, including competition from motor vehicles using the internal combustion engine; the railroad industry's low labor productivity as compared to manufacturing industries; and excessive government regulation of rates, which discouraged the necessary capital investment.[96] One other factor cited by insiders is important in assessing Alexanderson's role as an engineer-inventor of electric railway motors. This was the failure of the engineers involved in the design and manufacture of apparatus for electric traction to arrive at a consensus on a standardized system.[97]

The lack of agreement among electric-traction engineers was demonstrated by what amounted to a battle of the systems between DC and AC traction systems, which still remained unresolved as late as the 1920s.[98] The advocates of DC traction proved remarkably resourceful in devising technical improvements that maintained parity, if not outright superiority. In the United States, the battle was mainly waged between an evolving DC traction system and a single-phase AC traction system. Polyphase traction systems were being tried in Europe, but they were not seen as a feasible alternative in the United States at the time.[99]

The battle of the traction systems in the United States was not confined to a competition between the dominant electrical manufacturers, GE and Westinghouse, although it sometimes was subject to that interpretation.[100] GE did win the contract for the electrification of the New York Central Railroad, a project that served as a paradigm for DC interurbans, while Westinghouse engineers were responsible for such major AC traction projects as the New Haven Railroad.[101] However, a group of AC enthusiasts at GE, including Steinmetz and Alexanderson, undertook to create a competitive single-phase AC system for interurban railroads. Their effort led to a more subtle but very illuminating battle within the AC-traction fraternity, with a division along corporate lines. A spirited competition developed between the designers of a single-phase, series-wound AC traction motor developed at Westinghouse and the designers of a single-phase repulsion motor favored at GE. A modified repulsion motor designed by Alexanderson eventually became GE's principal entry in the competition.

The debate over the relative merits of the series- and repulsion-type AC motors for railroad traction antedated Alexanderson's active involvement. In 1901 Benjamin Lamme of Westinghouse designed a single-phase, series-wound motor suitable for railway use, and a contract was obtained to use the motor in the electrification of the Washington, Baltimore, and Annapolis Railroad (WB&A). Lamme discussed the new motor in an AIEE paper presented in 1902 before the WB&A system became operational. William S. Murray, a leading proponent of AC traction, later characterized Lamme's paper as the "entering wedge of the single-phase motor" for train propulsion.[102] In his paper, Lamme explained that his AC motor was similar in design to the standard DC traction motor except that it had a laminated field core. Among the several advantages of AC for traction, he mentioned the absence of rheostat losses, the elimination of electrolysis problems, and the better speed control achieved by the use of an induction regulator.[103]

As was frequently the case at AIEE meetings during the early twentieth century, the discussion took the form of a Westinghouse-GE debate. Steinmetz commented on Lamme's paper and expressed disappointment over Lamme's selection of a series-type motor. Steinmetz recalled his own failure to overcome such serious defects of these motors as severe sparking at the commutator and low power factor at high speeds. He predicted that some form of the repulsion motor invented by Elihu Thomson would prove more suitable for this application.[104] Lamme responded that he had rejected the repulsion motor because his calculations had indicated that it would be larger and heavier than the series motor for the same output power. He declined to give details on the construction of his series motor, for what he termed "good reasons."[105]

At professional meetings, the GE engineers continued to advocate the repulsion motor for railroad traction. Steinmetz presented a paper on the theory of the repulsion motor at an AIEE meeting held in 1904. He pointed out that it had a better power factor at low speed than did the series AC motor and also returned power to the line when operated in reverse.[106] At the same meeting, Walter I. Slichter presented a paper that gave the results of tests of a 175-horsepower (hp) repulsion motor constructed by GE. The motor was designed to operate at 1,500 volts, and Slichter stressed that the high-voltage capability gave it an advantage over the series motor, which could only be run at comparatively low voltage without excessive commutator sparking. During the discussion, Lamme again defended the series motor as being more compact, although he admitted that a step-down transformer was necessary between the high-voltage line and the series motor.[107] The low voltage limitation of the series motor was attributed to the voltage induced in the armature winding when it was short-circuited by the brushes during commutation.[108]

Steinmetz gave another paper on the theory of single-phase motors at the International Electrical Congress held in Saint Louis, Missouri, in 1904. In addition to the series and repulsion motors, he discussed the compensated, or Eickemeyer, motor, which tended to behave like a repulsion motor at low speed and like a series motor at high speed.[109] During the Saint Louis meeting, Steinmetz also presented an abstract of a paper on the compensated repulsion motor contributed by Ernst Danielson, the Swedish engineer whose influence on Alexanderson already has been mentioned. Danielson described the compensated repulsion motor as the most modern motor design for AC traction.[110]

An important encounter in the battle of the traction systems followed the decision in 1905 to undertake the electrification of the New York, New Haven, and Hartford Railroad. William S. Murray, a former Westinghouse employee, was appointed electrical engineer of the New Haven line in April 1905, and his recommendation of a single-phase propulsion system was adopted. Murray became perhaps the most articulate and persistent defender of AC traction and later the leading advocate of the "superpower" movement.[111] He characterized the New Haven's departure from the DC traction system used by the New York Central as being the "first gun in the battle of the systems."[112] Westinghouse received the contract for the first phase of the New Haven electrification, which called for the locomotives to be powered by large, series-type AC motors. The first section began commercial service in the summer of 1907 although the entire project was not completed until 1917.[113]

The immediate goal of the GE engineering team that Alexanderson joined in April 1906 was to design and develop a single-phase traction system that would be competitive with the New Haven system of the Westinghouse Company. Alexanderson's principal task was to design a single-phase motor with performance characteristics as good as or better than those of the 250-hp series motor designed by Lamme for installation on the New Haven locomotives. Pending the outcome of the battle of the systems, GE needed to be prepared to bid on later New Haven contracts or on other AC traction projects that might be undertaken.

Alexanderson's early efforts to acquire information about existing practice in AC traction are an illuminating example of technology transfer, a process that is one of the major themes in the history of technology.[114] The fact that GE employed many engineers who, like Alexanderson, had professional contacts in one or more European countries and fluency in foreign languages frequently proved valuable to the company in that it facilitated the transfer of technology across national boundaries. Alexanderson began his work on the traction project by studying an in-house memorandum entitled "Present Development of Alternating-Current Railway Motors."[115] By early June 1906, he had worked out a preliminary design of a 250-hp AC motor suitable for traction service. Shortly afterward, he embarked on a two-month trip to Europe, having promised Edwin Rice that he would collect information on the progress of electric railroads in Germany and Sweden.[116]

During his European trip, Alexanderson visited his old acquaintance

Ernst Danielson and learned the results of ongoing electric traction experiments in Sweden. The Swedish tests established the feasibility of using up to 18,000 volts on overhead trolleys. They also demonstrated that both the series and compensated AC motors gave satisfactory performance in electric locomotives.[117] While in Europe, Alexanderson also studied the traction motor designed by Marius Latour, a French engineer retained as a consultant by GE. As a result of the information brought back by Alexanderson, it was deemed unnecessary to bring Latour to Schenectady to consult on the traction project. Soon after Alexanderson's return to the United States, he reported directly to Rice on what he had learned in Europe.[118]

In November 1906, Alexanderson submitted a progress report to William Potter, managing engineer of GE's Railway Engineering Department, stating that he had completed an analysis of the repulsion motor intended for use as a traction motor. Alexanderson had concluded that the motor should be capable of approximately the same starting torque and efficiency as the series motor developed for the New Haven Railroad. He mentioned that he and Steinmetz were in agreement that a repulsion-type motor should be very successful. Alexanderson recommended that GE construct four repulsion motors and one series motor so that the engineers could gain experience with them and do comparative tests.[119]

Both Potter and Alexanderson spoke on the topic of electrification of steam railroads at a meeting of the Schenectady branch of the AIEE early in 1907. Potter spoke on some of the essential features that should be considered, including the issue of whether AC or DC should be used. Alexanderson made a presentation on the AC traction motor.[120] He made a trip to the Midwest to investigate the equipment being used on several interurban electric railroads. His report to Potter mentioned the Bloomington, Pontiac, and Joliet; the Toledo and Chicago; and the Warren and Jamestown railroads, all of which had installed AC propulsion systems.[121] Alexanderson's information gathering also included the acquisition of a new book on AC motors by A. S. McAllister and an issue of *Scientific American* containing an article on the opening of electric traction service on the New York Central and the New Haven Railroads.[122]

The traction motor project generated a substantial number of patent applications by Alexanderson. These covered design features of the series-repulsion motor such as winding arrangements, starting methods, speed control, and ventilation. Of his 46 patent applications filed during the years 1906–10, 39 were related to electric traction, with the majority being on commutator AC motors of the series-repulsion type.

The GE engineers subjected the new traction motor to comprehensive testing during May 1907. Rice, Steinmetz, and Potter were among the interested observers as the operating mode of the Alexanderson motor was adjusted in stepwise fashion: it acted as a repulsion motor when starting but acted as a series motor at high speed. A report on the tests prepared for GE's Patent Department stated that the desired "black" (sparkless) commutation had been obtained at full load and over a wide range of speeds.[123] The almost sparkless commutation led Alexanderson to suggest that the preventive or resistance leads required with the Westinghouse series-wound traction motor might not be needed at all with the series-repulsion motor.[124]

Steinmetz participated actively in the AC traction project at GE and undertook a detailed theoretical analysis of the commutation of single-phase traction motors. He concluded that the Alexanderson series-repulsion motor was capable of commutation far superior to that of the series motor over its full range of speeds. Steinmetz mentioned the new traction motor designed by Alexanderson during a discussion at an AIEE meeting in 1907. He noted that the principal difficulty with the AC traction motor had been commutation, but that Alexanderson now had solved the problem and soon would present a paper explaining how he had done it.[125]

Frank Sprague, the patriarch of urban electric traction in the United States, gave his personal views on the battle of the interurban traction systems during an AIEE presentation in 1907. He noted the "apparent division of the engineering world into two camps, alternating and direct current." Comparing the DC system of the New York Central and the AC system of the New Haven, Sprague praised the New York Central electrification as "epoch-making in forwarding the march of electrification" with a locomotive that was the "acme of simplicity never hitherto attained." He cited improvements in the design of DC motors during the past two years which would permit the use of much higher voltages. Sprague argued that the AC locomotive was at a considerable disadvantage because its total weight was unavoidably at least twice that of a DC locomotive with equivalent power. He suggested that the best situation for AC traction would be on long lines with low traffic density.[126] The discussants, including Steinmetz and Potter from GE and Lewis B. Stillwell from Westinghouse, questioned Sprague's conclusions and contended that it was still too early to decide on the better method or to standardize electric traction motors.[127] An economic comparison of the competing systems published the same year tended to support Sprague's position. It indicated that although construction costs of the DC and AC systems were similar, the operating costs of a 1,200-volt DC traction system were substantially lower than those of a comparable AC system.[128]

As Steinmetz had promised, at an AIEE meeting in January 1908 Alexanderson presented a paper on the GE series-repulsion traction motor. He reviewed the recent history of single-phase railway motors and noted that they were all one of two general types, series or repulsion. The motor he had designed was neither a series nor a repulsion motor, he explained, but rather a synthesis that exploited the best features of each. He stressed its great improvement in commutation by comparison with the series motor and also pointed to his motor's high starting torque. To demonstrate the theoretical possibility of achieving spark-free commutation by combining certain features of the series and repulsion motors, Alexanderson included in his paper a discussion of the general theory of commutation in repulsion motors. He identified four magnetic field components that influenced commutation and described how the relative phase and amplitude of the components could be manipulated in the design of a motor to satisfy the conditions necessary for almost perfect commutation. He explained that he had achieved these conditions through the use of a fractional-pitch winding on the armature, a proper ratio between the rotor and stator winding to compensate for leakage flux, and the use of a portion of the field winding as a reactance coil in series with the armature winding. He listed other advantages over the series motor, including operation at higher frequencies, which would enable the weight of motors and transformers to be reduced, and an improved heat characteristic due to the absence of high-resistance spark-preventive leads.[129]

Alexanderson's paper stimulated a long discussion that took up more pages in the AIEE *Transactions* than did the paper itself. The discussion displayed the usual GE-Westinghouse competition and transcended the still-undecided issue of DC versus AC traction systems. Lewis Stillwell of Westinghouse welcomed the announcement of a new traction motor that promised to improve performance and/or reduce cost. He termed the elimination of resistance leads a feature of "marked originality and of much practical value" but suspected that a price had been paid to achieve this gain. He observed that the central issue was whether the series-repulsion motor would prove superior to its competitors in actual service and that this would depend on more factors than just improved commutation. Benjamin Lamme, designer of the Westinghouse series traction motor for the New Haven Railroad, expressed doubt that the new GE motor could accomplish anything that could not be done by a well-designed series motor. He defended the New Haven type of traction motor and questioned whether the motor designed by Alexanderson "changes the broad problem in any way."[130]

The GE engineers gave an equally spirited defense of the series-repulsion traction motor. William Potter stated that the essence of the motor was its sparkless commutation, which in turn made possible other design improvements, resulting in increased reliability, reduced maintenance, and greater load capacity. He reported that a recent test by GE of a series-repulsion motor of the same physical dimensions as the Westinghouse motor for the New Haven line had delivered a 50 percent higher starting torque and good commutation to a speed of 75 miles per hour (mph). Steinmetz commented that the two major difficulties with early single-phase motors had been the low power factor and "hopelessly bad commutation." He credited his former employer, Rudolf Eickemeyer, with having solved the power factor problem by means of a compensating winding, and asserted that Alexanderson now had solved the second problem, poor commutation. Steinmetz reviewed the various methods—such as adding poles, using high-resistance leads, and using high-inductance leads—that had been tried in hopes of reducing commutator sparking. He credited Alexanderson with having taken a more logical approach to overcoming the root cause through a careful investigation of the magnetic field distributions. Steinmetz concluded by asserting that the Alexanderson series-repulsion motor marked the end of "the period of youth" in the development of AC railway motors.[131]

Despite the GE engineers' claims regarding its superiority over the series motor, the series-repulsion motor failed to stop a trend away from AC traction during the next several years. William Potter noted the trend in favor of DC traction in a paper published in 1910 comparing the economics of AC and DC systems. He reported that the AC system was proving to be inferior in first cost and in maintenance costs and had been "practically abandoned" on new railway electrification projects, while the advantages of the 1,200-volt DC system were such that he would not be surprised if AC traction were completely discarded in the United States. Potter concluded that railroad electrification was "really a problem in economic engineering and not simply a technical problem."[132] His pessimism about the future of AC traction was confirmed by events, as many interurban electric railroads, including the Washington, Baltimore and Annapolis, changed from AC to DC traction systems.[133] In 1915, Potter and an assistant published a review article on electric railroads which included further evidence that the battle of the traction systems was being won by DC. They reported that the single-track mileage of DC traction in the United States had surpassed the AC mileage by 1912 and was more than twice the AC mileage by 1914.[134]

Although AC traction obviously failed to live up to the sanguine expectations of its proponents, it did have some advantages, which forestalled a total victory for DC in the battle between systems. The

Alexanderson series-repulsion motor finally got an opportunity to demonstrate its capabilities in direct competition with the Westinghouse series motor. In 1913, Alexanderson wrote that a GE locomotive powered by series-repulsion motors had been operating on the New Canaan branch of the New Haven Railroad for nearly two years and had outperformed two of the Westinghouse locomotives. He reported that the commutator brush life on the GE motors had been about 25,000 hours, as compared to an average of 6,000 hours for the Westinghouse series-type motors, and mentioned that a more powerful AC locomotive using four 400-hp series-repulsion motors was being tested by GE.[135] In April 1915, Alexanderson informed Steinmetz that the GE locomotive had so impressed the New Haven officials that GE had been asked to bid on a contract for several units of similar design. Alexanderson also remarked that some leading European manufacturers had adopted the series-repulsion traction motor and that he now was convinced that successful AC locomotives would be made.[136]

Alexanderson found further vindication in several reports published during 1915 that Westinghouse had decided to use the series-repulsion motor in the electrification of the Philadelphia-Paoli line. He wrote Steinmetz that the Westinghouse motor appeared to be a virtual copy of the ones installed on the GE locomotive used on the New Haven Railroad. The series-repulsion motor now was being called a great step forward, Alexanderson noted, and it also had been adopted by Germany's Allgemeine Elektrizitäts-Gesellschaft (AEG). He mentioned that he had been told that the new Westinghouse motor probably infringed his patents on the series-repulsion motor, and concluded by suggesting that Steinmetz might find these developments of interest in view of Steinmetz's forecast, 15 years earlier, that the repulsion motor would provide the final solution of the single-phase traction problem.[137]

When an article on the electrification of the Philadelphia-Paoli line appeared in the *Electric Railway Journal,* Alexanderson promptly wrote to the editor that the motors described were not new but were similar to those in two GE locomotives used by the New Haven. He asserted that operating experience with the two GE locomotives had fully confirmed his expectations regarding the virtues of the series-repulsion motor, which he had expressed in an AIEE paper presented in 1908. An editorial in the journal had stressed the advantage of eliminating the high-resistance leads that had been an undesirable feature of the series AC motor. The editor noted that this enabled the series-repulsion motor to be designed with an armature similar to that of the proven DC traction motor.[138]

The article that had attracted Alexanderson's attention described the series-repulsion motor and the Philadelphia-Paoli system in considerable detail.[139] The Philadelphia-Paoli line was a 20-mile-long route that was part of the Pennsylvania Railroad system. George Gibbs, a consulting engineer, had recommended the electrification of this line in 1912, and electric operation began in September 1915. The decision to use AC rather than DC on the Philadelphia-Paoli apparently was based on the expectation that electrification of the Pennsylvania Railroad might later be extended as far west as Pittsburgh. For longer lines with lower traffic density than existed near New York City, AC traction offered economic advantages that were convincing to Gibbs and the executives of the Pennsylvania Railroads. The Philadelphia-Paoli electrification also was regarded as a milestone in that, for the first time, the Pennsylvania Railroad decided to purchase electric energy from a private utility instead of generating its own power.[140]

The Shunt-Repulsion Motor

During the process of developing the series-repulsion motor for electric traction, Alexanderson determined that the shunt-wound repulsion motor also might possess useful properties. The dialectic between opposites in the inventive mind again became manifest as he undertook an investigation of the use of the shunt-repulsion motor as a variable-speed motor in industrial applications where the shunt-wound DC motor had been dominant. In June 1907 he contacted Edwin Rice about this idea, and Rice encouraged him to determine the practical value of the variable-speed repulsion motor as quickly as possible.[141] Alexanderson also notified Sidney B. Paine, a leading GE sales engineer, of the possibilities of the shunt-repulsion motor. Paine had already achieved a reputation for his important role in the introduction of GE's AC motors into the textile industry beginning in the mid-1890s.[142] However, the AC motors used in textile factories were designed to run at constant speed, and the DC shunt-wound motor had remained the only available choice in applications requiring an adjustable speed.

In an AIEE paper in 1909, Alexanderson explained the theory of the variable-speed repulsion motor and gave the results of tests on an experimental motor. He described two methods he had employed to vary the speed, a speed directly proportional to the supply frequency and inversely proportional to the number of poles in the motor.[143] In the first method, he explained, the normal equilibrium of voltages induced in the armature could be upset by impressing an adjustable voltage across the brushes connected to the armature. He showed that this applied voltage resulted in a new speed, which could be higher or lower than the synchronous speed depending on the polarity of the applied voltage. He reported that he had obtained a speed range of up to three to one by means of this method of "armature control," using the voltage

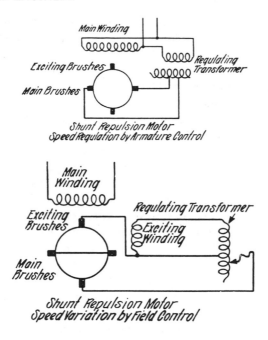

Figure 2.2. Circuit diagrams of the two methods that Alexanderson used to vary the speed of shunt repulsion motors. From Ernst F. W. Alexanderson, "Repulsion Motor with Variable-Speed, Shunt Characteristics," *Transactions of the American Institute of Electrical Engineers,* 28 (1909): 515.

from a speed-regulating transformer applied to the brushes. For example, a small 3-hp motor with a synchronous speed of 1,800 rpm had been operated at speeds ranging from 900 to 2,700 rpm.[144]

Alexanderson called the second speed-control method "field control." In this method, the speed variation was obtained by applying an adjustable voltage across a second set of brushes, called the "exciting brushes," with the main brushes being short-circuited. He stated that field control permitted a smaller speed range, of the order of 1.5 to 1, but that this method had been found especially useful for cases where the torque decreased with motor speed. During the discussion of his paper, it was brought out that his speed-control methods could be applied readily to constant-speed repulsion motors already on the market if the appropriate auxiliary apparatus were available.[145] Shortly after his paper was presented, GE announced that it was prepared to sell adjustable-speed repulsion motors in several sizes.[146]

The Westinghouse-versus-GE debates during the early twentieth century over the relative merits of certain types of AC motors suggest both the existence and the long persistence of machine-design traditions or technological styles. Alexanderson's preoccupation with repulsion motors over an extended period placed him within a corporate tradition dating back to Elihu Thomson's invention of a repulsion motor: a tradition continued by Steinmetz. This may be contrasted to the long dominance of the series-type AC motor at Westinghouse under Lamme's leadership. Within an engineering design tradition, there may exist qualitative differences in individuals' styles. For example, Alexanderson's approach to design seems to have differed considerably from Steinmetz's. In Alexanderson's case, there is less evidence of a reliance on mathematical analysis or the use of ideal models. Alexanderson alluded to his own design style during a discussion at an AIEE meeting in 1912, when he commented that if the designer had in his mind a clear physical understanding of the relationships among key parameters, he could anticipate the effect of changes in the parameters without necessarily using formulas or curves.[147]

Despite his disappointment in what seemed to be the declining fortunes of AC traction after around 1910, Alexanderson retained a strong interest in traction systems and had later opportunities to contribute to the field. In a reflective speech given near the end of his career, he mentioned the acute discouragement he had felt when AC traction had temporarily been almost eclipsed by DC, and expressed his belief that the economic analysis that had led GE's management to discontinue further development of AC traction had been shortsighted. However, he speculated, the decline of work on AC traction might have benefited him in the long run by turning his attention to other challenges. He noted that his early experience in traction engineering had helped convince him that the best hedge against the frustrations that seemed to overtake many inventors was to have many interests.[148] One of his interests was the high-frequency alternator, which again began to occupy a portion of his time during 1907. The fortunes of the Alexanderson radio alternator began to ascend even as those of his traction motor became problematic.

Chapter Three

High-Frequency Alternators
and Wireless Politics

When he resumed work on high-frequency alternators in 1907, Alexanderson had not yet acquired the broad perspective of a communication system designer. The project that had culminated in the dual-disk alternator used in Fessenden's December 1906 Brant Rock–Plymouth wireless telephony demonstration had been something of a diversion from Alexanderson's work on AC power and electric traction. General Electric still showed little interest in entering the field of wireless engineering beyond the level that the National Electric Signaling Company was willing to support. One important consequence of the continuing interaction with Fessenden was that it enabled Alexanderson to learn both the technical characteristics of the rapidly evolving wireless systems and the active role of the federal government and military services as patron, consumer, and regulator of the new communication technology. The lessons learned during the process of completing the development of reliable, commercial-quality, high-frequency alternators prepared Alexanderson for the opportunity that came in 1910 to join Steinmetz's new Consulting Engineering Department, which encouraged a broader technological system perspective among its members.

The Telephone Alternator Project

A continuing search by AT&T engineers for a suitable device that would amplify telephone signals carried by wire led them to a consideration of high-frequency alternators and to Alexanderson's resumption of work on an appropriate alternator. George Pickard's report to AT&T on NESCO's December 1906 wireless telephony demonstration stimulated a renewed interest at AT&T in the high-frequency alternator as a telephone amplifier. Actually, the telephone company had showed some interest in using an alternator as a repeater or amplifier many years before. John S. Stone, at the time an engineer employed by American Bell Telephone in Boston, proposed using a high-frequency alternator to amplify a telephonic current applied to the alternator field winding during a conference in June 1891. He anticipated that an amplified replica of the telephone signal might be obtained from the alternator armature, and in May 1892, he reported an experiment that he claimed had "proved the correctness of my premises." He did not pursue the investigation but did recommend that a machine be constructed "capable of giving very high frequencies."[1]

George Pickard, who had joined the engineering staff of AT&T in July 1902, proposed a variant of Stone's alternator repeater. Pickard suggested that amplification of telephone signals might be achieved by "unsymmetrically varying the permeability of the iron core" to "modulate the current induced in a coil wound on this core." His proposed apparatus included four coils, with two coils connected in opposition to an AC generator, the third coil connected to a rectifying commutator on the alternator shaft, and the fourth coil connected to the telephone signal to be amplified. He anticipated that the input signal would alter the magnetic permeability of the iron core and result in an amplified signal in the alternator output.[2] Hammond Hayes, who was then head of American Bell's Boston engineering staff, authorized an investigation of Pickard's telephone amplifier, but the results evidently were not encouraging.[3] At least the work by Stone and Pickard had set the stage for further investigations once improved high-frequency alternators became available.

The perceived need for a practical telephone amplifier became increasingly acute during the first decade of the twentieth century as AT&T extended its lines over greater distances. The severity of the need was alleviated somewhat by the loading-coil innovation that approximately doubled the distance of satisfactory communication by wire. However, there existed what amounted to a "presumptive anomaly" in the use of loaded lines, since telephony across the continent from New York to California did not seem possible by wire even with loading coils unless amplifiers were available.[4] The somewhat leisurely search for an amplifier changed to a crash program after AT&T promised in 1909 that it would inaugurate transcontinental telephony in time for the Panama-Pacific Exposition scheduled for 1914.[5]

In essence, amplification is a process through which a time-variant signal is magnified by means of energy extracted from a source such as a battery, a DC power supply, or a rotating machine. An important characteristic of a good amplifier is that it produces an accurate replica of an input signal without adding substantial distortion. Often, as in the case of telephone amplifiers, it is also important that the amplification be relatively constant over a considerable range of frequencies. Several alternative methods of amplification, including the high-frequency alternator, the magnetic amplifier, and the vacuum tube, came under investigation during the early twentieth century.[6]

In March 1907, soon after he received Pickard's report praising the

GE alternator used in the NESCO demonstration, Hayes wrote to Alexanderson to ask if he could meet with an AT&T engineer to discuss the design of an experimental high-frequency alternator. After a meeting with W. M. Gould, Alexanderson reported that Gould, on the assumption that the voltage induced in a secondary winding would be greater than an input voltage applied to a primary winding on the motor, had proposed using an induction motor driven above synchronous speed. Alexanderson wrote that he had questioned the feasibility of Gould's method and had instead recommended the use of a 40-kHz inductor alternator with a disk rotor and no rotating winding. Alexanderson also advocated the use of a laminated-iron core for the stationary armature winding.[7]

Alexanderson based his design proposal on experience with the high-frequency alternator he had designed earlier for Fessenden. The major modifications for the proposed telephone amplifier were to be the use of iron instead of wood for the armature core and the use of a single disk instead of the dual-disk rotor. Alexanderson received a $200 appropriation to begin the project and promptly advised GE's Patent Department to begin the process of obtaining a patent covering the new application of the high-frequency alternator. In an April 1907 memorandum addressed to Henry Geisenhoner of the Production Department, Alexanderson stated that the telephone alternator would use the same drive system and shaft arrangement as the NESCO alternator but would require a new rotor and armature design.[8]

As Alexanderson, now with the Railway Engineering Department, began the development of the experimental alternator for AT&T, his former colleagues in the AC Engineering Department continued to develop alternators for Fessenden. Apparently unaware that Alexanderson no longer was an active participant in the NESCO developmental project, Fessenden informed him in May 1907 of recent wireless experiments at Brant Rock and urged him to try to expedite work on a larger, 10-kw, alternator. Fessenden wrote that the high-frequency alternator had proved a great success and had enabled NESCO to do "all kinds of things" that could not be done by other means. He invited Alexanderson to visit the Brant Rock station at his earliest opportunity. By this time GE had completed two additional dual-disk alternators, which were similar to the one designed by Alexanderson and delivered to NESCO in 1906 except that they were equipped for gear rather than belt drive.[9]

Fessenden probably first learned of the new single-disk alternator from a letter he received from Alexanderson in June 1907. Alexanderson mentioned the problem encountered with the dual-disk machine that had failed to deliver the expected power level at higher

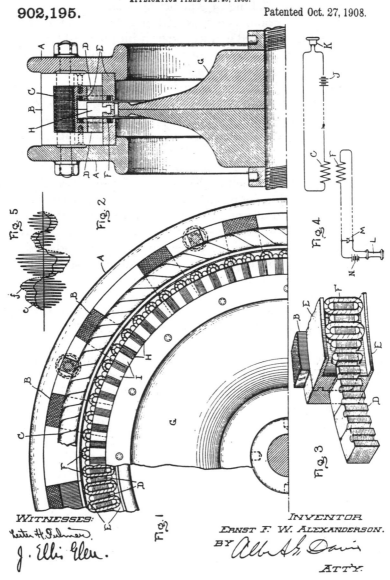

E. F. W. ALEXANDERSON.
TELEPHONE RELAY.
APPLICATION FILED JAN. 25, 1908.

902,195.

Patented Oct. 27, 1908.

WITNESSES:

INVENTOR
ERNST F. W. ALEXANDERSON.
BY
ATT'Y.

Figure 3.1. Patent drawing for Alexanderson's single-disk alternator designed to amplify telephone signals. The high-speed rotor is designated *G*, the field coil *C*, and the armature coil *F*. From Ernst F. W. Alexanderson, U.S. Patent no. 902,195, issued October 27, 1908.

speeds. He attributed the effect to elastic deformation of the two disks which had increased the air gap between the disks and armatures, and he reported that he now was at work on a new alternator with a single disk, which he felt would overcome the problem. He expressed his belief that it would be inadvisable to operate the dual-disk alternator at speeds above that required to generate 60 kHz but that a single-disk machine should be safe at considerably higher speeds. He did not mention that the new alternator was being built for AT&T, and Fessenden assumed that it was intended for NESCO. He responded by expressing his persistent doubt that an iron core could be used at such high frequencies and asked that the machine be designed so that both wood and iron cores could be tested and compared.[10]

Although now aware of Alexanderson's work on a single-disk alternator, Fessenden's assertive self-confidence and habit of premature optimism led him to continue supporting multiple-disk alternator development at GE, resulting in an almost comic duplication of effort. Only a week after he heard from Alexanderson about the single-disk alternator, Fessenden announced in a letter addressed to Caryl D. Haskins of the AC Engineering Department that he and his NESCO assistants had arrived at a solution to the problem of decreasing output at high speed. Fessenden described a proposed modification of the dual-disk alternator which, he claimed, would make it easy to build generators capable of producing 100 kw at frequencies of up to 150 kHz and more than 1 kw up to 300 kHz. Consequently, he stated, he thought it of doubtful value to proceed with construction of the single-disk machine: instead, GE should build three additional small units with wood-core armatures and a 10-kw alternator of similar design. With his letter he included a memorandum that revealed that the proposed modification was to use two separate armature windings, with one located adjacent to each of the two rotating disks. He predicted that the modified design would double the power output and also improve ventilation. Early in July 1907, Fessenden informed Alexanderson of the proposed modification and again encouraged him to expedite work on a 10-kw alternator since it was "absolutely certain that there will be no difficulty" in constructing it.[11]

As a result of Fessenden's new initiative, Conway Robinson of GE's AC Engineering Department began the task of designing a 10-kw alternator for NESCO in August 1907. Meanwhile, Alexanderson continued his work on the alternator for AT&T and did not become directly involved in the development of the 10-kw alternator for Fessenden. The compartmentalization at GE permitted the two projects to continue for some time before the duplication of effort was perceived by the interested parties. Robinson decided that it would be necessary to use several steel disks mounted on a single shaft to obtain the necessary output power of 10 kw. In October 1907, an order for parts to be used on Alexanderson's single-disk machine was mistakenly routed to the AC Engineering Department. This led Robinson to write to Alexanderson requesting more information unless Alexanderson wished to handle the order. Two weeks later, Robinson notified the production department that instructions on the order would come from Alexanderson, who was engaged in a special project with Fessenden.[12]

When Alexanderson visited the NESCO facility at Brant Rock in early September 1907, he rekindled Fessenden's interest in the experimental single-disk alternator. In a written summary of the visit, Alexanderson noted that Fessenden now expected all future wireless systems to employ high-frequency alternators and that a large variety of sizes and frequencies would be needed. Alexanderson mentioned that Fessenden wanted several armatures to be constructed, including both wood- and iron-cored types, for use with alternators that he already had received. He also had asked for an opportunity to experiment with the new single-disk alternator. Fessenden had finally become aware of the existence of two separate alternator projects in two different departments at GE. He instructed that both be continued for the time being, but added that if nothing had been done on the 10-kw alternator, it might be best to await the results of Alexanderson's experiments before undertaking construction of the new multiple-disk machine.[13] The advantage to Alexanderson of having two outside patrons for his project soon became apparent.

AT&T's interest in the alternator as a potential telephone amplifier ended abruptly owing to a major reorganization in which Hammond Hayes was dismissed from his position as chief engineer. Hayes, who had initiated the single-disk alternator project at GE, visited Fessenden at Brant Rock shortly after his dismissal and they evidently discussed the possibility of Hayes working for NESCO. Fessenden wrote to Hayes soon after the visit that he anticipated seeing more of him "if matters go as I hope." Hayes responded that he had a great deal of interest in Fessenden's work and would be glad to assist for the enjoyment but that he had no present intentions of entering a new line of work.[14] The Mechanical Department in Boston, which Hayes had directed for many years, was eliminated by the reorganization, although some of the Boston engineers were transferred to New York City. The resulting disruption not only ended AT&T's interest in Alexanderson's alternator but also delayed a proposed investigation of the de Forest audion as another potential telephone amplifier.[15]

Fessenden's enthusiastic support enabled Alexanderson to continue the single-disk alternator project despite the unanticipated loss of

AT&T as patron. In November 1907, Fessenden informed Alexanderson of the recent discovery that the daylight absorption of wireless signals was much lower for frequencies below 90 kHz. Fessenden stated that this discovery had convinced him that the future development of wireless communication favored the use of long waves that were well within the range of alternators. In December 1907, Fessenden came to Schenectady to observe tests of the single-disk alternator. Soon after this visit, he decided to cancel his order for the uncompleted 10-kHz multiple-disk alternator that had been designed by Robinson, but to continue his support of Alexanderson's project.[16]

Fessenden's decision to end his support of the Robinson project created apparent consternation within the AC Engineering Department and led to an effort to persuade GE's management to support the completion of the 10-kHz alternator. Robinson suggested that the company might be able to sell such alternators to colleges because of the increasing interest in high-frequency research. He also mentioned that Germany's AEG company had inquired about the availability of such a machine. He estimated that the alternator could be completed for an additional $250 and that it would deliver as much power as could be obtained from a dual-disk machine at 100 kHz.[17] Henry Reist, head of the AC Engineering Department, also urged that the Robinson alternator be completed, and recommended that some novel features of the design be covered by a patent. In a memorandum addressed to Francis C. Pratt, assistant to the chief engineer at GE, Edmund P. Edwards provided a summary of expenditures on the Robinson alternator project. Edwards asked whether it might "be possible to reconcile the two designs and avoid the present apparent conflict."[18]

In April 1908, Alexanderson and Edwards arranged to visit the NESCO station at Brant Rock to discuss the alternator issue with Fessenden, who reaffirmed his decision not to pay for the completion of the Robinson machine. Shortly afterward, Reist instructed Robinson to box the completed parts of the unfinished alternator.[19] Thus ended the Robinson alternator project. The personal relationship that had developed between Alexanderson and Fessenden since 1904 seems to have been the decisive element in the survival of Alexanderson's project and the demise of Robinson's. Thus the technological sequence of events that ultimately led to the formation in 1919 of the Radio Corporation of America depended on the slender thread of interpersonal relations between two engineer-inventors in 1908.

The Politics of Wireless in 1908

Fessenden engaged actively in the politics of wireless during 1908, and as an interested observer, Alexanderson became more knowledgeable about the complex political environment in which his alternator would be used. Fessenden lobbied against ratification by the U.S. Senate of an international agreement regulating wireless systems which had been negotiated at a conference in Berlin in 1906.[20] Theodore Roosevelt's administration supported confirmation of the agreement, which was scheduled to go into effect on July 1, 1908. Fessenden asked Elihu Thomson of GE to join him in opposing its ratification. Fessenden wrote to Thomson that it seemed that the U.S. Army and Navy were trying to gain control of wireless communication, although they had done little to advance its development. Thomson replied that he was in sympathy with the opponents of the agreement and would do all that he could to oppose its confirmation. He claimed that the movement to produce the agreement had originated in Germany and would be a "great victory for German diplomacy" if approved. He suggested that this would be the most persuasive argument in the U.S. Senate. Thomson subsequently submitted a brief against ratification to Senator Henry C. Lodge of the Foreign Relations Committee. In his brief, Thomson objected to the proposed limitation of commercial wireless to certain wavelengths on the grounds that it was unrealistic, since it would probably limit commercial wireless operations to nighttime. He argued that the state of the wireless art still was too new for the adoption of such restrictions.[21]

Fessenden also submitted a long brief against ratification. It included a "brief history" of wireless, a summary of NESCO's contributions, and "objections to the Berlin Convention." He alleged that army and navy officers were principal advocates of confirmation. "They care nothing for the commercial side, they can see in [wireless] only an instrument of warfare," he charged. He contended that restrictive legislation such as the agreement in question was premature, since "the art is young, full of promise and, if allowed to develop in a natural unrestricted way, will produce results that today seem beyond belief."[22] Senator Lodge later notified Elihu Thomson that the Foreign Relations Committee had held hearings and had decided to "do nothing in regard to [the agreement] at present." Lodge expressed his own "grave doubts as to the wisdom of its ratification."[23]

The Roosevelt administration then tried a different strategy, endorsing legislation proposed by the secretary of the navy which would make it unlawful to interfere with wireless stations transmitting official messages. James H. Hayden, a lawyer retained by NESCO, interpreted this as a clever effort to bypass the Senate Foreign Relations Committee. He assessed the proposed regulations as "more far-reaching and more harmful" than those contained in the Berlin agreement.[24] NESCO came out in support of a bill that had originated in the U.S.

House of Representatives and that reserved the band between 375 and 425 meters for governmental use but did not authorize the U.S. Navy to regulate the industry. This bill passed just before Congress adjourned in 1908, but it was vetoed by the president. The Senate declined to act on the Berlin agreement in 1908, but the agreement was ratified in 1912.[25]

Fessenden prepared a comprehensive paper on the history, state of the art, and politics of wireless communication for presentation at the annual convention of the American Institute of Electrical Engineers in the summer of 1908. The paper served several functions. Not only was it a tutorial on wireless systems for electrical engineers, many of whom were unfamiliar with the principles and practice of wireless, but it also warned members of the electrical engineering profession that other branches of the electrical industry might soon face the danger of government interference which the wireless industry was encountering. The first part of the paper was almost a bibliographic essay on the key inventions and discoveries in wireless communication, with numerous citations of patents and technical papers. Fessenden claimed much of the credit for having initiated and developed the sustained-oscillation, or continuous-wave, system that was beginning to supplant the damped-oscillation, or spark, system. He discussed the various methods of generating continuous wireless waves that had been tried, and concluded that the high-frequency alternators had several advantages over the alternatives available. He explicitly mentioned the contributions of Alexanderson, "to whose efforts the success of this type of generator is largely due."[26]

In the last section of his paper, Fessenden emphasized that "the difficulties in the development of wireless telegraphy have not been wholly of a technical nature." He attributed the slow development of commercial wireless to such causes as the difficulty of obtaining transmitter site permits at suitable locations, premature attempts at regulation, the U.S. Navy's efforts to gain control of all coastal stations used in ship communication, and the American policy of permitting the military services to use patented inventions without paying compensation. Fessenden included a detailed analysis of the rules and regulations incorporated in the Berlin agreement and explained why he found so many of them ill-advised. He argued that the restrictions would tend to prevent the introduction of improved methods. For example, the requirement that only syntonized systems be used would hinder or prevent the use of the heterodyne system that he had invented.[27]

The policy issues raised by Fessenden concerning what he saw as the inhibiting effects of military intervention and premature governmental regulation on the development of wireless have some broader impli-

cations for the history of technology. Comparison of the roughly con-
temporary development of interurban electric traction systems and
wireless communication systems suggests that regulatory policies were
an important constraint on the relative success of the two systems. Alexanderson, who was deeply involved in both systems, learned that there
were significant external influences on technological change. In the
case of wireless, he observed that Fessenden's attitude toward military
involvement depended on whether the military services were seen in
the role of influencing government policy on commercial wireless or as
patron and potential customer. The foundations of an increasingly
strong military-industrial alliance in electrical technology were being
laid during the first decade of the twentieth century. Because Alexanderson, unlike Fessenden, was not an entrepreneurial engineer, he
was less ambivalent in his relations with military technologists. Their
support would soon become a significant determinant in his career development.

The Completion and Impact of the Alexanderson
100-kHz Alternator

While Fessenden engaged in wireless politics, Alexanderson and his
associates at GE continued the development of a single-disk high-
frequency alternator. By 1908, the scope of the project had expanded
considerably beyond the original design of an experimental rotary telephone amplifier. Alexanderson designed a modified form of the machine, one that could be used at either 25 kHz or 100 kHz and could
accommodate three different rotors, including one with laminated
teeth. The last-named rotor was for use when the alternator was tested
as an amplifier. He filed a patent application in January 1908 covering
design features of the amplifier version of the alternator. In the amplifier mode, telephonic currents produced by a microphone were applied to the stationary field winding of the alternator. The rotating steel
disk modulated the signal induced in the stationary armature winding
as the laminated rotor teeth passed between the two windings. The key
claim in Alexanderson's patent was that after demodulation of the alternator output, the audible sounds would be "of similar quality to, but
of greater amount than, those that would be heard if a telephone receiver were substituted for the field winding of the alternator."[28]

The alternator did produce amplification as predicted, but the necessity of using laminated teeth at the rotor's periphery limited the
maximum safe speed to less than two-thirds that of a rotor without
lamination.[29] Although the rationale for the amplifier experiment was
diminished after the 1907 reorganization at AT&T, the experiment did
have an important consequence. Alexanderson's dissatisfaction with

the rotor with laminations led to one of his most significant inventions, the magnetic amplifier with no moving parts.[30]

In May 1908, Alexanderson wrote a memorandum for GE's Production Department outlining the steps needed to complete the single-disk alternator project. He mentioned that further tests would be run on the laminated-tooth rotor that already had been constructed, and went on to say that the 100-kHz rotor would be finished and various combinations of rotors and armatures would be tested. At about the same time, he informed Fessenden that slots were being cut in the rim of the 100-kHz rotor but that the work was very slow owing to the hardness of the steel. He enclosed a sketch that showed 300 slots cut in a 12-inch-diameter disk. After the first test of the new rotor in July 1908, Alexanderson notified Fessenden that comparative tests of wood and iron as armature cores had demonstrated conclusively that iron was "very much better." The machine had been operated up to a frequency of 60 kHz, Alexanderson continued, and the only difficulty encountered had been heating due to air friction. He recommended that all future efforts be devoted to perfecting alternators with iron-core armatures, since the use of such armatures would increase the output power by at least a factor of five. He noted that to determine the ultimate limit of high-frequency power for such machines, additional data on air friction and bearing friction would be required. He obtained data for graphs of friction losses and armature voltage for a range of air-gap lengths and for frequencies up to 62.5 kHz during the period from June 27 to July 13, 1908.[31]

The air-friction problem at higher rotor speeds proved more challenging than anticipated, but a suitable technological solution emerged as the tests continued. Alexanderson initially believed that the problem had been overcome by filling the slots with nonmagnetic solder, and this did indeed prove satisfactory up to a speed of about 12,000 rpm (60 kHz). However, he soon discovered that the solder had insufficient mechanical strength to withstand the stresses produced by higher speeds; it was simply thrown out of the slots. Fessenden suggested that tin or type metal be tried as an alternative to solder or that copper wire be soldered into the slots using solder with a higher percentage of tin. Finally, Alexanderson tried filling the slots with U-shaped wire of phosphor-bronze, and in December 1908 he notified Fessenden that this had solved the problem. In the same letter, he described a fortuitous thermal effect that tended to make the rotor shaft self-centering: when heated, the thrust bearings expanded and moved the rotor in a direction that tended to reduce bearing friction. Alexanderson told Fessenden that the alternator now was operating at its rated speed of 20,000 rpm (100 kHz) for long periods.[32]

The air-friction investigation also led Alexanderson to make an important contribution to the engineering science of high speed machines. Using the data for air-friction loss measured on the 100-kHz alternator, he developed an empirical formula for the air friction loss on a rotating disk as a function of the disk diameter and velocity. By plotting the loss on a logarithmic graph, he found that the power required to overcome air friction was directly proportional to the 2.7th power of the peripheral velocity and to the square of the disk diameter. For the 12-inch rotor used in the 1908 tests, the air-friction loss at 20,000 rpm was 5 kw, as compared to the useful power output of about 2 kw. The rapid increase of the loss with speed, as well as the reduced safety margin in mechanical stresses, led him to conclude that speeds greater than about 20,000 rpm should not be used in high-frequency alternators of this design.[33]

Edmund Edwards of GE played an active role in the promotion and marketing of the Alexanderson alternator beginning in 1908. In October of that year he wrote a memorandum that gave a very optimistic assessment of the alternator's market potential. He predicted an annual business with NESCO of at least $75,000 if the alternator design proved satisfactory, and he noted that Fessenden considered the alternator to be "*the* machine for small telegraph and telephone work." The United Wireless Telegraph Company was eager to get one of the new alternators, Edwards continued, and they would need 200 of them annually if the price were acceptable. He concluded by noting that the wireless industry was still in its infancy and that it was not yet possible to forecast accurately how much business might come to GE once wireless had become a "thoroughly established commercial industry."[34] Edwards urged NESCO to begin commercialization of the alternator-based wireless system. After visiting the United Wireless Telegraph Company and meeting with representatives of the navy and the army, he wrote to Fessenden encouraging him to cooperate with the military services, since they undoubtedly would go elsewhere for apparatus if he refused. Edwards also contacted Hay Walker, Jr., the president of NESCO, stating that he regarded the Fessenden system as superior to any other and urging that it be promoted more aggressively than NESCO was doing. Edwards mentioned that the recent *Republic* disaster had stimulated a great interest in wireless and that GE now was receiving many opportunities to bid on wireless apparatus.[35]

The *Republic* episode that Edwards mentioned had made the public aware of the potential role of wireless in enhancing the safety of ocean travel. The *Republic,* a White Star Line steamship, had collided with the *Florida* near Nantucket on January 23, 1909. The passengers and crew of the sinking *Republic* were rescued thanks to a wireless distress signal that brought several vessels to the scene. A published account of the accident stated that the episode should serve to "convince the last doubter as to the immense utility of the wireless" and predicted that the wireless would decrease the risks of sea travel to a minimum. The *Florida* had not had a wireless transmitter, and the writer predicted that wireless equipment soon would be mandatory on passenger ships.[36]

GE shipped the first of the 100-kHz, 2-kw alternators to NESCO in June 1909 to be used in a planned demonstration for army and navy officers. Fessenden soon wrote to Alexanderson to express delight with the performance of the machine, calling it a "thoroughly practical piece of apparatus."[37] GE completed two more alternators of the same design for NESCO later in the year, and Fessenden turned one over to the Army Signal Corps to be used in their own experiments. In December 1909, Edwards informed Fessenden that the price for a single 2-kw alternator would be $1,931, not including the DeLaval gear, and that the unit price for an order for six alternators would be $1,542.[38] By this time, after approximately five years of developmental engineering, the single-disk 100-kHz alternator had reached a marketable stage.

With Fessenden's encouragement, Alexanderson prepared a paper on the high-frequency alternator for presentation at the annual AIEE convention in 1909. After receiving an advance draft of the paper, Fessenden responded with a discussion of the history of such machines and concluded that most of the credit for the development of practical high-frequency alternators belonged to Alexanderson and his associates at GE. Fessenden predicted that the AIEE paper would create a sensation, especially in Europe, and that within five or six years alternators would supplant other methods of producing wireless electromagnetic waves.[39]

In his paper, Alexanderson summarized his work on high-frequency alternators since 1904 and discussed the mechanical and electrical characteristics of the single-disk 100-kHz machine. He reported that the rotor was made of chrome-nickel steel with an elastic limit of 200,000 pounds per square inch (psi). The maximum stress at the rated speed of 20,000 rpm was 30,000 psi, giving a safety factor of 6.7. He stated that the precision of construction had made temporary operation possible with an air gap of only 0.004 inch between the disk and the armature, although the clearance normally was 0.015 inch. A threaded adjustment permitted varying the clearance in increments of 0.001 inch. The disk had 300 slots cut in its rim, with a spacing between adjacent slots of 0.125 inch, and the slots were filled with a nonmagnetic material to reduce air-friction losses. Alexanderson devoted considerable space in his paper to a discussion of the flexible rotor shaft and bearings, and included a derivation of two "critical speeds" at which mechanical resonance occurred. He described a forced-oil lubrication system that had proved necessary because of the heat produced in the bearings and noted that he had added a set of "middle bearings" near the disk to prevent excessive vibration as the rotor passed through its critical speeds. He concluded that it now seemed probable that frequencies even higher than 100 kHz might be generated by rotary machinery.[40]

During the discussion, Arthur E. Kennelly of Harvard asked Alexanderson if he had noticed whether an electrical shock could be felt at 100 kHz, since it was well known that no shock was felt at a sufficiently high frequency. Alexanderson replied that a shock could be felt; he had observed no difference between the shock felt at 100 kHz and that produced by a direct current at the same voltage.[41] Kennelly evidently was not convinced, and he arranged for a series of experiments which he and Alexanderson reported in a joint paper published the following year.

Using the new alternator as a source, Kennelly and Alexanderson investigated human subjects' physiological tolerance of high-frequency currents over a frequency range from 11 kHz to 100 kHz. They pointed out in the paper that such measurements had not been feasible previously but that with the Alexanderson alternator, the frequency, voltage, and current could be manipulated and measured readily. Contrary to Alexanderson's original impression, they observed a substantial increase in the magnitude of current that could be tolerated as the frequency was increased. In the test arrangement, they submerged a subject's hands in a saline solution that contained electrodes connected to the output of the alternator. They defined the tolerance current as "the limiting current strength which the subject could take through his arms and body, without marked discomfort or distress." They found that a man could tolerate approximately 500 milliamperes (mA) at 100 kHz, but only about 30 mA at 11 kHz or 5 mA at 60 Hz. They attributed the difference to "reduced nervous sensibility at high frequencies."[42] Interestingly, Kennelly had been involved in measuring and comparing the effects of direct and alternating currents on human and animal subjects while employed by the Edison Electric Light Company 20 years earlier. His earlier experiments had been conducted in the context of the so-called "battle of the systems," when the allegedly greater danger of AC was stressed by advocates of the Edison DC system.[43]

Alexanderson and members of GE's Production Department continued to make improvements in the design of the alternator and the machine tools used to produce it. In February 1910, after trying a lighter-weight disk and shorter shaft, Alexanderson wrote to Fessenden that the results were "far superior" to those obtained previously. Alexanderson reported that the modified disk had a rim that was only 0.1 inch thick and that slots now were cut by a special precision machine that could cut any number of slots in a circular disk. The slots now were being filled by riveted aluminum fillers to reduce the effect of air friction.[44]

Alexanderson undertook a more substantial modification of the alternator, doubling the output frequency at a given speed, when Fessenden decided that the navy might need a 200-kHz alternator. In August 1909, Fessenden had written to J. R. Werth of GE that some navy officers had expressed an interest in installing alternators on all vessels so that a watch officer could telephone any ship in a fleet by wireless. He concluded that a 200-kHz alternator would be more suitable for shipboard use than would a 100-kHz unit because the 200-kHz unit would require a shorter antenna. In March 1910, Alexanderson informed Fessenden of a possible way of doubling the output frequency by doubling the number of slots cut in the rotor and making changes in the armature winding. He provided Fessenden with air-friction data that had led to the decision that a speed greater than 20,000 rpm was undesirable "even if this were possible for mechanical reasons." Early in June 1910, Alexanderson reported a successful test of the 200-kHz alternator.[45]

In March 1909, Fessenden contacted Edmund Edwards of GE to ask that the company begin development of a larger alternator capable of delivering up to 35 kw at a frequency of 50 kHz. Fessenden agreed to provide up to $2,500 to support the project. In July 1910, Alexanderson informed NESCO that the large alternator was undergoing its first test and had been operated at a speed of 3,000 rpm. He stated that the preliminary results indicated that the machine should deliver up to 40 kw at its rated speed. He mentioned that, before much more work was done, a decision was needed on how the machine was to be driven. He recommended using a 100-hp electric motor with a DeLaval gear but noted that a direct turbine drive might be employed if the alternator were to be used only at the NESCO station at Brant Rock.[46]

In February 1910, Fessenden approached AT&T about a possible buyout of NESCO. He wrote to Theodore Vail, who had become head of the telephone company during the reorganization in 1907, that NESCO would consider selling out to a company that had the necessary business organization to exploit the NESCO wireless system fully. Fessenden asserted that it would not be feasible for AT&T to acquire a comparable system from a competitor because NESCO controlled key patents that were necessary for commercial success. He included a cost analysis comparing wire and wireless communication, which indicated that the first cost of a wireless circuit was only about one-third the annual cost of operating a wire circuit. He claimed that wireless telephony had been carried out at a distance of over 400 miles and that a more powerful alternator being developed should extend the distance to 3,000 miles. Fessenden acknowledged that considerable developmental work remained "before the system can supplant the present telegraph and telephone lines," but he felt that future work would mainly concern details of commercial operation which would require "an elaborate engineering and business organization."[47] A summary of documents in AT&T files on Fessenden's system covering the period 1906–12 indicates that there was continuing interest, along with skepticism that the system would achieve commercial success before the fundamental patents expired.[48]

Technologists who invent and preside over the developmental phase of a technological system frequently seem to experience difficulty during the transition to the commercialization phase, in which somewhat different skills and considerations become paramount. Fessenden fell victim to this difficulty when he lost the confidence of his Pittsburgh financial backers and his position as general manager of NESCO late in 1910. The president of NESCO, Hay Walker, Jr., offered Fessenden a position that would leave him in charge of experimental work but without responsibility for business matters. He declined the offer and instead initiated litigation that drove the company into receivership, effectively leaving the wireless field to others. The effective demise of NESCO as a market for the high-frequency alternators soon led GE to look elsewhere.[49]

Long afterward, Alexanderson reflected on his early association with Fessenden and concluded that it had been a "productive partnership." Alexanderson expressed doubt that his own involvement in radio would have gone beyond the development of apparatus for Fessenden if the latter had remained active in the field. Alexanderson called Fessenden the "American Father of radio," who had been an inspiration to many and whose impact had been so great that work on his system continued after his own withdrawal.[50] By 1910 the development of high-frequency alternators at GE had reached a stage such that the loss of the principal outside patron failed to halt further development of the alternator-based technology under Alexanderson's leadership.

Alexanderson's inventive productivity was impressive during this period, and he was rewarded by seeing his annual salary increase from $2,900 in 1907 to $4,000 in 1910.[51] During the period 1903–10, he applied for 68 patents that ultimately were issued, and in 1907 he achieved his all-time high, with 17 applications. Only 6 of the 68 applications were related to the high-frequency alternator. The remaining successful applications were in the field of AC power machinery and related apparatus. This disparity probably was more reflective of GE's initial disinterest in building a strong patent position in high-frequency technology than of the relative time and energy that Alexanderson devoted to the two fields of invention.

ALEXANDERSON became a naturalized citizen of the United States on January 27, 1908. His certificate of naturalization described him as being 5 feet 9 inches tall, with fair complexion, brown eyes, and blond hair.[52]

Another major event in Alexanderson's personal life occurred on February 27, 1909, when he married Edith B. Lewin, whom he had met at GE, where she worked as a secretary. Edith was the daughter of Creighton E. Lewin and Minnie Porter Lewin of Rome, New York. (The Lewins had changed their name from Llewellyn after coming to New York from Wales.) An Alexanderson family tradition holds that Edith was an effective advocate of her future husband at GE for some time prior to their marriage.[53] In a letter written in May 1909, Amelie Alexanderson, the inventor's mother, expressed her delight at having received a portrait of Edith along with a detailed description of the new couple's home in Schenectady. According to this letter, Edith had provided more information than the elder woman could expect to get from her son in a year.[54]

In July 1909, Amelie Alexanderson wrote to Edith thanking her for her diligence in writing and remarking that Edith's account of Alexanderson's work had been "highly interesting." Alexanderson's mother went on to say that it seemed that the development of AC now was in an exciting period and at a turning point. In an apparent reference to the controversial battle between DC and AC traction systems, she expressed the hope that AC would prove victorious. She mentioned that she had heard of the "oppressive heat" in New York in the summer and was glad to hear that her son and Edith were going to the St. Lawrence River, where he was to read professional papers. This was a reference to the two papers that Alexanderson delivered at an AIEE meeting held in Frontenac, New York, in June 1909. In another letter, written in November 1909, Alexanderson's mother referred to Edith's "undutiful husband," who had failed to follow her good example by keeping his parents informed on the progress of his work—which she described as being "so very interesting to us."[55] Alexanderson's busy schedule caused him and his new wife to postpone a planned trip to Sweden until 1910.

The documentary evidence on Alexanderson's nontechnical activities during his early years at GE is limited. He is known to have been

an active member of the "Agora," a club of GE engineers who met fortnightly from October to April each year. The members presented papers on nonengineering topics such as occult phenomena, German
politics, scientific management, modern athletics, and the "insane desire for regulation."[56] In 1911 Alexanderson gave a talk to the Agora
on the subject of "woman and labor." He contended that the family was
the fundamental unit in the life of a nation and that there would be
less social disruption if there were government by and for families
rather than by and for the people. He expressed the view that female
workers were not a social necessity in factories, but stated that he favored equal pay for equal work for women who worked outside the
home.[57]

Chapter Four

Corporate Consultant: The Magnetic Amplifier, Applied Electronics, and the Phase Converter

Alexanderson began a significant new phase of his career at GE in July 1910, when he left the Railway Engineering Department to become a charter member of the Consulting Engineering Department. Charles Steinmetz, who had conceived the new department and who headed it, had invited Alexanderson and a few other senior engineers at GE who shared his philosophy and broad interests to join. The Consulting Engineering Department represented an imaginative approach to the creation of an environment within the corporation which enabled a few elite engineers to select challenging problems and be provided with the resources needed to solve them. It was designed to appeal to creative engineers who did not aspire to be administrators or to become too narrowly specialized. It provided an alternative opportunity for those engineers who might otherwise have been tempted to leave the company to become independent consultants or engineer-entrepreneurs. The Consulting Engineering Department was, in essence, complementary to the GE Research Laboratory, which Steinmetz also had been instrumental in founding some years earlier but had declined to head.[1] The differences between the new department and the Research Laboratory illuminate the much-discussed issue of the nature of science and technology and their interaction. GE's internal consulting engineer tradition survived the passing of Steinmetz, and several reorganizations. Alexanderson became the leading proponent of the continuing need for such a department at GE after World War I.[2]

Steinmetz gave his views on the mission and the organization of the Consulting Engineering Department in an internal memorandum in 1912 in which he stated that the department had been established to advise and consult with other GE engineering and commercial departments on "unusual phenomena, designs, troubles . . . either on request of the interested party, or voluntarily where such subjects come to our attention." He added that the department also was to advise and assist customers "and other outsiders" on proposals for new work or trouble with GE apparatus or systems, and "to study phenomena and problems of interest in electrical engineering, and therewith of interest to the Company, and their industrial applications." Steinmetz emphasized that it was essential for the new department's members to keep well informed on important engineering and development work, not only at the Schenectady works but also at those in Pittsfield and Lynn, so that they could disseminate information throughout the company. This would enable the department to "make all the development work, which is going on within the Company, available to the fullest extent to all who might possibly benefit by it." The department itself would undertake projects that required facilities or a combination of knowledge or experience beyond that available in the more specialized departments.[3]

In contrast to the "autocratic organization" of other departments, where the department head assigned work, Steinmetz stated that "each consulting engineer must be responsible for his work, the methods used and the results derived, and receive the fullest personal credit and recognition of his work, and to a large extent even choose his work." The principal limitation in selecting projects was that "the subject matter must be useful to the Company," but Steinmetz interpreted this to include fundamental investigations of phenomena such as corona or transients "which [are] not directly of special use to the Company, but [are] of value to the industry at large, in extending the application of electricity." Because of the "practically complete independence" of the department's engineers, Steinmetz had seen the need for some administrative and financial mechanism to "keep the department from disintegrating into a number of isolated units." He reported that he had delegated these functions to J. LeRoy Hayden, and that "this will enable us to devote all our time and energy to engineering." Steinmetz stressed the need for frugality of time and money. For example, he recommended that the consulting engineer, instead of ordering expensive apparatus, first investigate whether suitable facilities might not already be available somewhere within the company.[4]

Steinmetz outlined several methods of facilitating information transfer both within and outside the department. "No matter how important some work is," he wrote, "it is valueless if not brought to the attention of those who may utilize it, in such a manner as to have it used." Methods used within the department included frequent personal discussions among the engineers, periodic joint meetings in which one member would have the floor, and monthly written reports on each project. Each of the monthly reports was to include the project's history, its objectives and methods, and the results achieved and anticipated. Steinmetz felt that the process of writing up the reports was itself valuable, because it stimulated a reflective review that might provide information or insights that might be overlooked otherwise. To promote the dissemination of information outside the company, Steinmetz stressed the importance of writing articles and of reading papers at engineering society meetings. He suggested that this would be an effective way to enhance the engineer's reputation, and that it would be an asset both to the engineer and to GE.[5]

Since there were "many more important subjects of investigation" than the consulting engineers could handle individually, Steinmetz encouraged each of them to "develop and educate assistants" who could work with minimal supervision and who might ultimately become consulting engineers.[6] A procedure was worked out to identify and train promising recruits for the Consulting Engineering Department. The department began an extension course for young engineers who had spent at least a year in the Test Department. Those completing the course then were assigned to assist a consulting engineer on a specific project. An engineer who had gone through this screening process might then be invited to join the Consulting Engineering Department when a vacancy or new position occurred.[7]

The diversity of projects that attracted the new department's engineers is indicated by a report Alexanderson prepared in October 1911. He listed 22 investigations in which he had direct interest. Fifteen were related to traction motors, variable-speed industrial motors, or phase converters. The remaining 7 were related to high-frequency alternators, telephony, and miscellaneous projects such as a hydraulic-gas engine and a constant-voltage generator for an automobile lighting system.[8] It must have seemed an almost ideal institutional arrangement for an engineer-inventor whose interests transcended the boundaries of more specialized engineering departments.

As Steinmetz had indicated, members of the Consulting Engineering Department were expected to facilitate the dissemination of information and the transfer of technology. Alexanderson took advantage of his European connections and his occasional trips to Europe to carry this mission beyond the confines of the United States. Soon after he joined the new department, he embarked on his third trip to Europe since coming to the United States. His itinerary included two weeks in Sweden, a week in Germany, a week in France, and a week in England.[9] Except for a brief visit with his parents in Sweden, he spent his time gathering information on recent European developments in AC traction and power machinery. While in England, he learned of a new type of induction motor, known as the Hunt-Sandycroft motor, which was being manufactured by the Sandycroft Foundry Company. Alexanderson conferred with an engineer of the British Thomson-Houston Company, a subsidiary of GE, who recommended that GE acquire

Hunt's patent rights. After returning to Schenectady in January 1911, Alexanderson prepared a report that included the theory of the motor and an assessment of its commercial potential. During his trip he had also observed the operation of single-phase electric railroads in Sweden and Germany and visited several electrical manufacturing companies.[10]

The Alexanderson High-Frequency Alternator: Growing Outside Interest and Some Research Applications

Evidence of an increasing outside interest in the Alexanderson high-frequency alternator began to accumulate during 1911. A German engineer-inventor, Rudolph Goldschmidt, published a description of an alternator that he claimed would prove superior to the arc and spark methods of generating wireless signals.[11] Shortly after Goldschmidt's article appeared, Alexanderson received an inquiry from Germany's Telefunken Company concerning the status of the high-frequency alternators being developed at GE. He attributed the inquiry to Goldschmidt's article, although he may have discussed the GE work on alternators when he was in Germany a few weeks earlier. Alexanderson commented that he believed his pending patent applications would enable GE to establish priority of invention in high-frequency machines. In an internal memorandum, he mentioned the urgent Telefunken inquiry and suggested that work be resumed on the 35-kw alternator. He noted that nothing had been done on the machine since the previous summer owing to the lack of a suitable drive motor.[12]

Alexanderson read with interest an article in the *Electrical World* which defended American contributions to wireless engineering. The author singled out the Alexanderson alternator as an example that had succeeded despite being regarded as an "ignis fatuous," a will-o'-the-wisp, by foreign inventors. He credited Alexanderson with having placed continuous-wave wireless in such a position that it could be engineered with the "certainty of 60 cycle engineering."[13] Alexanderson cited the article as evidence that the advantages of alternators in wireless communication finally were being recognized. He again pointed out that the 35-kw alternator had been sitting idle for months, and raised the question of whether GE should continue to restrict itself to supplying NESCO, since other firms, such as Telefunken, seemed eager to purchase high-frequency alternators.[14]

Major George O. Squier of the U.S. Army Signal Corps used a 100-kHz Alexanderson alternator obtained from NESCO in a series of duplex telephony experiments reported at a meeting of the American Institute of Electrical Engineers in June 1911. Squier had an exceptional background, which enabled him to carry out original research and to be receptive to introducing the latest technology into military communications. After graduating from West Point in 1887, he earned a doctorate in physics from Johns Hopkins University in 1893. After transferring to the Signal Corps in 1899, he was instrumental in the first army experiments with airplanes and served on a cable-laying ship in the Pacific. He became chief signal officer in 1916, and he promoted the mass production of airplanes and electronic tubes for use in the war.[15] In his 1911 AIEE paper, Squier reported that he had combined the techniques of wire and wireless communication in an experimental telephone circuit in Washington, D.C. He had successfully operated a two-channel system using the Alexanderson alternator to generate the high-frequency carrier. Squier suggested that the use of wires as guides provided a solution to the persistent problem of interference between signals sent through the air and might make possible a new range of services over existing telephone lines.[16]

Frank B. Jewett, an AT&T engineer, commented on Squier's paper and expressed reservations about the commercial potential of the proposed multiplex, or wired wireless, system. Jewett stated that he had observed Squier's experiments earlier in the year and could confirm the results reported. However, he had "not been able to determine that the research, beautiful though it may be from a physical standpoint, possesses any great commercial value or possibilities." Among the reasons for his pessimistic assessment were an increase in signal attenuation at the higher frequencies, and the fact that existing toll lines with loading coils would need to be modified for use with the new system. Another discussant, Samuel G. McMeen, reminded Jewett that Bell's telephone had been just a "beautiful laboratory apparatus" before it was developed. Alexanderson, who was at the meeting, commented that he had done some experiments in which the alternator output was of considerable greater volume than the input. He also suggested that an alternator of somewhat lower frequency might be used in multiplex telephony without excessive distortion due to attenuation.[17] Despite Jewett's doubts about the multiplex application, the experiments reported by Squier had demonstrated the utility of the Alexanderson alternator as a reliable and flexible high-frequency source of energy.

Alexanderson's success in modifying the alternator so that it would generate frequencies of up to 200 kHz opened new opportunities for research on the high-frequency properties of materials, as well as on communication. Alexanderson himself began an experimental investigation of the high-frequency properties of iron. In November 1911, he informed his friend Arthur Kennelly of Harvard that the 200-kHz alternator now was "available for scientific investigation." Kennelly responded enthusiastically that he would like to have the opportunity to do experiments with "such a wonderful machine."[18] Lee de Forest, the inventor of the electronic tube known as the audion, and a pioneer wireless experimenter, requested data on the 200-kHz alternator and credited Alexanderson with being "more of an authority on the subject [of radio alternators] than anyone else in the country." Alexanderson suggested that de Forest probably could purchase an alternator from GE if he wished. In August 1911, Edmund Edwards of GE notified the president of NESCO that GE intended to begin selling the high-frequency alternators to customers other than NESCO.[19]

Alexanderson used the alternator effectively as an instrument to expand the knowledge base of the engineering science of materials. At an AIEE meeting in November 1911, he reported the results of an extensive series of tests on the high-frequency properties of iron. He stated that the principal objectives of the experiments had been to obtain data to be used in the design of high-frequency alternators and to determine the feasibility of using iron as the core of high-frequency transformers. His data on iron loss as a function of frequency led him to conclude that the common belief "that iron does not respond to high frequencies is entirely without foundation." He reported that the magnetic permeability of iron was the same at 200 kHz as at 60 Hz if the skin effect were taken into account. To demonstrate the feasibility of using iron in transformers at high frequencies, he included in his paper design calculations of a 5-kw transformer at five frequencies ranging from 60 Hz to 200 kHz. His calculations indicated that the core dimensions and weight decreased as the design frequency was increased. For example, the 60-Hz transformer had a weight of 19.5 kilograms (kg) and an efficiency of 90 percent, while the 200-kHz unit had a weight of only 1.5 kg and an efficiency of 98.3 percent.[20]

Alexanderson's mentor Charles Steinmetz, who had achieved his own professional reputation for a systematic investigation of the properties of iron at AC power frequencies two decades earlier, discussed the significance of Alexanderson's findings. Steinmetz stated that the paper was important for two reasons. First, it reported the successful development of a 200-kHz alternator, whereas Steinmetz had been "very proud when we succeeded in generating 10,000 cycles." Second, Alexanderson had provided quantitative data on the behavior of iron at "these extremely high frequencies which are far beyond any machine frequency ever produced." Steinmetz observed that the experiments had shown iron to be about 180 times as effective as air as a transformer core at 200 kHz.[21]

It may be recalled that the question of whether iron cores would be used to advantage at high frequencies had been a source of disagreement between Fessenden and Alexanderson during the early development of the high-frequency alternator. Consequently, it must have been satisfying to Alexanderson when Fessenden wrote to congratulate him on his AIEE paper and to credit him with having provided the "first definite and dependable knowledge in regard to this subject." Those working with high frequencies might now "go ahead with confidence and use iron without fear," Fessenden added.[22] Alexanderson's paper also stimulated an editorial in *Electrical World* in which it was noted that the frequency record again had been raised. The editor noted that the use of 200 kHz would reduce the necessary physical length of a resonant antenna to about 375 meters and make it simpler to load an antenna to resonance.[23]

The Invention of the Magnetic Amplifier

Early in 1912, Alexanderson successfully tested one of his most original and important inventions, the magnetic amplifier. The device was destined to become a key element in the Alexanderson system of international wireless communication. The invention served as a solution to a critical problem in wireless technology: the need to control or modulate several kilowatts of power generated by the larger high-frequency alternators.[24] The need for such a device is well illustrated by a Telefunken engineer's abortive effort in 1912 to control 7 kw of wireless power by using 72 microphones.[25] Alexanderson's invention of a ferromagnetic control system with no moving parts grew out of his experiments with an alternator as a telephone amplifier, which had begun in 1907. Although the rotary amplifier worked, its maximum safe speed and therefore its maximum frequency were limited by the need to employ laminated teeth on the rotor. The discovery that iron could be used in high-frequency transformers gave Alexanderson the final key to the invention of an alternative amplifier that did not require a rotor.

Alexanderson reported his first successful test of the magnetic amplifier in a memorandum dated January 19, 1912, and addressed to Albert Davis of GE's Patent Department. He attached a data sheet entitled "Magnetic Amplifier," showing a graph of output current as a function of a control current. He informed Davis that he had just completed an experiment in the laboratory which seemed to open new possibilities in wireless communication. According to Alexanderson's memorandum, the magnetic amplifier was simpler and more efficient than the alternator amplifier. The new amplifier had already produced a power gain of 34, and substantial improvements in the design were anticipated. The principal application Alexanderson could foresee for it was in the control of large energy levels in wireless telephony. GE had built a 35-kw alternator that wireless experts believed capable of transoceanic communication, and the magnetic amplifier would provide

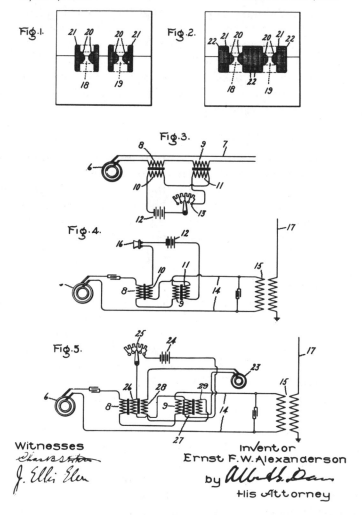

E. F. W. ALEXANDERSON.
CONTROLLING ALTERNATING CURRENTS.
APPLICATION FILED DEC. 7, 1912.

1,206,643.   Patented Nov. 28, 1916.

Figure 4.1. Patent drawing for magnetic amplifiers used to control the output of radio alternators. The drawing designated *Fig. 4* indicates the magnetic amplifier terminals by *8, 9, 10,* and *11,* while the microphone is *16* and the transmitting antenna is *17.* From Ernst F. W. Alexanderson, U.S. Patent no. 1,206,643, issued November 28, 1916.

the control capacity necessary for such a level of power.[26] An interesting sidelight of the first test of the magnetic amplifier was Alexanderson's adaptation of an old transformer for the crucial experiment. His habit of modifying old apparatus for new ends is said to have given him a reputation as being a "master salvager."[27]

The magnetic amplifier functioned by means of the dynamic interaction between two circuits carrying different frequencies and linked by an iron core. Although the device was structurally similar to a power transformer, it worked on a different principle, which did not depend on inductive energy transfer between a primary and secondary winding. A low-frequency current such as that provided by a telephone microphone was applied to a control or excitation winding. The second winding was connected in series or in parallel with the high-frequency alternator. The magnetic permeability of the iron core varied in a nonlinear fashion that resulted in amplification. The amplification factor was proportional to the ratio between the high frequency and the control frequency. Alexanderson later used a mechanical analogy to explain how the magnetic amplifier worked. He described it as being like a balanced throttle valve in a steam line, designed so that a small effort could control a large power flow.[28]

Shortly after his first test of the magnetic amplifier, Alexanderson requested an additional appropriation to support wireless telephone experiments. He pointed out that this seemed to be a very promising field and that others now were active in its development. If GE intended to maintain its leadership in the new technology, he argued, it would need to continue its research. In a written summary of the high-frequency alternator developmental program in February 1912, he noted that two distinct types of alternator had been developed. One was the type represented by the 100-kHz alternator; it was well covered by patents. The second type had been developed as a telephone amplifier and used a rotor with laminations. According to Alexanderson, this latter type of alternator probably would remain "on the shelf" if the magnetic amplifier proved as successful as he anticipated in controlling alternators of the first type.[29]

The catastrophic sinking of the *Titanic* in April 1912 made wireless communication front-page news and served as an indirect stimulus to Alexanderson's high-frequency alternator developmental program at GE. The *Titanic* struck an iceberg during a transatlantic crossing and sank within three hours, with a loss of more than 1,500 passengers. The disaster might have been even greater had a wireless distress signal not been received by an operator on the *Carpathia*, which rushed to the scene and picked up approximately 700 survivors. Unfortunately, other ships in the vicinity either were not equipped with wireless receivers or had no operators on duty. The episode brought renewed demands for maintenance of a constant wireless watch on large ships and for immediate regulation of wireless to reduce interference.[30]

THE PUBLIC tragedy of the *Titanic* almost coincided with a personal tragedy in Alexanderson's life: his wife, Edith, died on March 26, 1912, only a few days after the birth of their second daughter. Their first child, Amelie, had been born on September 23, 1910, and Edith, named for her mother, was born on March 13, 1912. A documentary gap in Alexanderson's professional papers during the months from March through August 1912 provides indirect evidence of the impact of his loss. Left with the sole responsibility of caring for two infant daughters, he retained a full-time nurse until he remarried in 1914.[31]

The Ion Controller and the Invention of a Selective Tuning System

An important new influence entered Alexanderson's life in the fall of 1912 when he met a young wireless enthusiast, John H. Hammond, Jr. In effect, Hammond, a client-inventor who dreamed of a global wireless communication system, filled the vacuum left by Fessenden. The son of a distinguished and wealthy mining engineer, Hammond had graduated from Yale in 1910 and established a private laboratory in Gloucester, Massachusetts, with financial support from his father. Hammond hired Benjamin F. Miessner, a former naval wireless operator, as a research assistant and also employed Fritz Lowenstein, a former assistant to Nikola Tesla, as a part-time consultant. In the summer of 1912, Hammond attended an international wireless conference in England and used the opportunity to learn about the latest developments in European wireless.[32] Soon after his return to the United States, he decided to purchase two of the Alexanderson 100-kHz alternators for planned experiments on the wireless control of boats and torpedoes. He instructed Alexanderson that the alternators should be capable of a field excitation of at least 1 kHz, and mentioned that he had seen a Goldschmidt alternator that could be used as a telephone amplifier at frequencies of up to 40 kHz.[33]

Alexanderson responded that experiments with the magnetic amplifier had caused him to abandon the idea of using a rotor with laminated teeth; moreover, he doubted that such rotors would be practical for frequencies of 100 kHz or higher. In October 1912, Alexanderson visited Hammond's Gloucester laboratory to discuss the proposed application of the alternators. The visit proved more fruitful than he could have anticipated, since it led to the invention of a selective tuning system for wireless receivers and to the initiation of a broad program of electronics research at GE.[34]

Hammond's entrepreneurial ambition was stimulated by his conversation with Alexanderson. A few days after their meeting, Hammond wrote to Louis W. Austin of the National Bureau of Standards that he believed that now was the "psychological time" to organize a company for the manufacture of wireless apparatus. Hammond noted that there was no American manufacturing company capable of supplying the U.S. government with needed wireless equipment, and added that there was a growing recognition of the primary importance of wireless communication for military purposes as well as a general adoption of wireless by commercial shipping interests. He went on to say that he was quite interested in arriving at some sort of cooperative arrangement with a company having the facilities and reputation of GE.[35] Hammond sent Alexanderson a copy of his letter to Austin and noted that he had learned from "inside sources" that the government planned to construct at least six high-power transmitting stations. He encouraged Alexanderson to pursue the possibility of an agreement under which GE would manufacture wireless equipment and develop new apparatus in cooperation with Hammond's laboratory. He insisted that the matter was urgent and that "time is precious now." Alexanderson replied that he had talked with several executives at GE and had found them favorably inclined toward the proposed cooperative agreement. He later reported that GE engineer-administrator Edwin Rice, Jr., who was named president of GE in 1913, had returned from a trip to Europe and would be consulted about an agreement between GE and Hammond.[36] Apparently the agreement did not materialize.

Magnetic and electronic amplifiers were among the topics discussed by Hammond and Alexanderson during their meeting in October 1912. Alexanderson summarized his magnetic amplifier experiments, while Hammond disclosed the results of experiments that Fritz Lowenstein had done with an electronic device they called an "ion controller." The device actually was a version of the so-called audion invented by Lee de Forest several years earlier. It contained three electrodes, one of which served as a source of electrons when heated in incandescence. The magnitude of current that reached a second, cold electrode was affected by the voltage applied to a third electrode, which became known as a grid. Lowenstein was among the first to discover that a telephonic signal applied to a grid produced an amplified replica in the cold electrode, or plate, circuit when the DC voltages applied to the three electrodes were properly chosen.[37]

In 1911, Lowenstein and Hammond had conducted tests using the ion controller as an amplifier of long-distance telephone conversations between Lowenstein's laboratory in New York City and Hammond's laboratory in Gloucester. They had used coded messages to indicate the degree and quality of amplification without revealing their experiments to AT&T. For example, when Lowenstein spoke of varnishing an imaginary box, it served to inform Hammond that the amplifier was being used. Lowenstein wrote to Hammond that he had "fairly jumped in delight" at the success of their first amplified call.[38]

Alexanderson had already perceived the need for a more selective and sensitive receiver of the wireless signals produced by the high-frequency alternator, and he immediately recognized a possible application of the ion controller in such a receiver. Within a few days after he returned to Schenectady from Gloucester, he wrote to Hammond to describe how several ion controllers might be used in a cascade arrangement, with each provided with a resonant circuit tuned to a single frequency. Alexanderson enclosed a sketch showing a wireless receiver with six ion controllers separating tuned circuits and connected in the cascade arrangement between the receiving antenna and a telephone. This marked the genesis of Alexanderson's interest in electronic amplifiers and of his invention of selective tuning by geometric progression, which culminated in another important wireless patent.[39]

The convergence of two streams of amplifier research at the Gloucester meeting is quite striking. It soon led to further diffusion of the technology and an accelerated rate of development as Hammond and Alexanderson brought the magnetic and electronic amplifiers to the attention of others. In December 1912, Hammond wrote to Alexanderson that George W. Pierce of Harvard, another of Hammond's consultants, already was at work on the magnetic amplifier. Hammond reported that Pierce had been able to adapt it to wireless telephony and suggested that Alexanderson and Pierce cooperate on further development. In turn, Alexanderson, by serving as a conduit of important technical information into GE, again provided an example of the benefits that Steinmetz had expected from the outside contacts of members of the Consulting Engineering Department. He arranged with Hammond to have an ion controller delivered to GE, where it led to the development of a more predictable high-vacuum electronic amplifier for use in the Alexanderson radio system.[40]

The Emergence of Electronics Research at GE

The early history of electronics research and development at GE provides an illuminating case of the interaction of science and engineering in a corporate environment. In the process of carrying out this research and development, Alexanderson of the Consulting Engineering Department and Irving Langmuir of the GE Research Laboratory became good friends and assumed leading roles in the creation of a GE tradition in electronics science and engineering. Langmuir had graduated from the Columbia School of Mines and had taken a doctorate in chemistry in Germany. He then taught for three years at the Stevens Institute of Technology in Hoboken, New Jersey, before coming to the GE Research Laboratory in 1909. His initial research was an investigation of gaseous chemistry phenomena in incandescent lamps.[41]

The possibility of using the ion controller in a wireless receiver with selective tuning led Alexanderson to seek Langmuir's help in improving the device. Langmuir's interest—and skill—in investigating thermionic and high-vacuum phenomena allowed him to identify deficiencies in the ion controller and correct them. Alexanderson, motivated by a critical problem in engineering, stimulated Langmuir and others at the GE Research Laboratory to shift their emphasis from the science of incandescent lamps to that of incandescent electronic devices. Laurence A. Hawkins, who had just become the Research Laboratory's executive engineer, introduced Langmuir to Alexanderson in January 1913.[42] Almost immediately, Alexanderson wrote to John Hammond, Jr., to request that he send what Alexanderson now called an "incandescent detector" so that the Research Laboratory could begin work on it immediately. When the requested device arrived early in February, Alexanderson notified Hawkins that he was sending it to the laboratory along with notes on some modifications proposed by Hammond. Alexanderson added that he was very eager to try the device as a high-frequency amplifier.[43]

In February 1913, Alexanderson wrote a memorandum to GE's Patent Department concerning his proposed method of selective tuning using the electronic triode; he sent copies to the GE Research Laboratory's director, Willis Whitney, and to Langmuir and Hawkins. Alexanderson pointed out that for the method to work properly, he needed a device that would enable one high-frequency current to control another with no reverse interaction. He speculated that the incandescent detector might be such a device. A few days later, Hammond wrote to Alexanderson that the device seemed to be less sensitive with an extremely high vacuum. Hammond suggested that the Research Laboratory investigators test varying degrees of vacuum, a suggestion that Alexanderson forwarded immediately to Hawkins.[44]

By late February 1913, Alexanderson completed an internal technical report on the theory of selective tuning. He stated that the most important improvement to be anticipated in the wireless art was the attainment of undisturbed communication, by means of systems that were highly selective in frequency. He added that selectivity could be improved in a geometric progression depending on the number of cascaded stages, provided that a high-frequency amplifier was available to isolate successive circuits. The report included a mathematical analysis with selectivity curves that showed the effect of adding stages. Alexanderson sent a copy of the report to Steinmetz and wrote in the cover letter that the incandescent detector in its present stage of development was "too sluggish" to function at high frequencies but that he expected the GE Research Laboratory to overcome this defect.[45]

Irving Langmuir's laboratory notebooks document the shift of his research emphasis from incandescent lamps to incandescent electronics and also provide insight into the dialectic between scientists and engineers at GE during this period. By coincidence, Langmuir began a systematic investigation of the so-called "Edison effect" the day before Alexanderson's first visit to Hammond's laboratory in October 1912. To observe this effect, an extra electrode was installed inside the glass envelope of an incandescent lamp. A current that increased in magnitude as the voltage applied to the lamp filament was increased could then be detected by a galvanometer connected to the supplemental electrode.[46] Langmuir observed and recorded an interesting secondary phenomenon whereby a blue glow in the experimental lamp was extinguished when a finger touched the outside of the bulb. During November 1912 he accumulated experimental data that included filament temperature and current and Edison effect current.[47]

Langmuir compared his results with those predicted by a thermionic emission theory proposed by Owen Richardson, a British physicist. Langmuir had discussed the theory with Richardson at a meeting of the American Physical Society the previous month and again when Richardson had visited the Research Laboratory in November. Langmuir discovered that Richardson's equation accurately predicted the current emitted from a heated filament but that a space charge that formed near the filament tended to reduce the Edison effect current below the magnitude predicted by Richardson's theory.[48] On November 22, 1913, Langmuir entered in his notebook an outline of a "new theory of Edison current" which took into account the space charge effect.[49]

When the incandescent detector from Hammond's laboratory arrived at the GE Research Laboratory in February 1913, Langmuir soon realized that it was similar enough to the Edison effect lamps that he could employ what he had already learned, including his modification of Richardson's theory. However, the direction of his research changed because of what Alexanderson told him about the potential use of thermionic devices in wireless technology. Langmuir mentioned in his laboratory notebook that Alexanderson had told him that mercury-vapor rectifiers seemed too sluggish to use at 100 kHz but that he himself anticipated that high-vacuum thermionic devices should work well at even higher frequencies. Alexanderson had experimented with mercury-vapor devices as modulators of the 100-kHz alternator and applied for a patent on the technique in June 1911.[50]

Langmuir added in his notebook that "the whole field of pure negative electron discharge is a very important one" as he embarked on the design of experimental high-vacuum triodes that would exhibit maximum sensitivity in wireless applications. He questioned Alexanderson about Hammond's claim that the ion controller was more sluggish than a crystal detector at high frequencies. Langmuir recorded that initial low-frequency tests of the first thermionic triode made at GE had revealed it to be more sensitive than the device received from Hammond and that it produced little distortion over part of its operating range. He observed that "this together with instantaneous response" should make the new vacuum device an ideal telephone amplifier.[51]

Although the first electronic amplifier designed by Langmuir was available for testing by early April 1913, the crucial test of whether it would work in high-frequency receivers was delayed because the necessary components for the tuned circuits were not immediately available. Hammond offered to make his laboratory facilities available to test the GE amplifiers, but Langmuir decided to try the triode using an Alexanderson alternator as a source before going to Gloucester. On May 12, 1913, Langmuir recorded in his notebook that he and William C. White, who had made the triode following Langmuir's design, had just tested it at a frequency of about 75 kHz and that "the results were exactly as I had expected."[52] White had graduated in electrical engineering from Columbia the previous year and was assigned as an assistant to Langmuir in 1913. He had been active in amateur radio for several years and was highly skilled in the fabrication and testing of electronic tubes and wireless circuits.[53] Two days after the test of the triodes Alexanderson informed GE's Patent Department that the Langmuir amplifier worked at 100 kHz and probably would work at even higher frequencies. He mentioned that the high-vacuum tube might be adapted for transmitting purposes if it were developed with an increased power capacity.[54]

After a successful test of Alexanderson's selective tuning circuit with two tuned amplifiers in the cascade arrangement, Langmuir recorded in a notebook entry dated May 18, 1913, that "we obtained the most striking results," that there was no noticeable sluggishness, and that "tuning by geometrical progression according to Alexanderson's scheme is an accomplished fact." A few days later, Langmuir and Alexanderson visited Hammond at his Gloucester laboratory. Langmuir recorded that Hammond had been very enthusiastic after their demonstration that the GE amplifier responded perfectly at high frequencies. Langmuir mentioned that George Pierce, a Harvard professor and author of a book on wireless, had come to witness the tests and had informed them that he had been using a condenser in series with the grid of an audion. Langmuir wrote that he understood at once that the sluggish behavior observed by Pierce was due to the condenser rather than an intrinsic property of the audion.[55]

In June 1913, Alexanderson wrote to Hammond that further improvements in the high-vacuum amplifier had been made. According to Alexanderson, Langmuir had obtained an incremental change of 5 mA in the plate current, corresponding to a change of 10 millivolts (mV) in the grid voltage. Langmuir also wrote to Hammond stating that the results were 10 times better than he had anticipated initially. Hammond then wrote Alexanderson that he had become convinced that it was not feasible for an individual or a small laboratory to build audions, since "the enormous facilities which you have in your Schenectady plant" were needed. Hammond promised to place a large order for the improved Langmuir amplifiers. Alexanderson forwarded a copy of Hammond's letter to Willis Whitney at the GE Research Laboratory and noted that it seemed likely that a considerable business would develop for the GE vacuum-tube amplifiers.[56] Alexanderson filed a patent application in October 1913 for the selective tuning system. In the patent specification, he cited two applications filed by Langmuir the same month covering the high-vacuum electronic amplifier that operated with a "pure electron discharge."[57]

Electronics research at the GE Research Laboratory developed momentum during 1913 as Langmuir and White were joined in the effort by Albert W. Hull and Saul Dushman. Hull joined the laboratory's staff in the summer of 1913 after teaching physics for five years. He became one of GE's most prolific inventors of electronic devices, and his inventions included the dynatron, the magnetron, and the thyratron. Dushman came to GE with a doctorate in physical chemistry from the University of Toronto and also became quite expert in high-vacuum techniques. His success in constructing a vacuum triode of unprecedented power-handling capability led Alexanderson to propose new applications.[58] In a memorandum to GE's Patent Department in December 1913, Alexanderson noted that Dushman's new tube had a rating of 500 watts and 20 kilovolts (kv). Alexanderson suggested that it now should be feasible to connect several of these tubes in parallel to control high levels of power. For example, he estimated that 10 of the tubes could be employed to control 100 kw if each tube were active 10 percent of the time, assuming an average power dissipation per tube of about 250 watts.[59]

By late 1913, the electronics group at GE began to employ a new nomenclature for the high-vacuum tubes. The term *kenotron* was initially selected as a generic designation for high-vacuum thermionic tubes after Langmuir and Dushman consulted with a professor of Greek at Union College. The word was derived from the Greek *kenos*, meaning empty. The term *pliotron* was coined to designate the three-element, or triode, vacuum tube; and *kenotron* gradually became limited to the two-element, or diode, vacuum tube. The new vocabulary served to emphasize that there was a difference in kind between the new tubes and the relatively low-vacuum de Forest audion. An additional incentive for the adoption of Greek-derived names was that the term *audion* mixed Latin and Greek roots and was thus a hybrid that language purists found distasteful.[60]

Alexanderson continued to work closely with Irving Langmuir and the GE Research Laboratory on wireless applications of the electronic tubes. In February 1914 he addressed a memorandum to Langmuir concerning the design of radio receivers and high-frequency circuit components. He included calculations of the theoretical voltage gain of a two-stage tuned amplifier using the Langmuir pliotrons. An interesting feature of Alexanderson's analysis was his use of an equivalent circuit for the pliotron: he represented it schematically as a voltage source in series with a 1-million-ohm resistor.[61] This may have been the earliest example of the use of a linear equivalent circuit for an amplifier tube in the analysis of active electronic networks. Perhaps it was Alexanderson's power engineering perspective that led him by analogy to adopt the same type of equivalent circuit that was often used to represent AC machines.

The rapid expansion of electronics research at GE during 1913 has been interpreted by George Wise as part of a significant change in corporate policy regarding research strategy at the GE Research Laboratory. Up to that time, the principal research focus had been directed toward the defense of the company's share of the market in established products such as incandescent lamps. The introduction of electronic tubes and related wireless apparatus provided an opportunity to move toward product and research diversification. Wise describes the shift from a strategy of defense to one of diversification as marking "a new epoch in American industrial research."[62] Alexanderson, Whitney, and Langmuir deserve considerable credit for this significant change.

An Investigation of Dielectric Hysteresis

Alexanderson made a significant professional affiliation early in 1913 when he became a member of the Institute of Radio Engineers. The IRE had been organized in May 1912, after the *Titanic* disaster. The 46 charter members had previously been affiliated either with the

Society of Wireless Telegraph Engineers, centered in Boston, or with a New York City group known as the Wireless Institute. Alexanderson became the 134th member of the IRE and was apparently the first from GE.[63] Membership in the IRE reinforced his growing commitment to the emerging profession that was changing its name from wireless engineering to radio engineering.

As one result of his membership, Alexanderson formed an important new connection with a leading member of the IRE, Alfred N. Goldsmith. Goldsmith had earned a doctorate from Columbia in 1911 and had been a member of the Wireless Institute. He was one of the three principals in the negotiations that resulted in the merger of the Wireless Institute and the Society of Wireless Telegraph Engineers. He became the long-time editor of the *Proceedings of the Institute of Radio Engineers* and, in that capacity, helped to provide a clearing house for the diffusion of the latest information on radio science and engineering. He encouraged Alexanderson and other radio innovators to publish their work in the *Proceedings*. It was in response to a request from Goldsmith that Alexanderson agreed to present a research paper before the IRE late in 1913.[64]

As the topic for his first IRE paper, Alexanderson chose the phenomenon of dielectric hysteresis at high frequencies. Hysteresis was an effect observed in magnetic or dielectric materials when they were subjected to a change in the magnetizing or polarizing force. The effect produced a small but important loss of energy in apparatus such as AC transformers with iron cores, or in capacitors subjected to an AC voltage. Dielectric hysteresis could often be neglected in the analysis of AC circuits at 60 hertz (Hz), but since it increased with frequency, it became more significant at radio frequencies.[65] Alexanderson's selection of dielectric hysteresis as a problem for investigation resulted from a conversation with Steinmetz. The striking parallel between Steinmetz's comprehensive investigation of magnetic materials and dielectric hysteresis at low frequencies during the early 1890s and Alexanderson's investigation of the properties of iron and of dielectric hysteresis at radio frequencies approximately 20 years later reveals not only the continuing intellectual influence of Steinmetz but also the continuity of research patterns at GE.[66]

Alexanderson's research on dielectric hysteresis had been assisted by Samuel P. Nixdorff, who had been assigned at Alexanderson's request to help with the measurements. Nixdorff, born in Baltimore, Maryland, in 1888, had attended the Baltimore Polytechnic Institute. He graduated in engineering from Cornell in 1910 and worked in GE's Test Department before becoming a trainee in the Consulting Engineering Department. He had prepared a technical report on a three-speed induction motor under Alexanderson's supervision in 1912. He became a long-time assistant to Alexanderson.[67]

Alexanderson presented his dielectric hysteresis paper at an IRE meeting in New York City in November 1913. He included considerable detail on how the experimental apparatus had been constructed. He explained that owing to unexpected difficulties with insulation loss and corona discharge, several changes had been necessary in the design of a high-frequency transformer that had been used in the measurement. Through the use of large spiral shields that served a dual purpose as a tuning capacitance, he had resolved the problems of providing electrostatic shielding of the transformer windings and tuning the circuit to resonance. With this design change, he had succeeded in constructing a suitable high-voltage transformer for use at 100 kHz. The experimenters placed each dielectric sample to be tested between two metal plates connected to the secondary winding of the transformer. The high-frequency alternator that served as a source generator had its output connected to the transformer's primary winding. With the secondary circuit tuned to resonance, the power dissipation in the dielectric sample could be determined accurately by taking the product of the voltage and the current, which were read directly from high-frequency meters.[68]

An unsuccessful effort to use asbestos tape to insulate the high-frequency transformer windings led Alexanderson to discover the anomalous behavior of asbestos at high frequencies. He found that asbestos exhibited a much greater dissipation at high frequencies than it did at 60 Hz, and in a typical example of his inventive style, he immediately thought of a practical application of the effect. He determined that at high frequencies asbestos had a power dissipation of about twenty times that of a sample of glass or porcelain. He asked Edward Hewlett of GE's Switching Department whether he thought this property of asbestos might not be useful for the fabrication of high-voltage insulators that would act to absorb transient surges that might otherwise propagate along power transmission lines and cause equipment damage. Alexanderson referred to the proposed asbestos protective device as a "hysteresis insulator." He later filed for a patent on this invention, demonstrating the opportunism of the inveterate inventor.[69]

In his IRE paper, Alexanderson stated that the fundamental goal of his dielectric hysteresis investigation was to obtain information that would enable the application of "systematic and scientific methods to the design of radio-frequency circuits." As Edwin Layton has emphasized, the linkage of such research to engineering design is a useful indicator of the difference between engineering and scientific research.[70] Alexanderson measured the high-frequency properties of a number of dielectric materials and presented the results graphically so that circuit designers would be able to predict accurately the dielectric loss for each element of a circuit. He stated that the power factor, defined as the ratio between the dissipation in watts and the volt-amperes absorbed, provided the "most important and characteristic information." He speculated about the "real nature" of the dielectric losses and about whether *dielectric hysteresis* was an appropriate term. After a discussion of some possible equivalent circuits for the analysis of dielectric losses, he concluded, "For practical purposes, the losses may be considered as if they were due to a true hysteresis, at a constant power factor, at all frequencies and all voltages."[71]

Alexanderson also addressed the question of a possible dielectric hysteresis in air, an issue he had raised with Steinmetz while the measurements were under way. He had informed Steinmetz that the test results seemed to show that air itself had a slight dielectric hysteresis at 100 kHz. Alexanderson acknowledged that he was hesitant to accept this interpretation and would welcome any thoughts that Steinmetz might have on the subject. In his note to Steinmetz, Alexanderson described the experimental setup and went through the analysis that had led him to conclude that there existed a slight residual loss that could not be attributed to corona discharge.[72] In his IRE paper, Alexanderson pointed out that the most perfect condenser with an air dielectric had measurable losses. Although a part of the loss might be attributed to radiation of energy into space or to incipient corona discharge, he continued, he had decided that an actual hysteresis effect in the air space between the condenser plates could not be discounted.[73]

Another discovery, which Alexanderson described as being "unexpected and striking," was that the corona loss and arc-over distance at high voltages were approximately the same at radio frequencies as at power frequencies. He noted that the measured corona loss at high frequencies agreed closely with the theory devised for power frequencies by Frank W. Peek, Jr., also a member of Steinmetz's Consulting Engineering Department. Alexanderson mentioned that Peek had helped check the high-frequency data against his theory. Alexanderson commented that his findings appeared to justify confidence in the use of Peek's theory to calculate corona loss for conditions where laboratory measurement might not be convenient. Sharing the cultural values of the Consulting Engineering Department, both Peek and Alexanderson employed a scientific methodology in much of their research, but the kind of technological knowledge needed in engineering design remained a constant touchstone.[74] In this instance, as was often the case throughout his professional career, Alexanderson's background in AC power engineering served as a valuable asset in interpreting high-frequency investigations.

A discussant of Alexanderson's IRE paper on dielectric hysteresis called it a milestone "in our emancipation from insulation difficulties," which should be "read by all radio engineers." He noted that past methods had been "empirical in the extreme" but that radio circuit design would become as systematic as the design of magnetic circuits of iron at power frequencies as a result of such research.[75]

As evidence of a growing interest at GE in the commercial potential of radio apparatus, an issue of the *General Electric Review* in 1913 featured a cover picture of the Alexanderson 100-kHz alternator and an article by Alexanderson on radio engineering. The editor predicted that Alexanderson's work would have "great consequences to the future of wireless telegraphy and telephony." The editor anticipated that the alternator would entirely supersede the spark method of high-frequency generation. He quoted Michael I. Pupin of Columbia University as having praised the Alexanderson alternator and having predicted that it soon would be in use on both sides of the Atlantic. In a paper published by the IRE in 1913, Pupin described the 100-kHz alternator as a "triumph of mechanical knowledge and engineering skill."[76]

In his *General Electric Review* paper, Alexanderson outlined what he expected would be future trends in radio engineering. He explained why he thought the alternator would supplant spark transmitters. Among his reasons were that the use of a continuous wave would increase antenna efficiency and that selective tuning would be more effective. He also noted that the alternator could be used for telephony, whereas the spark transmitter was suitable only for telegraphy. He predicted that radiotelephony would prove superior to radiotelegraphy for business purposes just as wire telephony had proved superior to wire telegraphy. He mentioned—without giving details—his magnetic amplifier, which could be used to modulate high-power radio alternators. The paper was reprinted under a different title in the *Scientific American Supplement* in May 1913.[77]

A modest market for the 100-kHz alternator developed during 1913. The Marconi Wireless Telegraph Company of America (a subsidiary of Britain's Marconi Wireless Telegraph and Signal Company), Columbia University, and the government of Japan were among the purchasers. In an internal memorandum written in October 1913, Alexanderson claimed that the alternators now were considered so valuable that their possession was being fought over. He recommended strongly that GE keep at least one alternator for use in developmental work, since the company would need to be in a position to fill orders from such important potential customers as Hammond and Marconi.[78]

Railroad Electrification and the Invention of the Phase Converter

Although no longer affiliated with GE's Railway Engineering Department after 1910, Alexanderson continued to take an active role in electric traction technology. The commercial outlook for railroad electrification appeared relatively favorable in the years just before World War I. An editorial published in *Electrical World* in 1913 called attention to the recent judicial reversal of an older ruling that damage due to smoke emitted by steam locomotives was not subject to legal liability. The basis for the reversal had been that it was now feasible to electrify railroads in urban areas and thereby eliminate the smoke pollution. The editor felt that the court decision would act as a strong incentive for accelerated electrification, although he noted that the court had neglected to reveal how the industry could obtain the capital needed for "immediate and wholesale electrification."[79] However, the long debate over the relative merits of DC and AC for electric traction continued to cloud the issue. This was alluded to in another editorial published in 1913, which commented on "the smoke arising from the battle of the systems" and expressed the hope that it would not "obscure the fundamental merits of electricity as a motive power."[80]

The authors of two papers presented at an AIEE meeting in May 1913 stressed the economic advantages of electric railroads over steam railroads. H. M. Hobart argued that the creation of large central power stations had so altered the situation that electrification projects that formerly would have been dismissed now could be "demonstrated to be highly attractive and from all standpoints commercially reasonable." He included an analysis of the costs of operating steam and electric passenger trains over a 100-mile route; it indicated that the electric train had a three-to-one advantage in terms of fuel cost per train mile.[81] Charles P. Kahler gave the results of an analysis of the cost of converting an existing steam train route in the western United States to electricity. He calculated that the conversion would cost about $8 million and that the annual return on the investment in conversion would be 12 percent. He concluded that even in such a case, with relatively low traffic density, it still would pay to electrify. A discussant of the papers suggested that Hobart and Kahler had been overly kind to steam rail proponents by assuming better results than were actually attained. He asserted that the time "has passed when the electrical engineer need adopt the position of apologist to the steam engineer."[82]

In 1914 the electric traction pioneer Frank Sprague called for the creation of an independent commission to search for ways to relieve the railroads of some of the burden of raising capital for electrification projects. He argued that the battle of the electric systems had tended to obscure the basic issue and that all interests should cooperate for the common good rather than contending on the basis of "sentiment and minor differences in transmission efficiencies." In view of what he termed the "radical differences in opinion" between leaders of the railroad industry and the electrical manufacturers, Sprague proposed the creation of a commission of "disinterested engineers" of wide experience to undertake a thorough study of the problem.[83] The editor of the *General Electric Review* commented, in an issue published in 1914 and devoted to electric traction, that technical considerations no longer were the paramount issue in the electrification of railroads: "the problem has resolved itself into one of finance."[84]

A review of existing railroad electrification projects in the United States published late in 1914 reported that there were 11 systems with a total of 286 electric locomotives and about 1,000 miles of track. The report stated that 6 of the 11 systems employed DC, 3 used single-phase AC, 1 used three-phase AC, and 1 used the split-phase system. The report also revealed that 6 of the systems used equipment supplied by GE, while the remaining five employed Westinghouse equipment. All of the GE installations used DC except for one three-phase system, while only 1 of the 5 Westinghouse systems used DC.[85]

ALEXANDERSON'S active involvement in railroad electrification and power engineering is evident from the distribution of his patents. Twenty-four of 42 successful patent applications that he filed during 1910–14 covered inventions in electric traction or power engineering. It was during this period that Alexanderson invented the phase converter, which led to the "split-phase" electric traction system. The phase converter made possible the successful operation of three-phase motors from a single-phase power supply. In its traction application, the phase converter enabled the use of efficient three-phase induction motors in electric locomotives fed from a single-phase trolley. This arrangement, known as the split-phase locomotive, combined the simplicity and rugged characteristics of the induction motor, the simplicity of single-phase power distribution, and the advantage of regenerative braking on steep grades. Alexanderson obtained more than a dozen patents on the phase converter and its applications.[86] Somewhat ironically, the split-phase traction system was introduced commercially by Westinghouse rather than by GE.

Shortly after he had filed his first patent applications on the phase converter, Alexanderson presented a technical paper on the invention at an AIEE meeting held in June 1911. In it, he pointed out that previous efforts to achieve balanced polyphase currents from a single-phase source had been ineffective. Noting that one method had used an induction motor between the supply line and a three-phase motor, he explained that its starting torque was low by comparison with that of a polyphase motor directly connected to a polyphase supply. He added that he had invented a way to overcome this deficiency by devising a connection that caused the phase converter to behave like a series transformer and to create a proper phase displacement. The use of his phase converter served to make the starting torque the same as it would be if the polyphase motor were fed from a polyphase power line.[87]

In explaining the theory of the phase converter, Alexanderson used reasoning and concepts quite similar to those he used to explain the results of his high-frequency dielectric hysteresis experiments. Instead of undertaking a "complete mathematical analysis," he stated, he had used a "synthetic method dealing only with input and output and losses." The synthetic method involved concepts such as "wattless volt-amperes," which made possible the simple calculation of the power factor or efficiency of AC circuits and machines and facilitated a comparison with data obtained directly from voltage and current readings. The method also could be depicted visually by phasor diagrams, which indicated relative magnitude and phase relationships among voltages, currents, and magnetic fluxes. Alexanderson included in his AIEE paper a calculation of the characteristic curves for a phase converter–motor combination which indicated close agreement with experimental data. In practical cases, he noted, one phase converter could supply several polyphase motors and would offer a substantial saving in weight and cost as compared to the alternative, a motor-generator set. He concluded that the phase converter should serve to reopen the prospect of using the more efficient polyphase induction motors instead of commutator motors with single-phase power sources.[88]

Alexanderson and his colleagues at GE initially tested the phase converter in electric traction by modifying an old single-phase locomotive to use a 200-hp phase converter between the single-phase source and polyphase drive motors. Alexanderson wrote that this trial had shown the split-phase locomotive system to be "thoroughly practical." He added that the test had convinced him that polyphase motors operated from single-phase distribution lines represented an attractive alternative, especially for heavy freight trains or for railroads operated in mountainous terrain.[89] After the successful operation of the modified locomotive, Alexanderson designed a more powerful split-phase locomotive that was built at GE. It utilized a 1,600-hp phase converter to drive six polyphase induction motors rated at 300 hp each.[90] However, GE did not introduce the split-phase traction system commercially.

The announcement in 1913 that Westinghouse had decided to use split-phase locomotives on the Bluefield section of the Norfolk and Western Railroad stimulated further discussion of the system in the technical press. Alexanderson ordered three copies of the issue of *Electrical World* which credited him with the invention of the system to be used by the Norfolk and Western.[91] An editorial in the same issue commented on the advantages of the phase converter technique, which made possible the use of the simple, compact, and rugged polyphase induction motor along with regenerative braking of the split-phase locomotive. The *Electric Railway Journal* also reported on this application of Alexanderson's invention and acknowledged that the Norfolk and Western contract had drawn renewed attention to the phase converter, which previously had seemed to be of only academic interest.[92]

Alexanderson described another application of the phase converter in a paper published in the *General Electric Review* in 1913, in which he explained how, rather than being used to drive polyphase motors from a single-phase source, the phase converter could be used in a power plant that was required to supply large single-phase loads as well as polyphase loads. He pointed out that unlike the older, slow-speed, salient-pole alternators, the high-speed turboelectric generators now used in modern power plants were unsuitable to supply large single-phase loads such as traction motors or electric furnaces unless special arrangements were made. The phase converter would constitute an effective solution of this problem, he suggested, for if properly designed, the phase converter was capable of "absorbing the unbalancing component of a power system and redistributing it so as to give a resultant balanced polyphase load."[93]

In 1915 phase-converter units designed by Alexanderson were installed in a Philadelphia Electric Company power plant that supplied a large traction load.[94] In 1916 Alexanderson and George H. Hill presented at an AIEE meeting held in Philadelphia a jointly authored paper on the theory and power plant utilization of the phase converter. Hill had graduated from Johns Hopkins University in 1895 and worked with the electric traction pioneer Frank Sprague until 1902, when he joined GE's Railway Engineering Department. Hill died of pneumonia a few months after he and Alexanderson presented the AIEE paper.[95]

Noting a trend toward the standardization of power plant design and the interconnection of power systems, Alexanderson and Hill urged that "general policy should be opposed to the establishment to any considerable extent of power systems having peculiar or special features making them unadaptable to efficient connection with other systems in the vicinity." Instead of building a special power plant to supply a large single-phase load, they continued, when mixed loads were anticipated it would be better to install equipment such as the phase converter in the powerhouse. They noted that all methods of phase conversion involved energy storage and exchange. For example, inductors or condensers were suitable for energy storage at high frequencies. But at power frequencies, energy storage was more readily achieved by using the inertia of rotating machinery and an arrangement such as the phase converter, which could transfer energy from one phase to another automatically.[96]

Alexanderson also contributed to the design of DC traction projects that were undertaken by GE during this period. In 1913, GE received the contract for a major DC railroad electrification project in Melbourne, Australia, when its proposal was recommended by the British consulting engineering firm Merz and McLellan. The contract called for the installation of 1,600 DC motors on 400 cars, and was described as "the largest single order ever placed for electric railway apparatus" up to that time. For the Melbourne system, Alexanderson designed a new 140-hp DC motor with improved ventilation that was provided by centrifugal fans built into the motor. The project was delayed due to World War I and did not become operational until 1919.[97]

Alexanderson also designed a regenerative electric braking system for DC motors which was introduced commercially by GE on a trunk line of the Chicago, Milwaukee, and St. Paul Railroad in 1915. The line extended 335 miles between Harlowton and Alberton, Montana, and crossed the continental divide, with its steep grades. This installation was the first to utilize regenerative braking with DC motors and the first to operate with a trolley voltage of 3,000 volts DC. The regenerative braking that served as an alternative to air brakes used the traction motors as generators on downhill grades. According to an article about this line published in the *General Electric Review* in 1916, a train descending a 2-percent grade generated about 60 percent of the energy required to ascend the same incline.[98]

During a discussion at an AIEE meeting in 1917, Alexanderson stated that regenerative braking had not been used earlier with DC traction systems because the fundamental requirements had not been understood; however, he had undertaken some experiments that had led to the discovery of some basic principles that enabled the successful design of a DC regenerative braking system. He mentioned two specific requirements: first, the traction dynamo should have a volt-ampere characteristic that ensured electrical stability; and second, the dynamo should have a speed-torque characteristic that ensured mechanical stability. In conclusion, after explaining how graphs of the machine characteristics could be used to determine the conditions necessary for both electrical and mechanical stability, he asserted that his analysis had been verified first at the GE factory, then on an experimental track, and finally on the line of the Chicago, Milwaukee, and St. Paul Railroad.[99]

In 1914, Alexanderson published a paper on the phenomenon of critical speed of railway trucks. This paper is an excellent example of his exceptional ability to be inspired by analogous reasoning and to apply insights gained from one area of technology to another. He had dealt with a somewhat similar problem in his earlier study of critical speeds of high-speed rotors in the high-frequency radio alternator. The reasoning process he employed in the two cases was quite similar, although the design implications were different. In his paper, Alexanderson stated that mechanical phenomena usually were so complex that the designer resorted to "cut and try" methods and used large safety factors. It was becoming apparent, he continued, that a "reasonable understanding of the fundamental physical principles" was needed. He wrote that the critical-speed phenomena encountered in rotating machinery were subject to the same general principles as those of an oscillating pendulum, but the critical speed of railway trucks behaved in a different manner. In critical speeds of the pendulum type, the mechanical resonance effect appeared at specific speeds but vanished at higher speeds. In contrast, in railroad trucks the effect was cumulative and became increasingly dangerous as the vehicle speed increased. He defined the critical speed of a rail truck as the speed at which the angle of reflection of a wheel with respect to the rail would be equal to the angle of incidence. The geometry he used in the analysis was similar to that of the reflection of an electromagnetic wave at an interface between two materials.[100]

Alexanderson derived theoretical equations for the critical speed and track shock in terms of weight, wheel base, and track parameters. One conclusion that resulted from his analysis was that the shock would be greater at any speed if the critical speed were reduced. He discussed some practical implications of his analysis by calculating critical speed and shock values for several alternative methods of suspending electric motors and for rigid and flexible connection between the truck and the railroad car. He calculated that the critical speed varied from 56 mph to 99 mph depending on the design used.[101]

War Begins in Europe: Its Impact on the Development of the Radio Alternator

A transition to a new phase of Alexanderson's professional life began in 1914 and accelerated with the outbreak of war in Europe, which

made the technology of transoceanic radio assume more than commercial importance. During the transition, Alexanderson became
more of a systems engineer and an administrator; his interests often
transcended departmental boundaries and even the GE corporate
boundary. His expanding horizons were indicated in a July 1914 memorandum to Steinmetz, in which Alexanderson stated that it was becoming almost impossible to determine how his salary should be
charged, since the problems he was addressing generally linked several
departments at GE. He commented that the benefits to departments
could not be measured accurately by the time he spent in each department but depended on his judgment on various engineering and
patent matters.[102]

When war began in Europe in August 1914, the need for a radio
system capable of reliable transatlantic communication became more
urgent and soon led to an accelerated rate of development of what became known as the Alexanderson system, which was based on the use
of powerful radio alternators. William White, an amateur radio operator affiliated with the GE Research Laboratory, was among the first
to learn that war had begun when on the evening of August 4 he monitored a radio transmission from the German station at Sayville, Long
Island, ordering all German ships to proceed at once to neutral
ports.[103]

A few days later, Alexanderson mentioned the war in a letter to John
Hammond, Jr., and asked how important it was to rush the development of a 50-kw radio alternator. Alexanderson noted that GE had decided to halt work on special developmental projects such as the 50-kw
alternator unless an outside customer would agree to purchase it upon
completion. Hammond replied that the war had made the prospects
for radio business so uncertain that he would not be able to place an
order at present. Alexanderson also addressed a memorandum on the
status of high-frequency alternator development to GE's president, Edwin Rice, Jr. Alexanderson pointed out that an alternator that should
be capable of transatlantic communication was about 80 percent completed and that various interested parties believed it would prove advantageous to GE to complete it.[104]

In September 1914, Alexanderson provided Steinmetz with a written
summary of the history and present status of radio development at GE.
Alexanderson noted that the National Electric Signaling Company had
paid GE for all developmental costs so long as Fessenden was in charge,
but that subsequent development had been slow. The 100-kHz alternator now was well established, Alexanderson continued, and several
units had been sold. He explained that a 35-kw alternator had never
been completed but that the experience gained from it was proving
valuable in undertaking a 50-kHz, 50-kw alternator, and he stated that

the 50-kw machine actually was the only high-frequency alternator for which GE had assumed the cost of development. He also pointed out that a broad patent position had been built up for GE which covered a complete system of radio based on the alternator. This patent coverage included his pending patent on a receiver with selective tuning which had been used to receive signals from as far away as San Francisco and Honolulu. He explained that this receiver used the Langmuir pliotrons and the heterodyne method of reception. Alexanderson described the receiver as being so sensitive that the 2-kw, 100-kHz alternator could be operated in Schenectady and detected in Pittsfield, Massachusetts, even without an antenna being connected. He predicted that the GE radio system would prove so superior to other systems that a profitable arrangement could be made with an operating company such as the Marconi Company.[105]

With regard to radio telephony, Alexanderson told Steinmetz that recent experiments pointed toward development of a complete system of long-distance telephony. He recalled that he had first tried using direct field control of an alternator for telephony but had abandoned that approach because it required a rotor with laminated teeth. He attributed the recent failure of a Goldschmidt alternator at a transmitting station in Tuckerton, New Jersey, to its having had a laminated rotor. Alexanderson added that he had come to the conclusion that the magnetic amplifier provided a superior alternative. He revealed that an improved magnetic amplifier, which used a by-product of the heat treatment of silicon steel as its core and worked well at frequencies up to 100 kHz, had now been developed at GE. The optimum radio system probably would use a combination of Langmuir pliotrons with the magnetic amplifier to control the alternator, he concluded. He predicted that this arrangement would make it possible to telephone across the Atlantic by radio.[106]

Alexanderson and his colleagues conducted radiotelephony experiments between the GE plants in Schenectady and Pittsfield beginning in August 1914. The experiments led Alexanderson to propose a system in which the radio link would be integrated with the existing wire-telephone network so that a subscriber might place calls from a home telephone which would be completed over a radio link. He continued to work actively on the design of receivers as well as on transmitting apparatus. In September 1914, he sent Irving Langmuir an analysis of a receiving antenna. Alexanderson noted that he had approached the analysis "like [that of] any other alternating-current circuit" that was "subject to well known laws." He expressed the hope that his analysis would be useful to Langmuir and others who shared an interest in solving the problem of radio interference.[107]

The potential strategic importance of transatlantic radio became evident to informed observers soon after the war began in Europe. In 1914, John V. L. Hogan wrote an article for *Electrical World* describing the German radio transmitter installation at Sayville, Long Island.[108] He reported that the German transatlantic telegraph cable had been cut by the British at a point near the Azores Islands, so that all the war news originating in Germany was reaching the United States through Sayville. According to Hogan, the receiver at Sayville, which used "special gas amplifiers" and the heterodyne method of reception, was receiving signals from a powerful transmitter located at Nauen, Germany, which had a 1,000-foot-tall antenna tower and a transmitter capable of 100 kw. He concluded by stating that direct radio transmission from Germany apparently was possible under propagation conditions that occurred almost nightly.[109] An editorial in the same issue mentioned the conflict between accounts of the war originating in London and Paris and those received from Germany. The editor wrote that there seemed no reason to doubt that the German version of the war news was being received directly by radio from Germany.[110]

Another article contributed by Hogan to *Electrical World* discussed a transoceanic transmitter located in Tuckerton, New Jersey, which used a Goldschmidt radio alternator. Hogan reported that he had been permitted to visit the station after it was taken over by the U.S. Navy on September 9, 1914. The alternator had been damaged and an arc transmitter from California had been installed until a new alternator could be obtained from Germany. According to Hogan, the Goldschmidt alternator had a laminated rotor and operated at a normal speed of 4,000 rpm to produce a fundamental frequency of 10 kHz, which was quadrupled by means of resonant circuits. Hogan credited the stations located in Tuckerton and at Eilvese, Germany, with being the first to maintain regular radio communication at such distances using waves generated by high-frequency alternators.[111]

In August 1914, American Marconi announced its intention of building a transatlantic radio station using a spark transmitter at New Brunswick, New Jersey.[112] However, perhaps influenced by the published reports of the German success in using alternators in transoceanic communication, the company had by late 1914 become increasingly disillusioned with the performance of spark transmitters and decided to contact GE as a potential supplier of high-power radio alternators. In December 1914, the chief engineer of American Marconi inquired whether GE could build a 200-kw radio alternator and, if so, what its upper frequency limit would be. He noted that the 50-kw alternator would be "rather too small" for reliable transatlantic work, although his company might find another use for it. Alexanderson replied that GE would be willing to prepare a bid on a 200-kw alternator but that he personally would prefer to furnish two 100-kw units that could be used separately or in parallel. He recommended that a complete alternator station be equipped with three 100-kw alternators with one in reserve; if necessary, all three could be used simultaneously to generate 300 kw. He mentioned that the 50-kw alternator would be ready for tests in about two weeks. GE agreed to sell the first 50-kw alternator to American Marconi for $20,000 and to supply duplicate units for $16,000 each.[113]

By 1914 Alexanderson's radio system based on the high-frequency alternator, the magnetic amplifier, and the Langmuir pliotron had thus reached a stage at which it was recognized as providing an attractive technological solution to the problem of creating the worldwide "imperial chain" envisioned by Marconi. The exigencies of war tended to eliminate barriers that might otherwise have delayed or prevented the successful introduction of the system. For example, Alexanderson and his GE associates were given access by the U.S. Navy to the giant antennas built by Marconi and Telefunken on the Atlantic coast which were needed to take full advantage of the alternator capability but were not available in Schenectady.

Alexanderson now was to leave the relatively comfortable niche of corporate problem-solver to spend a decade in an arena of power politics where his system became a pawn in the domestic issue of military versus civilian control of radio, and in disputes among nation-states over the control of international radio communication.

As MENTIONED above, Alexanderson suffered the untimely loss of his wife, mother of his two young daughters Amelie and Edith, in the spring of 1912. A happier event occurred on March 30, 1914, when he married Gertrude Robart from Boston. Gertrude enjoyed outdoor activities, including mountain climbing, and had met Alexanderson and his first wife during a family outing to Lake George, a location that became a favorite vacation spot for the Alexanderson family. Gertrude became the mother of Alexanderson's third daughter, also named Gertrude, born September 20, 1915, and his only son, Verner, born May 19, 1917.[114]

Chapter Five

Stentorian Alternator: The Creation
of the Alexanderson Radio System

During the war years, Alexanderson orchestrated the development of a transoceanic radio system based on the high-frequency alternator. His responsibilities expanded to include administrative duties, and he attempted, with some success, to influence communication policy decisions. During this period, the Alexanderson radio alternator entered its final phase of development with the completion of a 50-kw machine and the accelerated development of 200-kw units. The alternator, however, was only one element of a complex system that included arrays of towering antennas, modulators, speed regulators, high-voltage circuit components, and receiving installations. The Marconi transmitting station at New Brunswick, New Jersey, operated by the U.S. Navy after the United States entered the war, served as the proving ground of the system in much the same way that the Pearl Street station had for the Edison light and power system in the 1880s.[1] At New Brunswick, the separate elements of the Alexanderson system were melded into a reliable intercontinental communication technology that proved itself in maintaining regular contact with Europe during the final months of the war. The New Brunswick transmitting station became similar in many respects to a central power plant, although its unit of production was measured in words per minute rather than in kilowatt-hours.

Progress on the 50-kw Radio Alternator and the Start of the 200-kw Alternator Project

Early in 1915, developmental work on the 50-kw, 50-kHz alternator had advanced to the point that tentative plans were made to install the alternator at New Brunswick for tests under actual working conditions. Alexanderson informed Francis Pratt, assistant to GE president Edwin Rice, that although further minor changes might be needed, there was no doubt of ultimate success. In a February 1915 memorandum outlining the work remaining to be done on the alternator, he stressed the urgency of completing the machine since he had learned that the Marconi Company was very interested in the machine and was apt to place additional orders. He obtained data on the load characteristics of the existing antenna at New Brunswick and notified the chief engineer of

American Marconi that preliminary tests indicated that the alternator would deliver at least 50 kw to the antenna over a frequency range of 25–50 kHz.[2]

An interesting insight into the developmental process was given in a memorandum prepared by Alexanderson in April 1915. He reported that auxiliary apparatus was being developed under his supervision in cooperation with the factory foreman, and that there was a continual process of experiment but that little record of it was kept other than pencil sketches. The work had been carried out in such a manner that there was no recognized routine, he continued, although this would have to be changed once they went into production.[3] Alexanderson recalled on a later occasion that during development of the large alternators, he had worked closely with Bill Edwards, a skilled mechanic from the Turbine Department. According to Alexanderson, he and Edwards had spent weeks crawling over and under the alternator, with Alexanderson explaining the principles and depending on Edwards' skill to correct faults.[4]

Upon learning that Guglielmo Marconi himself intended to come to Schenectady to see the alternator, Alexanderson began corresponding directly with Marconi. He informed Marconi that a magnetic amplifier was being built which should enable control of the new alternator either for telegraphy or telephony. Alexanderson mentioned the discovery that the tuned circuits associated with the antenna would tend to hold the alternator at a constant speed. Marconi responded that he expected to arrive in the United States on April 24 on the *Lusitania*. He noted that he had become convinced that 150 to 200 kw of antenna energy would be required for commercial-quality operation at distances of the order of 2,500 miles.[5] The principal purpose of Marconi's trip to America was to testify in litigation between American Marconi and the Atlantic Communication Company, a subsidiary of Germany's Telefunken Company.[6]

Shortly before Marconi's visit, several representatives of his company came to Schenectady to witness the operation of the alternator. Alexanderson reported to Francis Pratt that they had seemed especially interested in the magnetic amplifier and had felt it would be of great importance in high-speed telegraphy. There seemed to be a consensus that the spark system was becoming a "dead issue" for future work, he added. He mentioned that one of Marconi's engineers had told him that the Marconi Company had to get radio alternators or get out of the business. Alexanderson concluded by noting that he and his associates were planning a demonstration of all the apparatus required for transatlantic radio.[7]

Alexanderson accepted an invitation to inspect the Marconi facilities at New Brunswick. In a memorandum prepared in early May 1915, after his visit, he reported that he had been permitted free access to all the apparatus and had been especially impressed with the antenna system, which employed thirteen 400-foot towers. He noted that Marconi preferred 200-kw alternators to the 50-kw alternator and that at least eighteen 200-kw machines would be needed to complete the planned communication network, which would include three American and six British stations. He wrote to Marconi that he knew of no reason that high-frequency alternators could not be built with a capacity of 200 kw or even more if necessary.[8]

Marconi's anticipated visit to Schenectady occurred on May 18, 1915.[9] The encounter with Marconi left a deep impression on Alexanderson's mind. More than 50 years later, he gave an animated account of the occasion when he and Marconi had stood together on a platform watching the alternator perform.[10] In early July 1915, representatives of GE and Marconi signed an agreement that provided for the continued development of the Alexanderson alternator system, with GE promising to restrict its sales to the Marconi Company. In return, Marconi agreed to purchase only alternators made by GE. The agreement included a provision that it could be terminated if Marconi failed to order alternators with an aggregate power of at least 5,000 kw in any single year. The estimated cost was given as $500 per kilowatt for the first 3,000 kw, $450 per kilowatt between 3,000 and 4,000 kw, and $400 per kilowatt for the next 1,000 kw. Alexanderson mentioned the agreement in an internal memorandum to Francis Pratt in which he stated that orders for equipment would be placed by Marconi as soon as the first alternator had been tested at New Brunswick. He noted that the design work would be started at once.[11]

A technical device that suggests the enthusiasm stimulated by the Marconi connection was mentioned in a laboratory notebook entry by Alexanderson's assistant, Samuel Nixdorff. On July 16, 1915, Nixdorff recorded that Alexanderson had proposed using an interrupter wheel for sending high-speed telegraph signals that would spell out the word *Marconi*. The device was to be used in an experiment to determine the maximum sending speed attainable with the alternator. A few days later, Nixdorff reported a test of the "Marconi wheel" at a rate of 600 words per minute with the signal being easily read.[12] In September 1915, Alexanderson notified Marconi that they had reached a sending rate of 1,500 words per minute without reaching the limit of the magnetic controller.[13]

Alexanderson had some misgivings about the relationship between GE and Marconi which probably were stimulated by his knowledge of earlier disputes between GE and Fessenden's company. He recom-

Figure 5.1. Guglielmo Marconi (on left) and Alexanderson in May 1915, when Marconi visited Schenectady to observe a radio alternator in operation. Courtesy of Verner Alexanderson.

mended to Francis Pratt that no apparatus be tested at the Marconi station until GE had the complete system operating, with all the accessories, and he stated that he wished to go on record that technical advice from the Marconi Company was neither needed nor wanted. Alexanderson urged that no arrangement be made that would permit an inference that the 200-kw alternator was designed under the direction of the Marconi Company or at their financial risk.[14]

In contrast to the increasingly close cooperation between GE and Marconi, a German radio authority, Jonathan Zenneck, was denied permission to see the Alexanderson alternator in operation. Zenneck and Ferdinand Braun had come to the United States to testify in behalf of the Telefunken Company in the litigation with Marconi.[15] In June

1915 Alexanderson responded to an inquiry from Zenneck that GE did not consider it advisable to allow him to see the radio apparatus at present but that he hoped that conditions would soon change so that they might exchange ideas. Rather ironically, the chief engineer of the National Electric Signaling Company, which had originally initiated the development of the Alexanderson radio alternator, also was refused permission to see the 50-kw alternator in 1915.[16]

The Alexanderson alternator also attracted a visit to Schenectady by a group of U.S. Navy officers in April 1915. Alexanderson informed Albert Davis of GE's Patent Department that the visitors expressed an interest in using the magnetic amplifier to control Poulsen-arc transmitters. Alexanderson recommended that the amplifier be put into production because it would place GE in a position to supply what might become an essential element of all high-power radio transmitting stations. He also informed GE president Edwin Rice of success in obtaining voice modulation of the 50-kw alternator. Alexanderson pointed out that this was sufficient power to permit transatlantic telephony under favorable conditions.[17]

Vacuum Tubes and an Alternator Speed Regulator

In April 1915, Irving Langmuir presented a paper on the work at GE on the high-vacuum tube and its radio applications at a meeting of the Institute of Radio Engineers. He reported that pliotrons had been used in radio receivers and had also been used to produce or control high-frequency power at levels of up to 1 kw. He credited Alexanderson with the invention of tuning by geometrical progression, and included a diagram of a two-stage amplifier that had given a "wonderfully high degree of selectivity." Langmuir stated that he and his associates at GE had conducted experimental two-way conversations using a combination of wire and radio links.[18] Lee de Forest commented that Langmuir's paper had been "one of the most interesting ever presented before the Institute." He added that such work was "a tribute to the exhaustive and scientific care with which such a resourceful corporation as the General Electric Company can attack any problem in which it may become interested."[19] Alexanderson also commented on Langmuir's paper, but later asked Alfred Goldsmith, editor of the IRE journal, not to publish his remarks—for reasons that Goldsmith could "well imagine."[20] The reasons presumably had to do with patent litigation between GE and AT&T over rights to vacuum-tube devices.[21]

Soon after the IRE meeting, Alfred Goldsmith wrote to Alexanderson that the IRE would be pleased to have him give a talk on the large alternator. Goldsmith added that he long had admired the "remarkable research work which had been done in the radio field by your company in Schenectady." Alexanderson responded that both he and Langmuir had been placed under a "gag rule" and would probably not be permitted to publish further papers for at least a year.[22]

Also in April 1915, Alexanderson demonstrated his exceptional ability to synthesize elements drawn from radioelectronics technology and from power technology in his invention of a speed regulator. The invention employed vacuum tubes and tuned circuits to achieve precise speed control of the radio alternator. In a memorandum prepared for GE's Patent Department, Alexanderson explained that the method depended on using rectified current to vary the speed of the alternator drive motor. He noted that extreme accuracy could be obtained by using tuned amplifiers and the principle of geometrical progression. A few days later he reported a successful test of the new speed regulator. He stressed that there was an urgent need to seek patent protection on the method because "outside parties" were working on the problem and GE should be able to show the "best possible case of diligence." He explained that the control circuit responded to a decrease in alternator speed by producing an increase in the field current of the DC drive motor, causing it to accelerate.[23] Alexanderson wrote to Albert Davis, "We have solved the problem of speed regulation for high frequency alternators for radio transmission." He stated that the regulator had held the speed to a variation of 0.05 per cent during telegraphic transmission.[24] In another memorandum, he wrote that the necessity for constant speed was "far greater" with the radio alternator than [with] any other electrical machine that had been used."[25]

The Impact of AT&T's 1915 Long-Distance Radio Experiments

Having purchased the rights to the de Forest audion in 1914, AT&T formed a radio development group early in 1915. In the fall of 1915 the group conducted a dramatic series of long-distance radiotelephony experiments with a vacuum-tube transmitter installed at the U.S. Navy's radio facility in Arlington, Virginia.[26] By early October, they reported reception of the Arlington signal in California and Hawaii. Soon afterward, AT&T disclosed one-way voice communication from Arlington to the Eiffel Tower in France.[27] The transmitter at Arlington used several hundred vacuum tubes operated in parallel for its final amplifier, which produced about 10 kw in the antenna.[28]

Alexanderson, who had called attention to the transoceanic capability of the 50-kw radio alternator several months earlier, felt some disappointment when he learned of AT&T's long-distance demonstrations.[29] He mentioned the newspaper stories about the Arlington tests in a memorandum to Francis Pratt requesting an appropriation to erect an antenna in Schenectady for use in radiotelephony experiments with the alternator. Alexanderson suggested that it might be worthwhile to radiate some strong signals, since he believed that GE was in a position to use a more powerful transmitter than any yet used in telephony. The request received immediate approval, and Alexanderson informed Steinmetz of plans to install a high-frequency transmission line between two buildings at the plant for use in the proposed radio experiments.[30] Saul Dushman of the GE Research Laboratory worked with Samuel Nixdorff on the design of a hybrid modulator using both vacuum tubes and the magnetic amplifier for use with the 50-kw alternator. Late in October 1915, voice modulation of the alternator was demonstrated for various GE observers, including Willis Whitney, Irving Langmuir, and Edwin Rice. Nixdorff recorded that Rice had "seemed quite pleased."[31]

Alexanderson discussed the various options open to GE in a memorandum to Francis Pratt soon after AT&T's Arlington experiments. He expressed his doubts that there would be any advantage gained by sending the alternator to be tested at the navy's Arlington station, since this would delay developmental work at GE. Pointing out that the newspaper reports of the transmission to Hawaii and Paris had already captured any sensational quality to be gained from a demonstration, he expressed his conviction that commercial service at such distances would require at least 10 times as much power as the AT&T engineers had used. Alexanderson added that the only demonstration of much value would be to proceed quickly to the commercial stage. He commented that he thought it would be to GE's advantage to have technical papers presented on the scientific aspects of their developmental work. He noted that such papers were appreciated by the IRE and would help retain the sympathy of the radio engineering profession and contribute to favorable business connections.[32]

Alexanderson again commented on the Arlington tests in a letter to a university professor, in which he stated that he had learned from a reliable source that a large battery of de Forest audions had been used as a transmitter. He commented that GE had had nothing to do with the Arlington experiment, although he suspected that the improvements in audions at the GE Research Laboratory had been infringed. He went on to say that GE was in the process of developing a radio system and was already able to control more power than the telephone engineers had used at Arlington.[33]

Progress on the Magnetic Amplifier

Alexanderson continued to urge a change in the GE policy that was delaying publication of information about the magnetic amplifier. In October 1915 he wrote to Francis Pratt that the need to manufacture and to publicize the magnetic amplifier was urgent. Alexanderson recommended that the new GE radio modulator be tested at the Marconi station at New Brunswick and that an IRE paper be given as soon as possible to establish GE's priority in the development. He argued that if GE did not manufacture the magnetic amplifier some other company would, with government backing, and GE would lose credit and the chance to sell the amplifiers at a profit.[34] He received the support of another GE engineer, Maxwell Day, who suggested that a first step should be an IRE paper by Alexanderson. Day continued that he thought it desirable to permit the U.S. Navy and the Federal Telegraph Company to borrow magnetic amplifiers for tests. He noted that several of the amplifiers might be sold to Federal if GE was willing to market them separately from the alternator. A few days later, Alexanderson wrote to Alfred Goldsmith that he would be able to present a paper on the magnetic amplifier at the IRE meeting in February 1916.[35]

Before Alexanderson's paper on the magnetic amplifier was read, Edmund Edwards expressed misgivings about releasing the paper until the "de Forest matter" had been cleared up. Alexanderson later wrote apologetically to Edwards that he had had to send the paper in without Edwards having seen it because Edwards had been out of town. Alexanderson stated that he had received the approval of the Patent Department. He also mentioned "the latest ruling that we should not infer that we have to get approval for publication."[36]

Samuel Nixdorff was Alexanderson's principal assistant in the development of the magnetic amplifier and was listed as joint author of the IRE paper on the subject. In a 1915 entry in his notebook, Nixdorff reviewed the history of the magnetic amplifier. He stated that the first two units had been built in 1912 for use with the 2-kw alternator, and a third had been completed in January 1915 for use with the 50-kw alternator. By July, he was at work on a fourth and a fifth version of the amplifier. After they received permission to prepare a paper, Nixdorff recorded that they were attempting to get experimental data to confirm Alexanderson's theory so that the method of design might be understood clearly and used in future amplifier design. At one point, he admitted that Alexanderson's analysis was not altogether clear to him but that he was entering it as a record of their work.[37]

Alexanderson delivered the magnetic amplifier paper at the IRE meeting in New York City in early February 1916. He explained that the amplifier had been developed as an accessory to the radio alternator in order to achieve both the advantage of control by means of a laminated magnetic circuit and the mechanical advantage of using a solid-steel rotor. He noted that the amplifier could be used either in series or in parallel with the alternator and that it served as an inductive reactance that was variable at voice frequencies. He included graphs of the amplifier characteristic, the alternator characteristic, and the combined amplifier-alternator characteristic. He stated that he and Nixdorff had obtained an amplification factor in the range of 100 to 350 with telephonic modulation. They had measured alternator power output of up to 72 kw, a level well above the nominal rating of the alternator. Alexanderson concluded that the magnetic amplifier could be applied in other ways, such as serving as a nonarcing key for high-speed telegraphy at rates of up to 1,500 words per minute. He noted that there seemed to be no limit to the level of power that might be controlled by the apparatus.[38]

In the discussion of the magnetic amplifier paper, Louis Cohen complimented Alexanderson for his "splendid research" on the high-frequency properties of iron. Cohen stated that the magnetic amplifier "unquestionably represents fine engineering skill in the conception of the method and the working out of the design."[39] Lee de Forest suggested that an appropriate distinction would be between ferric control, as used in the magnetic amplifier, and "sans-ferric" control, as used in vacuum-tube and Poulsen-arc transmitters. He noted that the vacuum-tube transmitter used at Arlington and designed by AT&T engineers cost about $10,000 per month to operate and could not compare to the Alexanderson system as yet. As the inventor of the audion, he stated, "I believe that I am entitled to express that sort of opinion of the audion if anyone is." He pointed out, however, that the advantage of the ferric system might change once more powerful vacuum tubes became available. De Forest contrasted the approach of the radio engineer, the telephone engineer, and the power engineer to the problem of generating radio-frequency power. He observed that the GE engineers had "tackled the problem from a power engineering standpoint."[40]

An editorial in the issue of *Electrical World* that contained a summary of the magnetic amplifier paper mentioned Stentor, the legendary Greek "with his brazen lungs." The editor noted that "the future electrical Stentor may have a 100 kw voice." He continued, "It is to be hoped that such tremendous utterance may be both musical and captivating, for the world would be a dreary globlet with 100 kw nasal voices perpetually shouting platitudes to the ethereal vault."[41]

Later in the year, Alexanderson explored the possibility that the GE Research Laboratory might be able to devise a way to make electrolytic iron suitable for use in magnetic amplifiers. He informed the laboratory's director Willis Whitney that the best results had been obtained using electrolytic iron made in Germany before the war. It would be very valuable, Alexanderson wrote, if the Research Laboratory could produce similar iron, possibly by depositing iron on mica with a process similar to that used to deposit copper on glass in condensers.[42] The contact between Alexanderson and Whitney provides further evidence of the frequent interaction between developmental engineering and activities pursued at the Research Laboratory.

The Multiple-Tuned Antenna and the Duplex Antenna

The transmitting antenna was both a highly visible component and a key element of the Alexanderson system and other long-distance systems. Antenna design required knowledge both of structural engineering (like that required for bridge design), and of radio engineering. The steel towers at Marconi's station were over 400 feet in height and were braced by cables anchored in concrete blocks.[43] During the winter of 1915–16, Alexanderson turned his attention to possible improvements in antenna design, and again his power-engineering perspective proved useful. His first significant antenna invention became known as the multiple-tuned antenna. In December 1915 he wrote of plans to modify the GE antenna at Schenectady, with the objective of creating a uniform potential gradient so that the highest possible input voltage could be used without causing electrostatic breakdown. A few weeks later he reported a successful test of the new arrangement, which promised to increase radiation efficiency substantially. He stated that tests already performed indicated improvement by a factor of about 2.5. Alexanderson explained that his method involved the use of several tuning coils connected to the antenna at intervals to provide separate paths to ground, thus reducing the effective ground resistance. He had already embodied the new idea in the design of a large antenna to be erected at the Schenectady works, he continued, and the innovation might also be of importance as a modification of the existing antenna of the Marconi station at New Brunswick. A few days later he reported that a further improvement had been achieved, resulting in an efficiency four times that of an antenna without multiple tuning.[44] Comparative tests of the effective radiation of ordinary and multiple-tuned antennas were made between Schenectady and Goldsmith's laboratory in New York City.[45]

As Alexanderson anticipated, the introduction of multiple tuning at the New Brunswick station resulted in a dramatic increase in antenna efficiency. His assistant William Brown, who participated in many of the early tests, later stated that multiple tuning had decreased the resistance of the New Brunswick antenna from 3.5 ohms to 0.5 ohms, increasing the efficiency by a factor of about seven.[46] From an analytical point of view, the antenna could be regarded as consisting of a small radiation resistance in series with a larger resistance. The energy radiated could be calculated by multiplying the square of the antenna current by the radiation resistance, while the larger resistance contributed to losses only. The effect of the multiple-tuning invention was to reduce the magnitude of the "wasteful resistance" without changing the radiation resistance. The advantage was somewhat analogous to that which the Edison parallel lighting system had over a series system.[47]

In March 1916 Alexanderson reported an important invention, which would permit the use of a single antenna for both sending and receiving. Double tuning would be used in the sending mode, he explained, while the antenna would behave like a magnetic loop in the receiving mode. An auxiliary transformer would be used to neutralize the effect of sending currents while at the same time causing the receiver to be sensitive to received currents. Later in the year he reported a successful test of simultaneous sending and receiving with two antennas mounted on a single tower; he and his assistants had been able to neutralize the transmitted signal so that weak signals from a remote station could be received at the same time. He went on to say that they planned to devise a commercial system of this design rather than using two separate antennas spaced some distance apart. He noted that such an arrangement would be very desirable for use on ships. The method became commonly known as duplex operation.[48]

In July 1916, Alexanderson informed Francis Pratt, assistant to GE's president, of an important experiment in duplex radio telephony. A GE executive had been called on an office telephone over a radio link to Alfred Goldsmith's laboratory at City College in New York City. Alexanderson stated that he believed this to have been the first demonstration of two-way radio conversation where the speaker could both speak and listen in the same way as over an ordinary wire telephone circuit. He noted that this was essentially the system that he and his GE colleagues planned to use in transoceanic telephony.[49]

Alfred Goldsmith, who was a professor at the City College of New York as well as editor of the *Proceedings of the Institute of Radio Engineers,* became an active participant in radio development at GE during 1916. Alexanderson visited Goldsmith's laboratory in New York City and arranged for cooperative transmission tests between there and Schenectady. Alexanderson was given a diagram of a new receiver recommended by Goldsmith in which a single vacuum tube acted simultaneously as a detector, an oscillator, and an amplifier. Samuel Nixdorff recorded in his notebook that it should be possible to use GE's oscillating pliotron in the same multipurpose circuit.[50] The importance to the work at GE of Goldsmith's access to the professional and academic communities was evident from the start. As editor of the IRE journal, he was in a position to pass on the latest information and to facilitate dissemination of the results of radioelectronics research at GE. In January 1917 Alexanderson recommended that the arrangement with Goldsmith be continued, since it had already proved very advantageous. He pointed out that Goldsmith had an extensive and accurate knowledge of the history and the literature of radio, and that conferences with him had been very helpful in developmental planning at GE. He noted that Goldsmith's laboratory facilities and skills had been at the disposal of GE, and that Goldsmith had made several inventions that had been assigned to GE. Goldsmith initially had been paid $1,000 per year for his services to GE, but Alexanderson suggested that the amount be raised to $1,600.[51]

Alexanderson prepared a technical paper on duplex telephony early in 1917, but it was not published until after the war. In the paper he noted the inconvenience of having to manipulate a switch to change from the transmitting to receiving mode. He explained that the duplex system he had devised eliminated this inconvenience. He reported that the system worked so well that signals from the West Coast could be received at the Schenectady station without interference from a transmitting antenna attached to the same mast: the receiving antenna had been protected from the strong local signal by means of a bridge arrangement similar in principle to a Wheatstone bridge. He suggested that "the corresponding equivalent in sound waves would be to have an ear which could be so adjusted that a person could stand close to a steam whistle without hearing the whistle but listen to a person speaking from a distance of a few hundred feet."[52]

The cause of the delay in publication of the paper was that the chief signal officer of the U.S. Army, George Squier, had asked that Alexanderson's paper be withdrawn because of the duplex method's possible military applications. Squier and Stanford Hooper of the navy had previously witnessed duplex operation during a visit to Schenectady, and Squier had asked that the work be kept secret. Alexanderson wrote immediately to Goldsmith, who already had a copy of his manuscript, and told him of Squier's request. He noted that Squier and Hooper had expressed "great interest" in the duplex system.[53]

A Plan to Complete the Radio Alternator System

By March 1916, Alexanderson had formulated a comprehensive plan for the allocation and coordination of the research and development needed to bring the radio alternator system to the commercial stage. He outlined the plan in a memorandum to Francis Pratt in which he pointed out that the problems covered such a wide range that there should be little difficulty in agreeing on an appropriate division of labor. It was very desirable to arrive at an understanding, Alexanderson continued, since the solutions ought not to be worked out independently one at a time because each would affect others. He listed the problems that would need to be solved to make the transoceanic radio "commercial and practical": modulation, amplification, detection, duplex operation, secrecy, prevention of interference, interconnection with wire telephony systems, and antenna design.[54] He was proposing a holistic approach to the creation of a complex technological system.

A meeting was called to discuss Alexanderson's plan and to solicit the ideas of others at GE. Among those attending were Willis Whitney, Irving Langmuir, and Chester W. Rice, the son of GE president Edwin Rice. They discussed the need to increase amplification for the transmitter and agreed that the GE Research Laboratory would try to develop vacuum tubes that could convert a whisper into a microphone into full modulation of the alternator. Amplification for receivers was regarded as already being sufficient.[55]

After the meeting, Alexanderson wrote a more detailed plan for the orchestration of the radio work. He proposed that the work be divided among five groups, with each being given primary responsibility for specific problem areas. The groups included the GE Research Laboratory, the GE Standardizing Laboratory, Goldsmith's laboratory at City College, an engineering group in building 28 at GE, and an engineering group in building 61. Goldsmith was to work on receiving antennas and methods for interconnection with wire systems and would also cooperate in making quantitative measurements of received signals. The Standardizing Laboratory, which had been created in 1896 with responsibility for developing and calibrating instruments, was to develop high-frequency instruments. The GE Research Laboratory was to develop vacuum-tube amplifiers, detectors, and oscillators. The engineers in building 28 were to work on modulators, magnetic amplifiers, and antenna measurements. The engineers in building 61 were to work on alternators, large magnetic amplifiers, tuning coils, and antenna ground resistance measurements.[56] GE's Standardizing Committee found Alexanderson's plan generally satisfactory, and in early April 1916 it approved an appropriation of $5,000 to be used for experimental work under his direction.[57]

The Search for Alternatives to the Vacuum-Tube Triode

Concern over the vulnerability of GE's vacuum-tube patents led to a search for practical alternatives. In November 1916 Alexanderson reported the successful operation of the radio alternator without the use of pliotrons in the modulator. He stated that the alternator had been controlled by means of a microphone and a magnetic amplifier alone. Not only was this important because of the pliotron patent situation,

Figure 5.2. A 10-horse team hauled a fully modern 50-kw Alexanderson radio alternator to the New Brunswick transmitting station in 1917. Courtesy of the Smithsonian Institution, Washington, D.C.

he noted, but also it would permit radiotelephony at moderate distances with minimum complexity and great reliability, characteristics that were especially desirable on ships. He suggested that this arrangement might prove to be the cheapest and most reliable radio system that could be made.[58]

The same concern was a principal motivation for experiments undertaken at about the same time by Albert Hull of the GE Research Laboratory. At Alexanderson's suggestion, Hull experimented with magnetic control of vacuum tubes as an alternative to the grid control used in pliotrons. Alexanderson informed Davis of some novel effects observed by Hull which depended on the spiral motion of electrons under the influence of magnetic fields. Alexanderson reported that, in the belief that amplification would likely be produced, he had suggested that a target electrode be placed where it would be barely grazed by the electrons.[59] This was the beginning of Hull's investigations that led to the development of the magnetron tube.

Testing the 50-kw Radio Alternator at New Brunswick

The 50-kw radio alternator finally was moved from Schenectady to the Marconi transmitting station in New Brunswick in February 1917. Before the move, Alexanderson wrote to one of the Marconi engineers that the total weight of the apparatus was about 22 tons, including 12 tons for the alternator alone. The engineer responded that it would be wise to plan to move the machine within the next two or three weeks, before the roads thawed.[60] Alexanderson and William Brown went to

New Brunswick to help with the installation, and Brown remained at the site on temporary assignment.[61] Brown had worked with Alexanderson on radio development and testing for about two years while assigned to the Testing Department. Alexanderson had recommended in January 1917 that Brown be transferred to the Consulting Engineering Department to continue his work in radio.[62]

The move to New Brunswick led to a revival of concern over the relationship between GE and Marconi. Francis Pratt wrote to GE president Edwin Rice that he wanted to make sure that it was clearly understood that the Marconi Company would not be able to claim equity in the alternator because of its cooperation in the tests. Pratt advocated that, if it were the only way to avoid misunderstanding, GE ought to insist on paying all expenses incurred in the tests.[63]

In April 1917 the entry of the United States into the war affected the radio activities at GE. Alfred Goldsmith notified Alexanderson of a presidential proclamation that placed the U.S. Navy in charge of all radio transmitters. Goldsmith reported that he was attempting to persuade the navy to permit his station to remain in operation as a Department of Commerce transmitter. He advised Alexanderson to discontinue operation of the Schenectady station until it was cleared with naval officials.[64] Official notification to discontinue all radio tests was received at GE on April 17, 1917.[65]

Alexanderson soon made a trip to Washington, D.C., to seek permission to continue radio experiments and to discuss GE work of interest to the military. He was invited to spend a day aboard the battleship *New York,* and used the opportunity to conduct duplex radio experiments. He reported that he had made up the needed apparatus from equipment found in the "scrap pile" in the ship's radio room. The experiment had worked very well, and he recorded that the naval officers felt that he had arrived at an entirely practical solution. He stated that the results had surpassed his own expectations, since he and the navy operators had been able to distinguish signals sent at almost the same wavelengths. In Washington, he also met with Gen. George Squier of the army Signal Corps.[66]

Shortly after his return from Washington, Alexanderson informed Edwin Rice that GE had received government permission to carry out experiments with the 50-kw radio alternator at New Brunswick. He also mentioned that the antenna at New Brunswick had been modified—in accordance with his multiple-tuning invention—by the installation of six tuned ground connections distributed along the original antenna. According to Alexanderson, he and his colleagues had used only 1.37 kw but had already demonstrated reduction to practice of the new high-power system that he termed "an important advance in the art." He concluded by stating that the New Brunswick station was now available for transatlantic service. Later in the month, he reported that they had reached the Panama Canal using a transmitter power of only 4 kw, and that they probably could cover the entire United States with 4 kw under ordinary circumstances.[67]

Alexanderson's optimistic reports on the performance of the alternator system may have prompted Edwin Rice's May 1917 decision to cancel the agreement giving the Marconi Company exclusive rights to the system. However, the official reason given by Rice was the Marconi Company's failure to place orders for the large alternators or to agree to share the expense of development.[68] Soon after the agreement was cancelled, Alexanderson asked that additional personnel be assigned to the radio project. In a memorandum to Francis Pratt, Alexanderson wrote that the GE engineers had "created a new art of communication that will be permanently useful." It was not sufficient to make apparatus, he continued; they must also train operators to use it. He explained that some of his assistants had been detailed to New Brunswick, making it unnecessary for him to be there and leaving him free to "cover much broader ground." However, replacements were needed in Schenectady for those assigned to duty at New Brunswick.[69]

In a memorandum addressed to Francis Pratt in July 1917, Alexanderson reviewed the overall significance of the New Brunswick tests and stated that he believed that the time had come for a determination of broad company policy. If GE did not take the financial initiative necessary to put a high-grade transoceanic system into service, the government or some other company soon would do so, he argued. He suggested that if GE wished to undertake the responsibility, an order should be placed immediately for the construction of two additional alternators, to be located in France and Great Britain. GE would thus have the system in production so as to be able to meet the anticipated demand from other nations after the war.[70]

Alexanderson again stressed the need for an early policy decision in a July 1917 letter to Edwin Rice written after a visit to New Brunswick. Alexanderson pointed out that the tests had been carried about as far as needed through his personal efforts and that the system was ready for service. He recommended that action be taken immediately to ensure that GE receive the full benefit of its considerable investment in time and money. He expressed certainty that the design features of the system could not be kept secret very long and that the improvements would soon be adopted by competitors, causing GE to lose all credit unless it acted promptly. He noted that the GE system did not require vacuum tubes when the magnetic amplifier was driven by a high-speed Wheatstone telegraph sender, and that this was an advantage from a patent perspective. He mentioned that the modified antenna at New Brunswick was the only one in existence that made such high-level radiation possible.[71]

With an eye to future contract negotiations, Alexanderson prepared data on the economics of the GE radio alternator system. He stated that it would hardly be worthwhile to undertake commercial production without a definite order of at least $1 million. He pointed out that there were 13 transatlantic cables in operation and that they were each operated at about 15 words per minute, with an investment of about $3 million per cable. He estimated the cost of each transatlantic radio link at about $500,000 with a message rate of 100 words per minute. Thus, he concluded, each radio link would carry as much information as six cables costing a total of $18 million. He noted that the addition of a multiple-tuned antenna and magnetic amplifier would multiply the value of a radio station by a factor of about five. He estimated that a charge per word of 5 cents was needed to pay for the use of cables, whereas 2.2 cents per word would be sufficient to pay for the use of a radio link that operated at 100 words per minute.[72]

Alexanderson stressed the unique opportunity that had been given to GE when it was granted access to the Marconi station at New Brunswick after the station was taken over by the U.S. Navy. In an August 1917 memorandum to Francis Pratt, Alexanderson pointed out that such an opportunity to have the use of a large station for no purpose other than experiment might never come again. He later reported that the New Brunswick station had attracted wide attention for its simplicity because the alternator delivered power to the antenna in the same way that an ordinary power generator delivered power to a transmission line. He noted that the wave produced by the high-frequency alternator was remarkable for its purity of tone and the relative absence of harmonic interference with other stations. He stated that the system made unprecedented secrecy and freedom from interference possible. In September 1917, Alexanderson reported a total expenditure on the radio system of $29,573 during the year to date, with $10,171 of the total going to production costs and the remainder to experimental costs. He stated that all the activities were interdependent and that their ultimate objective was to "develop a practical system of radio communication which represents the greatest possible advance in the art."[73]

Foreign Interest in the Alexanderson Alternator

The Alexanderson radio system at New Brunswick attracted the interest of several foreign governments during 1917. Edwin Rice was informed that the French, Italian, and British governments were all interested in acquiring radio alternators and that the Priority Board might determine which of the allies would gain the most benefit from getting the first machine. The Marconi Company attempted to prevent GE from quoting prices to the Italian government, but Rice stated that he could see no reason to comply other than courtesy to Sidney Steadman, Marconi's representative in negotiations with GE.[74]

Alexanderson attended a conference in Schenectady with members of a French scientific commission in July 1917. He reported that the French were eager to get a 200-kw station at the earliest possible date. He also inspected some radio receivers that had been brought over by the French commission and reported that they were quite similar to the receiver developed by Chester Rice at GE. However, Alexanderson observed that the pliotrons used by Rice had an amplification factor 10 times that of the tubes used in the French equipment. Alexanderson also mentioned that the French were using iron-core transformers designed along the lines he had outlined in a 1911 AIEE paper and "made from iron rolled in this country, obtained at my recommendation to Dr. Weintraub who acted as agent for the French Government."[75]

Edmund Edwards mentioned the French and Italian interest in a September 1917 letter to Anson W. Burchard, GE's representative in the intermittent negotiations with the Marconi Company. Edwards noted that there were three competing radio systems: the spark, the arc, and the alternator. He expressed his belief that the Alexanderson alternator system was the best so far developed and probably would be superior to the Poulsen-arc system at its highest level of development. He reported that this opinion was supported by the French scientific commissioner, who had recommended that his government install an Alexanderson alternator. Edwards added that the representatives of both the French and Italian governments had denigrated the achievements of the Marconi Company and had stated that it would never be permitted to operate in France. He pointed out that there were no Marconi stations in many countries and that most countries operated radio as a government monopoly. He concluded that the only markets open to Marconi other than the market for ship stations would be the market for fixed stations in Britain, Canada, and the United States. He recommended that GE offer to sell alternators to Marconi but retain the freedom to solicit other business.[76]

After a meeting with Sidney Steadman of Marconi, Edwards again wrote to Burchard, informing him that Steadman had wanted to purchase the exclusive right to sell the GE equipment. Edwards wrote that Steadman had claimed that the Marconi Company had a contract with the British government to equip a network of stations around the world. The Marconi Company was to install and operate the network throughout the British empire, and eventually would turn control over to the government. Edwards again expressed opposition to signing an exclusive contract but indicated that he would agree to an arrangement giving Marconi preferential delivery of equipment except when it was necessary to meet the needs of the U.S. military or allied governments during the war.[77]

Additional reasons for GE to decline to enter an exclusive agreement with Marconi were provided by a GE engineer, H. M. Hobart. Hobart wrote to Alexanderson from England in October 1917, advising him to try to persuade GE to retain complete control of the new radio system and develop it to the ultimate. Hobart wrote that when he returned, he would "satisfy [Alexanderson] of the importance of keeping *complete* control" (emphasis his). Subsequently, Edwards wrote to Burchard that he believed that Hobart had uncovered some reasons that made it unwise for GE to consummate a deal with Marconi. Edwards noted that he could readily believe that Hobart's reasons were good and that he would have cabled them if he had thought that a cable message would get through without trouble.[78]

Hobart's warning was clarified after his return to the United States. He reported that he had observed widespread discontent with the policies of the Marconi Company. He attributed the animosity to the company's tendency to minimize engineering progress and to concentrate instead on financial operations. Hobart asserted that Marconi had charged monopoly prices and had a policy of buying out potential competitors, thus stopping technological progress. He concluded by expressing his conviction that the Alexanderson system could be exploited in competition with Marconi with both commercial and technical success.[79] Some of the factors that would culminate in the formation of the Radio Corporation of America were already becoming evident in 1917.

The Invention of the Barrage Receiver

The perceived danger that Germany might manage to cut the transatlantic cables and try to jam transatlantic radio stimulated Alexanderson to conceive another invention, the "barrage" receiver. In October 1917, Alexanderson informed Albert Davis, a GE executive, of a conversation with Lieutenant Paternot of the French mission to the United States in which the likelihood of such action had been stressed. Alexanderson and Paternot had discussed the Germans' possible use of high-power transmitters to interfere deliberately with radio reception in France and Britain. Alexanderson reported that he had thought of a method to counteract the jamming by using two long multiple-tuned antennas. He explained that the signals received by the two antennas could be adjusted so that the relative phase difference was 90 degrees. He added that any source of quarter-phase energy could be used to produce a rotating magnetic field equivalent to that in an induction motor. For example, a signal received from the west would rotate clockwise. He proposed to use a local source to enable the total neutralization of an interfering signal. Once again his reasoning was guided by a power analogy. He had also employed the analogy of induction and synchronous motors in an earlier discussion of the possibilities of using phase modulation rather than amplitude modulation in radiotelephony.[80]

The barrage system was tested experimentally in Schenectady using two antennas separated by about a half mile and with a mobile receiver mounted on a trolley car. A signal station and a barrage station had been operated on the same wavelength. Alexanderson reported that it had been possible to separate the two signals between the antennas but not elsewhere.[81] In a memorandum to Albert Davis, Alexanderson described an "interallied radio system" based on the barrage principle. He stated that signals that originated in the United States would be received in France but not in Germany.[82] A test of the barrage method at New Brunswick was observed by Lieutenant Paternot and by a U.S. Navy representative. Alexanderson reported that after this demonstration the barrage apparatus had been packed and sent immediately to France. He stated that there was evidence that the Germans were preparing an offensive directed at transatlantic communication. He informed Edwin Rice that GE had received urgent requests to install barrage receiving systems at all the important navy radio stations.[83]

Alexanderson provided further details on the barrage-receiver invention in an IRE paper presented after the war. He noted that not only could the method be used to counteract deliberate interference in wartime but it could also be used to increase the number of usable channels in peacetime. Because of this duality, he noted, the term *barrage* was quite appropriate both because of the familiar military meaning but also because of "the original meaning of toll or stoppage prevention." He added that the antennas used in the first barrage receivers had consisted of two insulated wires extending 2 miles in each direction from the receiving station. The standard receiving set had been modified by the addition of an intensity coupler and a phase rotator, with the latter designed on the principle of the split-phase induction motor of the type he had introduced for electric traction. He explained that the position of the secondary coil of the phase rotator could be altered so that the voltage induced in it could have any needed phase relationship to the primary voltage, and that "a signal originating in any direction whatever may be neutralized by adjustment of the intensity couplers and phase rotators."[84]

He also reported that during the tests conducted at New Brunswick in 1918, he and his assistants had been able to neutralize completely the signal from the powerful New Brunswick transmitter while receiving a station in Wales that differed in wavelength by only 600 meters. They had been somewhat astonished to find that the barrage receiver could be used to pick up a weak signal from San Diego despite the fact that the San Diego station was nearly in line with the New Brunswick transmitter. He compared the phenomenon to having "a radio field glass of four miles in diameter; and for such dimensions, the focussing effect is sufficient, even at considerable distances, to produce an effective discrimination."[85]

A Project for Visual and Photographic Reception of Radio Signals

During 1917 one of Alexanderson's assistants, Charles A. Hoxie (1867–1941), was assigned to a special radio project that in retrospect may be seen as an early precursor of the radio facsimile and television development that Alexanderson took up after World War I. Hoxie already had some experience in both wireless and commercial photography when he came to work at GE in 1912. In addition to his experiments on visual and photographic radio reception at GE, he later invented a method to record sound on motion-picture film.[86]

Under Alexanderson's direction, Hoxie began work on a method of "visible radio signaling" early in 1917, a project sponsored by the U.S. Army Signal Corps. The initial objective was to develop a means to communicate directly with airplanes by radio where acoustical signals might be overcome by cockpit noise. Alexanderson wrote a memorandum for Albert Davis concerning the project in April 1917. Alexanderson reported that a radio alternator modulated by 60-Hz pulses had been used as the transmitter. He stated that the receiver had included a vibrating galvanometer tuned to 60 Hz. Alexanderson commented that the signal would be almost impossible to jam, and reported that he had heard that the British had lost a battle when the Germans had jammed British radio communication. A few days later, he reported that Hoxie had carried out a successful demonstration of visual radio signaling between a transmitter and a receiver located at Alexanderson's home.[87]

By July 1917, Hoxie had begun experiments on the photographic reception of radio signals. After observing these experiments, Alexanderson informed Albert Davis that it appeared that the same instruments might be adapted to the radio transmission of pictures. Alexanderson reported that a close examination of the photographs showed that at points with no signal, the record was a black line. A strong signal resulted in a white line. Alexanderson pointed out that by varying the signal deliberately it would be possible to obtain any degree of shading, from black to white. He explained how scanning would be achieved by means of a drum driven by a clockwork and a screw thread so that the exposed line would follow a curve that would cover the whole paper. The picture to be transmitted would be mounted on a drum, a spot would be exposed to light, and the reflected light would be focused by a lens on a light-sensitive selenium cell. The current produced would then be used to modulate the radio alternator. He suggested that such a service should be of value to newspapers and the police and would offer some advantage over the transmission of pictures by wire.[88]

By the end of the war, Hoxie's photographic recorder had been developed to the point that it could record telegraphic signals at a rate of 500 words per minute. The instrument proved capable of reading signals through severe interference that made speech reception difficult or impossible. Only a three-minute delay was required after reception until a permanent photographic record was available to be translated at the operators' leisure.[89]

The New Brunswick Station Begins Transatlantic Service

GE's experience in operating the New Brunswick station provided it with valuable economic as well as technical data. A memorandum in October 1917 stated that the station had been operated a total of 9,183 hours during the period from February to September at an estimated cost per month of $1,300. At about the same time, Alexanderson reported that the station was being held in readiness for use in an emergency and had a guard force of 32 men furnished by the U.S. Navy. He added that even if the telegraph cables were not cut, most of them were expected soon to become inoperative because it was not feasible to operate repair ships in submarine-infested waters.[90] The performance of the alternator at New Brunswick led to an urgent request from the navy for the delivery of one or more additional alternators at an early date. Edmund Edwards, a GE executive, responded that a 200-kw alternator and its associated apparatus should be ready for testing by January 1918. The same day, Edwards notified Anson W. Burchard, a GE vice-president, that he deemed it prudent to let the Marconi situation drift in view of the navy's request.[91]

As the development and testing of the radio alternator system continued, Alexanderson requested additional office space for his assistants. He explained that the rapid increase in radio work had tended to confine him to a desk and prevent him from personally monitoring all the laboratory work, and he added that he would like to establish a daily routine that would enable him to meet with his assistants in the

Figure 5.3. The 200-kw Alexanderson radio alternator and associated apparatus at the New Brunswick transmitting station in December 1918. Courtesy of the General Electric Company, Schenectady, N.Y.

office to discuss their work and give instructions. He stated that he and his assistants needed storage space for records and reports that eventually would serve as a basis for standardization of the alternator system.[92]

In January 1918, Alexanderson reported to Francis Pratt that the French had decided to build a large multiple-tuned antenna based on his design in France, at an approximate cost of $500,000. According to Alexanderson, the New Brunswick tests had been sufficiently impressive to convince the French of the need for such an antenna. Alexanderson and Nixdorff worked out a preliminary cost estimate for a 400-kw transmitter; it came to about $187,000, including $150,000 for two alternators, with the remainder covering the cost of such devices as magnetic amplifiers and tuning coils.[93]

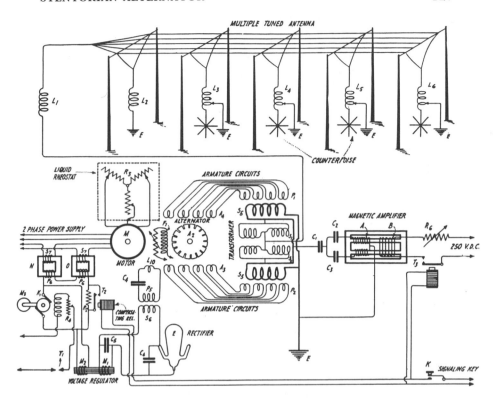

Figure 5.4. Diagram of transoceanic transmitting station at New Brunswick, N.J., including the Alexanderson radio alternator, the magnetic amplifier, and the multiple-tuned antenna. From Elmer E. Bucher, "The Alexanderson System for Radio Communication," *General Electric Review*, 23 (1920): 813–39.

The 50-kw alternator at New Brunswick began to be used for official radio traffic to Europe in February 1918. Alexanderson reported to Pratt that on February 8 the station had begun service from Washington to Rome, Italy, using a power of about 25 kw. A few days later he wrote that the station was in continuous use, handling traffic with both Italy and France. He stated that service would have been crippled without the availability of the alternator.[94]

By early March 1918, the first 200-kw Alexanderson radio alternator had been completed and tested at the Schenectady plant and was ready for installation at the New Brunswick station. A railroad strike delayed shipment of the alternator until early May. The J. G. White Company contracted to move the alternator from the railroad terminal to the transmitting site in New Brunswick and install it on its prepared base.

The move was no small undertaking, for the machine weighed about 30 tons.[95] By mid-June 1918 the alternator had been installed successfully and Alexanderson went to New Brunswick to participate in the first operational tests.[96]

Albert H. Taylor, who was engaged in radio research for the U.S. Navy, later recalled that the field intensity produced by the 200-kw alternator in the vicinity of the New Brunswick antenna was so strong that blue sparks were drawn from the bayonets of the navy sentries and sometimes "burned their fingers in a very annoying fashion." It was also necessary to ground automobiles and the nozzle of a gasoline hose at a nearby pump to prevent spark-ignited explosions. Taylor also remembered an occasion when Alexanderson had brought a GE receiver to the Belmar receiving station in New Jersey. While it was being tested by Taylor and Alexanderson using a 2,000-foot-long antenna wire, lightning had struck a tower and given Alexanderson a "pretty lively shock" through the headphone. According to Taylor, "Even this didn't cause him to quit the experiment."[97]

Although the 200-kw radio alternator was similar in principle to the earlier inductor alternators designed by Alexanderson, it was much larger in size and differed in details of design and in the auxiliary apparatus required. The rotor was a solid steel disk with a diameter of 64 inches and had several hundred radial slots milled around the rim. The rotor shaft was supported by main bearings located near the disk and by thrust bearings at each end. The bearings were lubricated by a forced oil system and also were water cooled. The water cooling system also removed heat from the stationary armature to a nearby cooling tower and pond. About 200 gallons of water per minute was pumped through the alternator cooling system. An air-gap equalizing mechanism was employed to maintain the air gap between the disk rotor and the armature on each side at about 0.035 inch during operation. The rotor was driven at speeds up to about 2,750 rpm by a two-phase induction motor and a double-helical step-up gear with a ratio of about three to one. Depending on the motor speed, the gear ratio, and the number of poles, the alternator generated frequencies in the range from about 10 kHz to 30 kHz. Once the frequency was selected, the speed variation had to be kept below 0.25 percent to maintain a constant power output. The Alexanderson speed regulator proved capable of maintaining the speed with less than 0.1 percent variation.[98]

Between June and September 1918, it was found necessary to modify the insulation of the transmitting antenna at New Brunswick to withstand the higher voltages produced by the 200-kw alternator. The speed regulation and keying system were tested, and a way was devised to change the operating wavelength quickly from 8,000 meters to

13,000 meters. Also, some changes were made in the design of the large tuning coils for the antenna.[99]

In July 1918, Alexanderson reported to Francis Pratt that the 200-kw alternator provided more power than the engineers knew what to do with. Alexanderson stated that the insulation limits of the antenna had been reached and that operation was somewhat handicapped by an inadequate power supply for the alternator-drive system. From a broad perspective, he continued, the foundations had been laid for a reconstruction of the radio art. He concluded that insofar as it was GE's policy to invest in the advancement of engineering art, the technical success had provided ample justification for GE's expenditures. Later the same month, he reported that the total development, production, and transportation costs for the 200-kw alternator system at New Brunswick had been approximately $200,000. He wrote to Pratt in August 1918 that the navy officers were now beginning to favor the alternator and had abandoned an earlier plan to install an arc unit at New Brunswick. Early in September, Alexanderson informed Edwin Rice that the 200-kw alternator had been used in tests between the United States and France and that the equipment was now at the disposal of the navy.[100]

Other nations in Europe began to express an interest in the Alexanderson radio system. In September 1918 Alexanderson informed GE vice-president Anson Burchard that GE had been approached by representatives of the Dutch and Swedish governments about the possibility of installing one of the stations. He predicted that "every sovereign government" would insist on having at least one powerful station after the war. He added that the interests that controlled the greatest number of such stations would control world communications.[101]

Soon afterward, Alexanderson reported to Edwin Rice that official tests of the New Brunswick transmitter had produced signals in France and Italy stronger than any heard before from America. Alexanderson stated that the tests of the superiority of the alternator system had convinced the U.S. Navy, and that he and Edmund Edwards had been asked to go to Washington to discuss bids on a proposed new navy transmitting station in Monroe, North Carolina. The station was to have four 200-kw alternators, and approximately $2.5 million was being appropriated for the project. Alexanderson also mentioned that the second 200-kw alternator that had been intended for installation at New Brunswick would be sent instead to a station in Marion, Massachusetts.[102] At about the same time, Stanford Hooper of the Radio Division of the navy's Bureau of Steam Engineering wrote to another navy officer about what he termed the "remarkable results" achieved by the 200-kw alternator. He continued that the alternator system seemed destined to supplant the arc system and "is the coming thing." Hooper also commented on the Alexanderson barrage receiver and described its performance as "almost unbelievable."[103]

The New Brunswick station used the call letters WII. During 1918, a navy radio operator at the station, H. R. Webster, composed a patriotic poem entitled "The Call of W Double I." Alexanderson sent a copy of the poem to a GE executive, evidently to seek approval for its publication.[104] It was eventually published, but not until after the war. A typical stanza read:

> It speaks and the whole world listens, And in the far-off city of
>   Rome
> A happy American sailor hears the call of his Home, Sweet Home.
> It comes to the listening German, As the Eagles' own warning cry,
> And he shudders as he hears that warning,
> The call of W double I.[105]

Late in October 1918, Alexanderson informed Edwin Rice of a dramatic event that had involved the alternator station in New Brunswick. He reported that on October 24 direct radio communication had been established with Germany. The New Brunswick station had transmitted a message to Germany from President Wilson calling for the abdication of the Kaiser. Alexanderson added that the antenna current had reached 563 amperes and had been limited only by the power supply available.[106] Alexanderson had been at the station at the time of the call, which was directed to the German station POZ at Nauen. He had been working with his assistant William Brown, testing the speed regulator of the 200-kw alternator, when the urgent request from Washington to use the alternator to send Wilson's message arrived.[107]

The end of the war followed soon afterward, on November 11, 1918. By coincidence, the bids on the proposed transmitting station in Monroe were opened the same day. Alexanderson expressed optimism that the end of the war would not affect construction of the new station and that GE would get the contract.[108] Stanford Hooper, however, wrote to George C. Sweet, a navy radio officer, that the navy was finding it difficult to decide between the alternator and the arc, although the alternator was "in popular favor." He added that although the navy had data indicating that the alternator was superior, "the advantage is not as great as some people may be led to believe."[109]

On November 18, 1918, Alexanderson received a message from Lieutenant Paternot of the French commission to the United States stating that the war was over but not radio development. Paternot expressed pleasure at having had the opportunity to witness "your marvelous achievement in high power radio science."[110]

Alexanderson's 1918 Analysis of GE's Policy Options in Radio

In April 1918 Alexanderson submitted an analysis of GE and the radio situation to Francis Pratt, assistant to the GE president. In it, he suggested that three considerations should be kept in mind in the formulation of GE policy: first, GE's ability to install a complete, operational system suitable for both long- and short-range communication; second, the available facilities at GE to manufacture and sell apparatus; and third, the legal and patent situation relating to the sale and operation of a system. Alexanderson asserted that he had been active in all three areas. He explained that he had tried to build a patent position that covered all phases of radio by working on the design of specific apparatus that would be needed in the event that GE decided to install and operate a system. He pointed out that the New Brunswick station was now in daily use for transatlantic service and required the lowest power ever used for such service. He added that a completely new solution of the transmitter problem had been achieved, one that was free of patent complications; and he stressed that GE could, if necessary, operate both a transmitter and a receiver without the use of vacuum tubes. He conceded that in doing so, the company would be infringing the Fessenden heterodyne patent, but stated that all long-distance radio systems were doing this. Alexanderson observed that the tube transmitter was superior for shortwave transmission over short distances but noted that GE was not in a position to exploit this because of the de Forest patents. However, he felt that GE had such a strong overall patent position that it should be able to come to terms on the use of vacuum tubes. He expressed his belief that the records of his early work on the use of vacuum tubes in radio transmission applications showed that his work antedated that done by other laboratories.[111]

Alexanderson suggested that the most natural association for GE would be with the Marconi Company, since it already had an organization and since Edward J. Nally, vice-president of American Marconi, was also head of the Pan American Communication Company, which represented both the Marconi and Federal Telegraph interests. Alexanderson speculated that Federal Telegraph had become involved in the Pan American Company because of its South American rights rather than its communications technology. If GE wished to get into the radio field independently, he concluded, it should plan to install a pair of stations in Europe and in the United States. He recommended that for legal reasons, GE should install a pair of receiving stations that did not require the use of vacuum tubes.[112]

Alexanderson continued his discussion of policy options in a May 1918 letter to Anson Burchard, in which he wrote that the development of radio now had reached a point where financial policy was more of a determining factor than was engineering. He argued that the proper attitude was not to wait but to decide if it was worth the cost to make a success.[113]

The Development of an Electric Ship Propulsion System at GE

Although during the war the principal focus of Alexanderson's activities was the development of his transoceanic radio system, he also contributed to the introduction of electric ship propulsion in the U.S. Navy. William Emmet, also a member of GE's Consulting Engineering Department, was the leading proponent of the innovation, and he took advantage of Alexanderson's virtuosity in the design of traction motors somewhat similar in type to those needed for ship propulsion. The two engineers had discussed the problem as early as 1905, although Emmet did not begin his promotional campaign until around 1909.[114]

During the next few years, Emmet published numerous papers and gave talks at every opportunity, in accordance with his philosophy that "the promotion of new things and methods, must always be to some extent one man jobs, and my definiteness and fixity of purpose was the important factor." In his view, "engineering invention is not simply a matter of the office and draughting [sic.] room." He contended that the engineer should have both a thorough knowledge and "a complete confidence in his plans, and then by some means he must impart that confidence to those who have the authority to make the knowledge useful."[115] Alexanderson shared the same philosophy as demonstrated in the case of the radio alternator and later innovations.

Emmet's campaign achieved its first substantial success in July 1911, when GE was awarded a contract to install an electric propulsion system in the *Jupiter,* a Navy collier. The *Jupiter* was one of three colliers being built at the same time, a circumstance that provided an opportunity to do comparative tests on three different methods of propulsion. The *Cyclops* was equipped with reciprocating steam engines, and the *Neptune* used the Melville-McAlpine geared turbine drive installed by Westinghouse.[116]

Emmet discussed the advantages of using electric motors for ship propulsion in an AIEE paper given in 1911. He pointed out that this method of propulsion enabled the turbines to be operated at their most efficient speed under different conditions, a feature that was especially important for warships.[117] During the discussion, Charles Steinmetz stated that the electric drive would increase a ship's maximum speed, cruising radius, and promptness of control. Another discussant suggested that the Hulmann locomotive was, in effect, one of Emmet's ships on a railroad track.[118]

An editorial published in the *General Electric Review* in 1912 stressed that the *Jupiter* project would not show off the advantages of electric drive as well as a large battleship would, but that the *Jupiter* was the only large vessel for which a contract could be obtained at the time. It was equipped with a Curtis turbine that drove an AC generator at 2,000 rpm, and a 2,750-hp induction motor was connected to each of two propeller shafts. The ship was designed to operate at 14 knots with the propellers driven at 110 rpm.[119]

The *Jupiter* was built at the Mare Island shipyard in California and was commissioned in September 1913. Emmet attended the sea trials off the California coast and reported that the turboelectric generator had delivered 6,300 hp at a ship speed of 14.78 knots. The *Jupiter*'s chief engineer, Samuel M. Robinson, was especially impressed with the fuel economy and submitted a very favorable report on the drive system. Robinson had become interested in electric propulsion after hearing a speech by Emmet at Annapolis and had been at the GE plant for a time to learn more about the system.[120] Robinson maintained his close ties with GE throughout his career in the navy and became one of Alexanderson's best friends. He later became the first director of the Bureau of Ships and was the first engineering officer in naval history to achieve four-star rank.[121] The success of the *Jupiter* led to a decision by the navy to adopt the electric drive for the battleship *New Mexico* and several other ships. After the war, the *Jupiter* was converted into an aircraft carrier and was renamed the *Langley*.[122]

Early in 1917 Emmet published an article in *Electrical World* which reported that the electric drive would be used in seven battleships and four battle cruisers that had recently been authorized. He stated that the battle cruisers presented "the most stupendous propulsion problem" ever encountered, since each ship would require a 180,000-hp system to achieve the design speed of 35 knots. He pointed out that only two or three large cities consumed this much power. In contrast, the battleships would only require 33,000-hp propulsion systems. He mentioned that opposition to the adoption of electric drive had been expressed by turbine inventors Charles Parsons, Charles G. Curtis, and others. Emmet suggested that the opposition resulted from the fact that experienced shipbuilders preferred to use "machinery they know how to build." He also noted that certain inventors had received royalties from installations that used the direct turbine drive, and that these royalties would be terminated with the advent of the electric drive which would be free of royalties because of licensing agreements.[123]

As Emmet indicated, the announcement that the navy planned to introduce the electric drive on a large scale provoked controversy. Schuyler S. Wheeler, head of the Crocker-Wheeler Company, launched a debate over the issue during 1917. Wheeler argued that in the absence of sufficient experience with the new system, the adoption of the electric drive was a dangerous experiment. He recommended the appointment of a select committee of engineers to investigate the matter. Frank Sprague, acting as a consultant to the navy, defended the navy's decision.[124]

Only two of the planned four high-speed battle cruisers were actually built, because of an armaments limitation agreement reached after the war. These two ships, the *Saratoga* and the *Lexington,* were converted to aircraft carriers. Each ship was propelled by eight electric motors described by Emmet as more than five times as large as any built previously.[125] He emphasized that a principal advantage of the electric system was that the ship would not be seriously affected by temporary loss of one or more of the motors or generators. For example, a speed of up to 19 knots was still possible with four motors and three generators out of service.[126]

ALEXANDERSON designed the propulsion motors used in the battleship *New Mexico* and received a number of patents on inventions related to ship propulsion.[127] He had originally become actively involved in August 1912, when he reported to GE's Patent Department that he might be able to obtain a broad patent in the field of ship propulsion. He explained that the method would use polyphase induction motors, with the speed to be controlled by changing the excitation of the generator. His patent application was filed the following April; it also covered a method of reversing the ship's direction by changing the motor connections.[128]

Early in 1913, Alexanderson sent Steinmetz a copy of a report on the use of a special squirrel-cage winding in a ship-propulsion motor. This described the type of motor that later was installed in the *New Mexico.* According to Alexanderson, it was originally intended for use in a 16,000-hp ocean liner that Emmet had proposed, but apparently nothing came of it.[129] Alexanderson continued to work on improvements in ship propulsion. These were covered by three patent applications filed in 1915–16.[130]

In a May 1918 memorandum to Emmet, Alexanderson discussed the electric propulsion of the *New Mexico,* the *California,* and the proposed battle cruisers. He stated that the most important new feature of the motors in the *New Mexico* was the use of a double squirrel-cage winding to obtain the required reversing torque. He also discussed the possible mistakes that might be made in operation, and means of protecting the propulsion equipment from human error. In December 1918, he spent a day aboard the *New Mexico* observing sea trials of the propulsion system.[131]

Alexanderson discussed the design characteristics of the motors used in the *New Mexico* in a paper published in 1919. He credited his assistant Albert H. Mittag with having helped with the analysis that had formed the basis for the design. (During this period, Alexanderson is said to have referred to Mittag, Nixdorff, and William Brown as his "three musketeers."[132] In his paper, Alexanderson noted that the *New Mexico*'s propulsion system was "an electric power plant of considerable size," although it was more like a constant-current system than a modern constant-voltage plant.[133] He went on to say that the motors of the *New Mexico* differed from those commonly used in industry and from those used in the *Jupiter*. The induction motors used in the Jupiter had required starting rheostats and had been designed for approximately constant speed, whereas the *New Mexico*'s motors employed two squirrel-cage windings and did not require starting resistances. One of the windings had a high resistance; it was used in starting and to provide a high reverse torque when needed. The second winding had a low resistance, which facilitated high-efficiency operation during normal operation at full speed. Alexanderson stressed that the motors and generators had to be designed as composite units, and he used graphs with the propeller characteristics superimposed on the motor-generator speed-torque characteristics to illustrate various operating situations. To reverse the ship's direction, he noted, the motor connection, the generator excitation, and the turbine governor were all changed simultaneously. Combinations of propulsion motors and generators were available to deliver power levels of 32,000, 16,000, or 8,000 hp as needed. In conclusion, he noted that simplicity and reliability had been primary goals in the design.[134]

During the official trials of the *New Mexico,* in December 1918, it was found that the motors could be reversed at a speed of over 21 knots and the propellers brought to full astern within 70 seconds.[135] Samuel Robinson reported that the *New Mexico* was from 20 to 30 percent more economical of fuel than was the *Pennsylvania,* which used direct turbine drive. He stated that another advantage of the electric drive was that the propulsion machinery could be more dispersed and could be better protected.[136]

SOMETIME during the war, Alexanderson gave a talk entitled "Efficiency and Democracy" at a meeting of the Agora, or "Fortnightly Club," in Schenectady. He contended that the war had proved the fallacy of the economic theory that the world could not afford a war. He also questioned the doctrine that labor should not be wasted on non-productive tasks but insisted that the stress should be on finding worthy things for people to do with their lives. Despite Americans' "lamented extravagance," he continued, they seemed generally more happy and prosperous than Europeans. He suggested that those who preferred warships to steam yachts should so indicate on their "wishing list." Although they might not have the pleasure of operating the ships unless they were naval officers, he added, "we will all have the fun of making them." He asserted that "a sweat shop system where efficiency of production is developed scientifically for its own sake has obviously no social value."[137]

He went on to say that a change in values was needed so that the accumulation of capital was no longer a central motive. Money should be regarded as a "useful lubrication of the social machinery but not as the motive power," he suggested. He predicted that the time would come when the federal government would subsidize needed technological development such as the electrification of railroads. He conceded that some might see such ideas as socialistic but contended that they were "true democracy," in which it was recognized that ample capital was needed to keep the social machinery lubricated, even if provided by the government.[138]

At the end of 1918, as he approached his forty-first year, Alexanderson had good reason to be proud of his achievements as a professional engineer-inventor. On December 31, 1918, he filed his 154th patent application, entitled "System of Radio Communication."[139] He had averaged almost 14 patent applications per year during the preceding four-year period of intense development of his radio alternator system. He had crossed the threshold from being an inventor of elements for other people's systems to being the promoter and architect of a communication system that was soon to be given his name.[140] During the next year, he would become something of a public figure and symbol as he and his radio alternator became a focal point of a campaign to prevent British domination of international radio, a campaign that led to the creation of the Radio Corporation of America.

Chapter Six

"Alexanderson the Great": The Origins of the Radio Corporation of America

With the end of World War I in 1918, the Alexanderson transoceanic radio system was a technology ripe for commercial exploitation. The unresolved issue was who would control it. The U.S. Navy had at least temporary custody of the powerful station at New Brunswick where the only 200-kw Alexanderson alternator was located, and where the technical merits of the system had been amply demonstrated. Key naval officials and officers, including Secretary of the Navy Josephus Daniels and Stanford Hooper, the head of the Radio Division of the navy's Bureau of Engineering, hoped that the navy's tight wartime control of American wireless could be continued. They soon reluctantly came to realize that military control would not be feasible in the postwar political climate. Thereupon they turned to a new strategy, aimed at preventing the operation or ownership of wireless stations on American soil by foreign interests. The foreign interest that was perceived as the most threatening was British Marconi, which was resuming negotiations with GE aimed at obtaining sole rights to the Alexanderson system.

After a complex process of lobbying and negotiations among the several interested parties, a new corporation known as the Radio Corporation of America was formed in 1919. The new firm acquired the assets, including the technical and administrative personnel, of American Marconi, resulting in a suspicion in some circles that all foreign influence had not been eliminated. With strong encouragement from the navy's radio enthusiasts, GE executives played a major role in the formation of RCA, which as a result gained full access to GE's radio technology and productive facilities. RCA also acquired, on either a temporary or a permanent basis, the services of several GE radio specialists.

Alexanderson was installed as the first chief engineer at RCA while continuing to work on a shared-time basis for GE. His new position brought with it challenges he had not faced before, and he learned that technology involved more than patentable inventions and the development of exotic hardware. Among the more immediate and obvious difficulties he encountered was the divided allegiance of a staff that included the former chief engineer of American Marconi and several of his assistants. Alexanderson's primary responsibility was to preside over the world deployment of the Alexanderson radio alternator system, an urgent task if RCA were to establish credibility as the world's foremost international telecommunication firm. Much of his energy was necessarily devoted to negotiations at international conferences and to arranging appropriate sites for his alternator transmitters.

During his tenure as RCA's chief engineer, Alexanderson also reached the pinnacle of the radio engineering profession as president of the Institute of Radio Engineers. Despite the sometimes frenetic pace of these activities, he managed to continue making patentable inventions that addressed problems in transoceanic radio systems. He became especially interested in experiments with the transmission of pictures by radio, an interest that by the late 1920s led naturally to experiments with television. The management of RCA saw the need for an effective public relations campaign from the start to gain public support and to counter the skepticism of some in government that the company was not as patriotic as its name implied. Alexanderson's achievements were grist for the public relations mill, and he became something of a celebrity in an era when most celebrities were athletes or entertainers rather than engineers.

Professional Recognition from the Institute of Radio Engineers

Alexanderson was quick to perceive that the official end of hostilities had eliminated the restrictions on publication of technical papers. In a December 1918 memorandum to Albert Davis of GE's Patent Department, he noted that the chief signal officer of the army, Gen. George Squier, was presenting a paper on developments in aeronautics during the war. Alexanderson suggested that this opened the way for GE to get military approval to publish papers on topics that had been held up at Squier's request, such as Alexanderson's own paper on duplex radio. He also commented that he did not believe that any problem that the company had had with the Signal Corps could be attributed to Squier. He stated that Squier had been among the first to use a radio alternator and foresee its possibilities. Moreover, Squier had been the first government official to recognize the importance of the multiple-tuned antenna and had exerted his influence to permit the continuation of experiments at New Brunswick when the war started. According to Alexanderson, Squier had proposed building a transatlantic transmitting station for the army's use but had been overruled in an agreement between the secretary of war and the secretary of the navy.[1]

In January 1919, Alexanderson wrote to Alfred Goldsmith, editor of the *Proceedings of the Institute of Radio Engineers,* and expressed satisfaction that the military restrictions on publications had been removed. Alexanderson wrote that he now wished to publish his paper on duplex radio, which enabled operation of transmitter and receiver at one location, with some additions based on work done during the war.[2] A discussant of the paper, which was presented at an IRE meeting in April 1919, credited Alexanderson with having disclosed the method of balancing transmitting and receiving antennas to the navy early in the war, and thus having enabled a general investigation of such methods to be launched at the Washington Navy Yard.[3]

Alexanderson, who had declined an invitation to be nominated as vice-president of the IRE in 1915, wrote to Goldsmith in February 1919 that he hoped to assume a more active role in the Institute's affairs. Alexanderson commented that the IRE enjoyed a considerable advantage in being an international organization, and was probably in a better position than any other institution to undertake an initiative for the benefit of the radio art which would be independent of competing commercial and governmental interests.[4] Alexanderson served as vice-president of the IRE during 1920 and as president the following year.

In 1919 Alexanderson became the second recipient of the prestigious IRE Gold Medal. The award was intended to recognize an individual who was responsible for a major advance in the science or art of radio communication. To be eligible, the recipient had to have described the innovation in a publication of recognized standing, and the innovation had to be in actual use.[5] When he received the medal in New York City on May 17, 1919, Alexanderson briefly reviewed the history of radio. He credited Marconi with having been the first radio engineer, and Fessenden with having first conceived the possibility of bringing radio engineering into the realm of power engineering.[6]

Lee de Forest evidently complained that Alexanderson had not mentioned his contributions to radio, for Alexanderson wrote to de Forest that he had felt that it would have been inappropriate to discuss work other than that which had led directly to the developments for which the medal was given. Alexanderson went on to say that he thought it was only fair that credit be given to de Forest rather than to the highly advertised 16-story building. (Presumably this was a reference to AT&T, which had acquired the rights to de Forest's audion patents.) He concluded by saying that such a building might house an efficient organization for the design and manufacture of apparatus, "but no amount of advertising can convince me that sixteen story buildings make fundamental inventions."[7]

The Invention of the Magnetron

It was concern over GE's vulnerability to outside control of vacuum-tube patents that motivated Alexanderson to resume a search for alternatives to the vacuum tube. Although it had been found possible to modulate the Alexanderson alternator without the use of vacuum tubes, sensitive receivers still required the use of vacuum-tube amplifiers. In mid-January 1919, Alexanderson reported to Albert Davis of GE's Patent Department on recent experiments on the magnetic control of vacuum tubes. Alexanderson enclosed theoretical calculations and a tentative design of a vacuum tube with magnetic control. He stated that the operation depended on the tendency of electrons in a magnetic field to follow paths like those of comets, and he noted that there was a critical value of the magnetic field that would cause the electrons to graze the anode.[8]

Albert Hull, of the GE Research Laboratory, worked closely with Alexanderson on electronic devices that might circumvent the need to use conventional vacuum-tube triodes such as the GE pliotron. In a memorandum to Hull in February 1919, Alexanderson stated that a magnetic valve was needed for use in a receiver to make possible a practical demonstration of reception with two-element tubes. He went on to say that if Hull could make a two-element tube with negative resistance, it could probably be used in reception with approximately the same efficiency as the pliotron. Alexanderson mentioned that he was working on the design of a dynatron detector in anticipation of the availability of Hull's newly invented tube.[9] The dynatron exhibited negative resistance as a result of secondary emission from an electrode bombarded by electrons. Hull had already reported in a 1918 IRE paper that the dynatron could be used as an amplifier or oscillator over a wide range of frequencies.[10] Alexanderson soon informed Albert Davis of a successful demonstration of reception that had not required a triode. According to Alexanderson, Hull's dynatron had been employed in a circuit that promised to make GE independent of the de Forest and Armstrong triode patents.[11]

Alexanderson continued to work on the two-element tube with magnetic control. In April 1919, he reported further experimental tests of the device and stated that he had predicted its characteristics on the basis of a theoretical analysis. He termed this one of the rare instances in which a design based solely on theory had worked as expected on the first try. The analysis had indicated that the device should exhibit a negative resistance over a range of the current in a control winding. If so, it could be used as an amplifier, oscillator, or detector and be substituted for a triode for many purposes. Since the tube worked on a different principle than any developed previously, Alexanderson continued, an appropriate new name was needed. He mentioned that he had considered calling it a "comet valve" or a "boomerang valve" but had finally settled on "ballistic valve." His report included a characteristic curve showing the plate current as a function of the control current, and clearly demonstrating the negative resistance behavior.[12]

In May 1919, Alexanderson informed Albert Davis that the ballistic valve had been used successfully as a local oscillator to generate the signal needed for heterodyne reception of continuous waves. Alexanderson stated that the quality of reception had been equivalent to that seen when the triode was used as the oscillator, and that the magnetic valve had also been used to amplify telephonic signals. He pointed out that the ballistic valve, when combined with a synchronous-resistance detector, gave GE the practical equivalent of a receiving system covered by the patents of Fessenden, Vreeland, Armstrong, and de Forest.[13] The synchronous-resistance detector had been proposed by Alexanderson as an alternative to the heterodyne receiving method invented and patented by Fessenden.[14]

Although the initial motivation for the development of magnetic control of electronic tubes was to circumvent radio patents not held by GE, Alexanderson's continued interest in electric power technology stimulated him to conceive other possible applications. In a June 1919 memorandum to Albert Davis, he suggested that the magnetic valves might be used in the conversion of high-voltage direct current to alternating current in traction systems. Alexanderson pointed out that the magnetically controlled tubes were more rugged than triodes and would make possible the use of a gas such as argon, which was not feasible with electrostatic-grid control. He explained that gas-filled tubes would permit much larger currents than vacuum tubes did, and mentioned that it might ultimately prove desirable to use metallic envelopes rather than glass for power tubes. He stated that he had been thinking about the design of an electric railway with a DC distribution system and an electronic converter that would make it possible to control motor speed by altering the frequency of the converter output.[15] One of Alexanderson's young assistants, David C. Prince, had been assigned to work on the DC-AC converter. In mid-June 1919, Prince reported that an experiment was being done to see whether a 1-kw vacuum-tube converter was feasible; if so, the GE engineers might try to develop a 100-kw converter.[16]

In September 1919, Alexanderson recommended to Willis Whitney, head of the GE Research Laboratory, that Albert Hull be provided all necessary facilities to develop a commercial-quality magnetic tube. Alexanderson stated that the tubes already developed were adequate for use as detectors in order to resolve patent difficulties. He also mentioned that the devices had been tested as oscillators up to frequencies of 200 kHz and might be used to construct shortwave apparatus.[17] Hull's work continued, and by January 1920 Alexanderson had decided that the magnetic tubes might prove superior to the pliotrons, regardless of the patent situation, because of their lower cost and more rugged characteristics. He reported that he and his colleagues had achieved a power level of about 300 watts with one of Hull's tubes and hoped to reach 1 kw using a tube of larger dimensions. Alexanderson expressed his approval of a suggestion that the generic term *magnetron* be adopted for this most recent addition to the "tron family."[18]

Hull stressed the potential power applications of the magnetron in an AIEE paper published in 1921. He remarked that electrical engineers were accustomed to thinking of electron tubes as "interesting playthings not engineering tools." Such a view was mistaken, he continued, since electronic devices were not inherently small and had already evolved from milliampere to ampere ratings. He predicted that kiloampere ratings would soon be achieved. Hull characterized the magnetrons as the fourth method to have been developed for control of electron discharge. He stated that the diode and dynatron worked by getting electrons from metals by "boiling or splashing," while the triode and magnetron used electrostatic or magnetic control of electrons already emitted. He concluded that the principal applications of the magnetron would probably be in power engineering rather than in radio.[19]

Communication with President Wilson's Ship

President Wilson's participation in peace talks held in France during 1919 provided an opportunity for some interesting and well-publicized communications experiments involving the Alexanderson system at New Brunswick. In December 1918, a few days after Wilson's arrival in France on the *George Washington,* Alexanderson informed Willis Whitney that the New Brunswick station had been operated continuously in an effort to keep in touch with the ship during the voyage, and that the navy hoped to communicate by voice on the return trip. Since vacuum tubes would be used for some of the apparatus needed, Alexanderson asked that Whitney assign someone at the Research Laboratory who was skilled in high-vacuum technology to the project.[20]

By late February 1919, when President Wilson returned to the United States during a month-long interruption of the negotiations, arrangements had been made for one-way voice transmission from Washington, D.C., to the *George Washington.* On February 22, when the ship was still several hundred miles at sea, Secretary of the Navy Daniels completed a call to the president from Washington using the 200-kw alternator at New Brunswick. Alexanderson submitted a report to Edwin Rice, GE's president, on the event and its significance. He wrote that this test transmission marked a development that differed in an important way from those made during the war, and he pointed out that the experiment placed GE in a favorable position to obtain patent rights on principles requisite for long-distance telephony. He compared the importance of the innovation tested to that of the Pupin loading coil used in conventional wire telephony.[21]

Alexanderson provided Rice with a firsthand account of how the stage had been set for the demonstration at New Brunswick. He stated that when he had arrived in New Brunswick on February 18 there had seemed little likelihood of success, because the necessary connecting apparatus had just been installed and had not been tested. Alexanderson mentioned "an amusing incident" he credited with contributing to the demonstration's eventual success. In order to run tests on the system using outside control, the operators of the transmitter had invited sailors at New Brunswick to keep the system operating by talking to their girlfriends by telephone. This had made it possible for Alexanderson and his assistants to monitor the effect of the system on the girls' voices and to make necessary adjustments. He stated, however, that the complete system, including the alternator, had not been tried until the telephone connection was made to Secretary Daniels's office in Washington.[22]

In a February 1919 memorandum to Edmund Edwards, a GE executive, Alexanderson stated that the station at New Brunswick had become an important public utility, whether it was to be operated by the government or by the Marconi Company. He commented that so long as it was navy policy to further development of the radio art, he thought it appropriate to complete the work while the station was under the navy's control. Alexanderson credited the navy with having a farsighted policy and reported that he had initiated an accelerated effort that involved GE's Consulting, Production, and Switch Board Departments.[23]

President Wilson arrived back in France for a resumption of the peace conference on March 14, 1919. Alexanderson informed Edwin Rice that transoceanic telephony had been demonstrated when signals from New Brunswick were received on Wilson's ship in the harbor in Brest. Alexanderson added that he and his colleagues hoped to employ a duplex system on the return voyage. At about the same time, Alexanderson suggested to Stanford Hooper that it would be desirable to install duplex equipment on the *George Washington* so that President Wilson and Secretary Daniels might be able to converse both ways as conveniently as over an ordinary telephone.[24]

In 1920, John H. Payne of the GE Research Laboratory published an account of the radiotelephone experiments on the *George Washington*. He stated that in March 1919 the Research Laboratory had received a request from the navy to build and install on the ship a radiotelephone transmitter the president could use to communicate with Washington during his return from France. Payne reported that he and his co-workers had built a transmitter using pliotron tubes. The set had been designed to operate at a wavelength of 1,800 meters and had been installed on April 12, 1919. Payne revealed that the system had been tested during a round trip to the United States in May without the president on board, and had been used during Wilson's return voyage in July 1919.[25]

Further details on the two-way radiotelephone experiments were provided by Harold H. Beverage, an assistant to Alexanderson, in a technical paper published in 1920. Beverage credited Alexanderson with having proposed an arrangement of antennas and tuned circuits to achieve two-way communication from the *George Washington*. Beverage stated that he and Payne had been aboard the ship during two transatlantic crossings. He mentioned that in Brest they had been able to receive voice signals from New Brunswick intermittently, and on one occasion they had heard Alexanderson's assistant William Brown singing "America." Beverage reported that President Wilson had spoken with Assistant Secretary of the Navy Franklin Roosevelt on July 8, 1919, via a circuit between the ship and Washington, D.C. He stated that reception from New Brunswick had been generally good except when the ship was between 400 and 600 miles from port.[26]

Postwar Negotiations over Control of the Alexanderson System

During the period of the *George Washington* radiotelephony experiments, decisions were being made that culminated in the formation of RCA. Although Alexanderson was not one of the principals in the negotiations, he gathered much of the technical and economic data that formed the basis for the decisions. He took great pride in his role as the chief designer of the potential global communication system for which the New Brunswick station was the prototype, and he seemed less concerned about whether the system would be operated by the navy or by a private corporation than about whether it would function effectively. The creation of RCA is an exceptional instance of the interaction of technology with domestic and international politics. Three institutions had participated in the creation of the New Brunswick facility: the Marconi Company, GE, and the U.S. Navy. The same three institutions shared responsibility for the creation of RCA, although it proved expedient to deemphasize Marconi's role.[27]

Stanford Hooper, head of the navy's Radio Division, became one of the principal proponents of a policy of excluding foreign interests from control of radio facilities in America. Three days before the signing of the Armistice, Hooper wrote a memorandum to the director of naval communications, Capt. David W. Todd, recommending that the issue of government ownership of radio stations be placed on the agenda of the peace conference. Hooper also suggested that a former director of naval communications, Adm. William Bullard, be released from his current assignment to participate in activities related to radio.[28] Bullard was in fact appointed to succeed Todd a few weeks later and returned to the United States in March 1919.[29] The following day, Hooper wrote a fellow officer, George C. Sweet, that he expected British Marconi to oppose government ownership "as they are very high handed due to their strength." However, Hooper continued, "the American Marconi Company is very progressive now and very pro America and [is] cooperating with us in all our plans." In late January 1919, Hooper informed Sweet that the bill for government ownership of radio had failed to pass and that public opinion was opposed to government ownership of public utilities.[30]

Meanwhile the negotiations between GE and the Marconi Company over the Alexanderson alternator resumed. In a January 1919 memorandum to Anson Burchard, the GE representative, Alexanderson wrote that it might be helpful to have information on the engineering aspects of a possible cooperative arrangement. He asserted that the engineering reasons for such an arrangement were quite strong and should be weighed carefully against any legal or political objections. If GE were to have the opportunity to install the alternator at all the Marconi stations, he continued, it would serve to establish GE as the unchallenged leader in the field. He suggested that the system would prove comparable in value to the existing submarine cable system. Alexanderson also pointed out that GE would be at a serious disadvantage without access to the antennas of the Marconi stations. He noted that from Marconi's perspective, there were three alternatives to his alternator: the Poulsen arc, the spark system used at Carnarvon, and the German alternator of the type used at Nauen. He argued that the Poulsen arc was incompatible with the Marconi antennas, the spark transmitter was unreliable, and the German alternator was less efficient than his alternator.[31] Shortly thereafter, he recommended to Edwin Rice that GE establish a basic price of $100,000 for the 200-kw alternator, not including the power supply and other auxiliary apparatus.[32]

In late January 1919, Alexanderson attended an IRE meeting held in Washington, D.C., at which he had an opportunity to talk with the chief engineer of American Marconi, Roy Weagant.[33] Alexanderson informed Albert Davis that Weagant had told him that the Marconi Company was very impressed by the GE alternator system and wanted to use it for all transoceanic stations. Alexanderson mentioned that Commander Hooper had announced that the radio field was again open for private enterprise and that the government would encourage further development.[34] The IRE meeting in question took place soon after the government ownership bill had failed to pass in Congress, and approximately one month before the first radiotelephone contact with President Wilson on the returning *George Washington.*

By late March 1919, the negotiations between GE and Marconi had reached a point at which Alexanderson evidently assumed that an agreement was imminent. He provided Weagant with data on the New Brunswick station and with recommendations for equipment that would be needed for new installations. The data included the station's power consumption and reliability of service. According to Alexanderson, there had been no equipment failure since the 200-kw alternator had gone into service, and the station had operated under control from Washington 93 percent of the time during February. If an arc unit were used for the same service, he continued, the power consumption would be "absurdly high" by comparison with that of the alternator system. He proposed four station configurations, using the New Brunswick station as the standard, rated at 100 percent (equivalent to a single 200-kw alternator). Alexanderson suggested that other standard stations be rated at 70, 140, and 200 percent. He estimated that the cost of a 70-percent station would be $125,000 for the alternator and $100,000 for the antenna towers. He computed the total cost for a 200-percent station at $590,000, which included $250,000 for alternators and $340,000 for the towers. Since the estimated cost of supplying power to such a station was $27,500 per year, he calculated that the cost of power per word transmitted would be 0.377 cents at a rate of 7.3 million words per year.[35]

On March 26, 1919, Anson Burchard reported to GE's management that American Marconi was prepared to place an order at once for 14 200-kw Alexanderson alternators and that British Marconi would want at least 10. He mentioned that the Marconi companies planned to use the alternators only at stations where the demand would be at least 500,000 words per year.[36]

A few days later, Stanford Hooper received a letter that gave him the opportunity for which he had been waiting to intervene in the negotiations. Hooper had developed a strong aversion to the existence of foreign-owned radio transmitters on American soil as early as 1913, when he was fleet radio officer and the German station at Sayville had interfered with naval communications. After his appointment as head of the navy's Radio Division in 1915, Hooper was in a position to determine or influence policy, and apparently missed no opportunity to pursue the task of "Americanizing American radio." He later claimed to have pressured American Marconi to employ only American citizens as radio operators and executives in order to get orders for radio apparatus from the navy.[37]

Officially, Hooper received word of the GE-Marconi contract negotiations in the form of a letter forwarded to him from the office of the assistant secretary of the navy, Franklin Roosevelt. The letter was from Owen D. Young, a GE vice-president; it was dated March 29, 1919, and had been written in response to the navy's request that Alexanderson be assigned to develop an improved speed-control system for a German-built alternator located at Sayville. Young objected to the proposal on the grounds that GE preferred not to engage in developing one element of a non-GE system. In the letter, Young also gave some details of the impending agreement with the Marconi Company. Hooper revealed several years later that he had already known about the negotiations but had needed some official notification in order to act. He stated that he had been informed of the negotiations several weeks earlier by Edmund Edwards of GE, who had been motivated by a "deep sense of patriotic duty."[38]

According to Hooper, he and E. H. Loftin of the navy's Radio Division had already discussed the question of what American company was best situated financially and technically to launch an "ideal American radio corporation." They had narrowed the field to two prospects, GE and AT&T. Choosing GE seemed to them to be best because it would enable them to block the sale of the best existing high-power system to a foreign company, an outcome consistent with Hooper's goal of establishing a virtual Monroe Doctrine of radio. The almost simultaneous arrival of Admiral Bullard from France and Young's letter to Roosevelt served Hooper's purpose perfectly. Hooper and Sweet quickly informed Bullard of what they had learned about the impending sale of the Alexanderson alternator system. Not surprisingly, they found him receptive to their plan to create an American-controlled radio company.[39]

In April 1919, Admiral Bullard arranged a meeting with GE executives at which he and Commander Hooper urged that GE not enter the proposed contractual agreement with the Marconi Company. GE was represented at the meeting by Edwin Rice, Owen Young, Albert Davis, Charles W. Stone, and Edmund Edwards. Several years later, Young disclosed that he had been informed privately by Bullard during the meeting that President Wilson had expressed his concern that American interests in radio communication be protected. Bullard had conferred with Secretary of the Navy Daniels in Paris and had either talked with Wilson directly or been informed of his view on radio by an aide. The precise nature of the president's recommendations to Bullard remains a mystery, although in response to critics of the process by which RCA was formed, Young later cited Wilson's involvement.[40]

The GE executives yielded to Bullard and Hooper's persuasion and agreed to suspend negotiations for the sale of the alternator system. Later in the month, Young traveled to Washington for further talks with Bullard and Hooper on what should be done. The naval officers encouraged GE to organize a new radio company that would install and operate a high-power radio system capable of worldwide communication. In response, Young asked that the navy give its official sanction to the proposed enterprise. Bullard and Hooper were willing to do this, and Commander Loftin was assigned the task of helping Albert Davis of GE to draft an appropriate document. The draft agreement specified that all employees of the new company in policy-making positions or operators of transmitting and receiving stations would be American citizens. It required that ownership and voting rights be at least 80 percent American, and gave the navy veto power over the sale, either by GE or by the new company, of radio apparatus capable of ranges of over 1,000 miles.[41]

Before a scheduled meeting with Secretary Daniels in May 1919, Hooper prepared a memorandum that summarized how the government would benefit if the terms of the draft agreement were to be implemented. He pointed out that the establishment of centralized control would help to solve the serious problem of radio interference. Since Congress had been unwilling to give this control to the navy, Hooper continued, the best alternative was to attempt to have it "placed in the safest possible hands." He predicted that there would be both domestic and foreign opposition, so that a parent organization was needed that was "absolutely free from foreign influence or domination" and strong in prestige, productive facilities, and financial backing. GE seemed to fill these requirements superbly, he stated, and besides, "its loyalty is unquestioned."[42]

The meeting with Daniels to consider the draft agreement between GE and the navy was held in May 1919 in Washington, D.C., with Roosevelt, Bullard, Hooper, Young, and Nally in attendance. Daniels, who still hoped for eventual government control of radio, declined to sign the document, at least until he had consulted certain members of the Cabinet and Congress. GE, with Hooper's informal encouragement, decided not to await the outcome of Daniels's inquiries and proceeded without official government sanction.[43]

An early consequence of GE's decision to expand its radio activities was the formation of a new department at GE, with Alexanderson as its head. The department was called the Radio Engineering Department, and officially began operations on May 1, 1919.[44] Alexanderson outlined the functions of the department in an April 1919 memorandum to Francis Pratt, in which he stated that the Radio Engineering Department would undertake general research and development, the manufacture of radio apparatus, and the design and engineering of radio stations which would be built on contract.[45] Most of the personnel who initially were assigned to the new department had already been working on radio or power electronics projects under Alexanderson's direction. His staff of assistants increased from 15 in March 1919 to 23 a short time after the Radio Engineering Department was established.[46] Many members of his staff eventually ended up with RCA.

The romantic and nationalistic qualities of the 200-kw Alexanderson alternator at New Brunswick served as the theme of a paper by GE publicist Charles M. Ripley, published in July 1919. He described the achievements of the station as a "romantic story of accomplishment, out-rivaling the imagination of a Jules Verne." He reviewed the use of the transmitter to send President Wilson's ultimatum to Germany, and speculated on "what romantic stories will someday be written of the struggle of American wireless engineers to outdo the Germans." He pointed out the contrast between the array of giant towers at the most powerful station in America and the mule-drawn canalboats on the nearby canal. Ripley listed some of the physical characteristics of the big radio alternator, and credited it with expressing "the will of the American people to a listening world." No device had ever been made that could equal the alternator in clear enunciation, he wrote: New Brunswick "doesn't slur its words." He concluded by asserting that radio technology had now reached a state such that only capital and business initiative were needed to establish a communications system that could duplicate cable telegraphy in speed, accuracy, and economy.[47]

By late July 1919, negotiations between GE and American Marconi had reached a point where Owen Young of GE and Edward J. Nally of American Marconi signed a statement of intent, with a final agreement being contingent on the outcome of further negotiations with British Marconi.[48] Nally and Albert Davis were chosen to go to England to try to work out the terms that would, in effect, make American Marconi into the nucleus of a new radio company, although with its "foreign" name, Marconi, erased.

Alexanderson was notified in late July 1919 that Davis was going to England to try to negotiate a deal with British Marconi and needed all available information to refute possible arguments that the Alexanderson system was no better than others or that specific elements of the system were not indispensable. Alexanderson prepared a fairly detailed brief outlining the virtues of his system and the deficiencies of others. He conceded that no single feature of his system was really indispensable but stressed that it was a complete transmitting and receiving system that did not "infringe the patents which heretofore dominated the radio situation." He described it as "the best system that has been developed up to the present time" and the only complete system under the control of one corporation. He argued that the Alexanderson alternator was superior to the French-designed Latour alternator and that, in any case, he had "invented that type alternator before Latour" and had American and British patents on it that antedated those of Latour.[49]

It was not only the patent situation that made the system valuable but also "the engineering and manufacturing organization which is behind it," Alexanderson continued. He credited a large number of individuals and departments at GE with having contributed to the success of the system. The mechanical design of the radio alternator reflected Emmet's experience with high-speed machine design, and groups headed by Erben, Farquhar, and Steenstrup had worked on problems related to manufacturing, while organizations headed by Whitney, Robinson, Hewlett, Laycock, and Peterson had contributed to the production of other elements of the system. Because of the diversity of skills that had been and would be needed, Alexanderson commented, the principal function of the Radio Engineering Department "will continue to be one of steering this whole development along consistently thought out lines." He ended his brief with the assertion that the development in such a short time of a complete radio system that could supersede those used previously proved GE's flexibility and placed the company in a "position to maintain the lead that we have acquired in the radio technique."[50]

Alexanderson anticipated that as many as 100 radio alternators might be needed in a worldwide radio system. In early August 1919, he prepared a memorandum stating that the 200-kw alternators could generate any wavelength between 10,000 meters and 20,000 meters. Everything now was standardized except for the power transformer, he noted. The preliminary plans were to utilize four groups of wavelengths in the United States, and this would necessitate 50 alternators for American business and the same number for foreign communications.[51]

In the course of thinking about the design of the world system, Alexanderson decided to prepare a paper on the subject for presentation at a joint meeting of the AIEE and the IRE which was scheduled for October 1919. When the completed manuscript was submitted for approval, he wrote in a cover letter that it seemed desirable that the paper be published, since it gave the technical basis for an international radio system. He pointed out that an international conference would soon be held to consider the regulation of the wavelengths used in radio, and that his paper would be helpful to the conference delegates.[52]

Alexanderson's paper, entitled "Transoceanic Radio Communication," was presented on October 1, 1919, in New York City. He devoted the first section to a discussion of whether the available technology of longwave radio was adequate to meet the anticipated needs. He observed that the task of satisfying these needs would not be easy, but that at least the subject of long-distance radio had "emerged from the cloud of mystery that used to surround it, and we are in position to treat the problem coolly and scientifically like any other problem in electrical engineering." He defined a "first class station" as one that radiated such power that its signals could be received throughout the world. There were currently only 5 such stations, he stated; 2 in the United States and 1 each in England, France, and Germany. Alexanderson pointed out that without some improvements, there would only be room for a total of 12 first-class stations in the most desirable band (with wavelengths between 10,000 meters and 20,000 meters). Since the average transmission rate for each station was about 20 words per minute, he concluded that 12 stations would not be sufficient to satisfy the demand. However, he explained, there were several promising ways to extend capacity, such as increasing the rate to perhaps 100 words per minute, improving receiver selectivity to accommodate more channels, and using more directional antennas. He estimated that a combination of these improvements might increase the information-handling capacity of transoceanic radio up to 200-fold.[53]

In the second section of his paper, Alexanderson discussed the characteristics of the 200-kw transmitter at the New Brunswick station and provided the first published account of the multiple-tuned antenna. He used equivalent electrical circuits to explain the theory of the antenna that had at New Brunswick proved its superiority to the original Marconi antenna. He drew an analogy between the multiple-tuned antenna and a high-voltage power transmission line with loads at intervals along it. Alexanderson also described the operation of the electronic speed-control system of the alternator he had invented. He noted that speed regulation had been more complicated because instead of a DC motor of the type used to power the earlier 50-kw alternator, an AC induction motor was used to drive the 200-kw alternator. However, he stated, the problem had been solved through the use of a saturable-core coil that served to adjust automatically the voltage applied to the induction motor as the load on the alternator changed. He explained that he used a resonant circuit to detect slight changes in speed and a vacuum-tube rectifier circuit to produce control current in the saturable-core coil.[54]

Alexanderson Becomes Chief Engineer at RCA

Edward Nally and Albert Davis successfully completed their negotiations with officials of British Marconi in early September 1919, opening the way for the final steps in the process of incorporating a new
American radio company.[55] In early October 1919, in a memorandum
to GE's president, Edwin Rice, Alexanderson discussed his own potential role in the new company. He remarked that it seemed he was expected to take the lead in formulating technical policies and in
influencing international relations in the development of the radio. In
order to function effectively, he added, he would need an official position in the radio company. He anticipated that his tasks would require
delicate judgment in the deployment of the system and in the control
of expenditures for research and development. He proposed that he
be paid a minimum combined salary of $20,000 per year, partially in
the form of stock dividends from the radio company, with the balance
from GE; this would contribute to his effectiveness in the new company
and would serve to safeguard his future and his peace of mind. Earlier
in the year, he had signed a three-year contract with GE which provided
for a salary of $9,500 for the first year, increasing to $11,000 in the
third year.[56]

The Radio Corporation of America was granted its official charter
by the state of Delaware on October 17, 1919, and a preliminary agreement by American Marconi to seek its stockholders' approval was
signed by Edward Nally and Owen Young on October 22, 1919.[57] The
following day, Young wrote a confidential letter to Nally informing him
of strong opposition from officials of the navy and the U.S. government
who believed that Marconi was taking over GE's radio interests. Young
explained that "very grave opposition" had arisen when Davis had notified the navy that Nally, a vice-president of American Marconi, was
to be the president and chief operating officer of RCA. Young added
that the agreement between GE and British Marconi which had been
viewed so favorably by the government was being seen differently since
it appeared that it would be executed by former members of the Marconi organization. "Such an unfortunate reaction must be prevented,"
he concluded, for "if the new company is to start right, that point of
view must be corrected."[58]

The agreement between RCA and American Marconi was approved
on November 20, 1919. The agreement specified the assets of American Marconi that were being acquired by RCA—including the New
Brunswick station, as well as other stations located on the East Coast,
in California, and in Hawaii.[59] The membership of RCA's first board
of directors included four executives from GE, including Edwin Rice,

Owen Young, and Albert Davis; and four from American Marconi. Young was selected as the first chairman of the RCA board, and Nally, despite the opposition aroused by his appointment, became RCA's first president.[60]

A cross-licensing agreement between RCA and GE, also signed on November 20, 1919, included trade restriction clauses that became controversial. The 25-year agreement gave RCA an exclusive license to use and sell radio devices that were covered by patents held by GE, or that might be invented in the future by GE employees. In addition, GE agreed that except for sales to the United States government or to complete existing orders or contracts, it would not sell patented radio devices directly for use in the United States or elsewhere. Under article 5 of the licensing agreement, RCA agreed to purchase 12 Alexanderson alternators and accessories at a price of $127,000 each, with payment to be in the form of 304,800 shares of preferred stock. GE promised to furnish technical information and assistance in the installation of the radio equipment.[61]

In late November 1919, Alexanderson reported that nine of the 200-kw radio alternators for RCA were already under construction and that orders for eight more were anticipated shortly. He recommended that the budget for GE's Radio Engineering Department be increased by 32 percent for the year 1920 owing to the need to build up the necessary organization to achieve the new plans. He reported that the department's expenditures during 1919 had been $108,680.[62] In a December 1919 memorandum to Albert Davis, Alexanderson commented on how the New Brunswick station might be used to best advantage after it was turned over to RCA. He stated that he had been told by David Sarnoff, a former Marconi employee who now was the commercial manager of RCA, that the station would be used for traffic to England. Alexanderson pointed out that this might well encourage Germany to build a transoceanic station in the United States. To forestall this possibility, he stated, he had come up with a plan to operate two alternators at New Brunswick, with one to be used for German communications and the other for British.[63]

Alexanderson's official position with RCA had still not been decided when he wrote to Edwin Rice in early December 1919 giving reasons why he should be designated chief engineer. Alexanderson stated that he understood that Roy Weagant, former chief engineer of American Marconi, would be given an independent position as consulting engineer with RCA and provided with research facilities.[64] Alexanderson went on to say that Nally had proposed that he also be called a consulting engineer but that he did not consider this adequate. If this were done, Alexanderson pointed out, RCA would have two consulting engineers of apparently equal rank and no chief engineer. He noted that this arrangement would place him in a difficult position, since he would have the duties of a chief engineer without the title, and he would be dealing with an engineering staff that had previously reported to Weagant. He explained that he would need to gain the confidence of the whole engineering organization, which would include both GE and former Marconi engineers, and blend them into a single group with a common purpose. There was a great likelihood that this effort would fail, he concluded, unless his authority was made clear from the start by giving him the title of chief engineer of RCA, even though half his salary was paid by GE and his employment contract remained with GE.[65] His new position became official on January 1, 1920.[66]

Alfred Goldsmith, editor of the IRE *Proceedings* and former radio engineering consultant to GE, became the director of RCA's Research Department. He had terminated his consulting contract with GE in May 1918 to become research director at the Marion, Massachusetts, laboratory of American Marconi.[67] In December 1919 he prepared a memorandum for Edward Nally outlining the goals of the Research Department for 1920. Goldsmith stated that he planned to concentrate on the "commercial research" needed to develop radio apparatus. He wrote that he did not think that purely scientific research that would produce results of doubtful practical value was justified. Among the projects to be investigated, he listed tests of reception from airplanes, high-power transmitting tubes, and low-power alternators; some of the research was to be done in cooperation with GE. He reported that his department already had a total of 27 employees, with an annual payroll of $60,000. He requested an appropriation of $98,000 for the coming year.[68]

Evidently, some communication problems occurred between the GE and RCA engineering staffs. Early in 1920, Alexanderson informed Goldsmith that each department at GE had its own design data and that information was exchanged only through personal contacts among the engineers. According to Alexanderson, it had been found virtually impossible to exchange such information through correspondence. He wrote that even when the engineers were working for the same company but in different buildings, difficulties had been experienced. He concluded that barriers could be overcome to some extent by holding frequent conferences of the engineers.[69] Alexanderson subsequently visited RCA's Research Department to discuss future plans and the coordination of GE's and RCA's efforts. Goldsmith reported that he and Alexanderson had "found themselves to be in agreement."[70] A technical committee that included Goldsmith and Alexanderson as members was established during 1920 to facilitate cooperation between the two companies and the development of new products.

Admiral Bullard and Commander Hooper continued to take an active interest in RCA and the politics of radio during 1920. In January, Bullard was appointed to attend meetings of RCA's stockholders and directors to "present and discuss informally the Government's views and interests concerning matters pertaining to radio communication."[71] In March 1920, Alexanderson wrote to Edward Nally, RCA's president, that he was quite disturbed by some critical comments about RCA's technical achievements which Bullard had made at a recent IRE meeting. Alexanderson, who agreed tacitly with Bullard, attributed the problem to inadequate staffing of the RCA engineering force. He recommended that volunteers should be recruited from GE.[72] Commander Hooper drafted a letter that was signed by his immediate superior and sent to both GE and AT&T in January 1920 strongly urging that an agreement be negotiated to settle the dispute over vacuum-tube patents. Such an agreement would be in the public interest, Hooper argued, and it would contribute to the safety of ships at sea by ensuring that the navy could obtain the tubes needed for radio apparatus without difficulty.[73] The requested cross-licensing agreement was signed in July 1920 and extended immediately to include RCA and Western Electric.[74]

In March 1920, Hooper wrote to Owen Young to complain that a *New York Times* article about RCA had neglected to mention the navy's role in the formation of the company. Hooper added, "The sooner a little General Electric Company spirit of patriotism is injected at the Radio Corporation, the better the latter will succeed."[75] Young replied that he would like to meet with Hooper and make his contribution to the organization of RCA a matter of record; he also asked whether Hooper would be interested in resigning from the Navy to join RCA in some capacity. Hooper responded that he doubted whether RCA was in a position to pay him enough, and that he would not consider leaving the navy for less than $18,000 or $20,000 per year. He pointed to the need for RCA to build a strong enough staff to gain public and government confidence. Young responded by defending RCA's policies as being prudent, and contended that the investing public had to be convinced that the management was conservative with its capital funds before a rapid development of radio could take place.[76]

RCA Begins Commercial Operation

Activities at RCA entered a new phase when the government turned over control of the New Brunswick station on February 29, 1920. The first commercial messages were sent to England the following day.[77] By early April, Alexanderson had enough data on traffic and revenue to report to Albert Davis that "investing in radio communication at this time should be very attractive from a financial point of view." According to Alexanderson, the New Brunswick station would produce an annual revenue of $504,000 if operated for 300 days per year at 12,000 words per day with an average rate of 14 cents per word. He stated that the rate actually being charged to the public for messages to England was 17 cents per word. He went on to say that the RCA station in Marion, Massachusetts, should generate an annual revenue of $864,000 on traffic with Germany at an average rate of 24 cents per word. Alexanderson pointed out that the proposed RCA station on Long Island, New York, would have about 12 times the message capacity of New Brunswick and would yield gross earnings of over $10 million per year at the rate currently charged for German messages. He estimated the total cost of a two-alternator transmitting station at $1,100,000, not including land cost and buildings. This figure included $526,000 for antennas and $254,000 for equipment supplied by GE. He calculated that the proposed Long Island station, which would have ten 200-kw alternators, would cost about $6,480,000.[78]

During 1920, the 50-kw alternator that had been used at New Brunswick since 1917 was moved to the Tuckerton, New Jersey, station, and four new 200-kw alternators were completed at GE. These four machines were installed at stations in New Brunswick; Marion, Massachusetts; Bolinas, California; and Kahuku, Hawaii. Each transmitting station was paired with a receiving station. Receiving sites on the East Coast were located at Belmar, New Jersey; Chatham, Massachusetts; and Riverhead, Long Island, New York.[79] At the start, the radio message charge to the public ranged from 17 cents per word for British traffic to 72 cents for Japanese traffic. The comparable charges for messages transmitted via submarine telegraph cables were 25 cents per word for British traffic and 96 cents per word for Japanese.[80]

Several articles dealing with the Alexanderson system and its significance were published during 1920. In February, the *Saturday Evening Post* published an article entitled "A New Day in Communications," which placed Alexanderson in the select company of Hertz and Marconi. According to the author, Alexanderson's radio inventions had enabled the United States to rise from the status of a minor power to that of a major power in world communications. After reviewing the circumstances that had led to the founding of the "strictly Yankee" corporation RCA, the author forecast that American radio apparatus would become standard throughout the world and would "enable us to send American news of a commercial and political nature to the peoples of other lands."[81] And in October 1920 the *General Electric Review* published a comprehensive paper by Elmer E. Bucher on the technical features of the Alexanderson system. Bucher had come to RCA from American Marconi. His paper contained many previously unpublished details on the construction and operation of the 200-kw alternator and the components of the system.[82]

The potentially revolutionary role of radio in the international communication of ideas was a theme of a paper Alexanderson presented at a joint meeting of the IRE and the New York Electrical Society in November 1920. He began the paper by comparing radio with older technologies, such as the steam engine, the telegraph, and the telephone. He noted that each of the earlier innovations had "introduced a new era in human affairs" by overcoming barriers of space and time, but that each had "certain serious limitations." For example, he pointed out, a submarine telegraph cable could be cut during wartime or "censored by its owners and controlled by military and naval power." By contrast, he continued, no one could prevent radio messages from arriving at their destination: "Radio makes the transmission of ideas from man to man and from nation to nation independent not only of any frail material carrier such as a wire, but above all it renders such communication independent of brute force that might be used to isolate one part of the world from another." He suggested that it was this aspect of radio that had inspired engineers, statesmen, and others in the United States to engage in the promotion and development of international radio.[83]

Another theme of Alexanderson's paper was the "close connection which now exists between electric power engineering and modern radio engineering," and how it had led to the development of the central station for radio. He commented that GE's entry into radio had evolved gradually and naturally from the company's power-engineering activities. Despite the fact that radio used frequencies of the order of 1,000 times greater than those used in power, he called it a "most remarkable fact . . . that *the generally established principles of the alternating current power technique could be applied to the radio technique almost without change*" (italics his). He reviewed the history of radio up to the time that the "two schools of engineering pursuing different aims, with widely different modes of thought" had confronted a common problem. He illustrated the contrasting modes of thought by stating that "the one had been thinking in terms of power factor, kilowatts, and phase displacement, the other in terms of wave length, decrements, and tuning." Alexanderson added that a third perspective came from electrophysicists who had been brought into contact with the radio problem and had given it a new impetus through their work on vacuum tubes.[84]

Alexanderson stressed that the same economic considerations that had caused the creation of large central electric power plants had stimulated the creation of powerful central radio plants such as the one being constructed on Long Island by RCA. Both types of plants served to "provide for the utilization of the plant investment and operating force to the utmost by shifting the equipment from one service to another and combining it to meet various demands." He explained that the peak demand of radio traffic to Europe and to South America or to the Pacific would vary widely depending on the time of day or the time of year. For example, in the winter less power would be required for communication with Europe but more power would be needed for communication with South America. For Alexanderson, who was well versed in the concepts of load diversity and load factor from power engineering, the parallel was an obvious one.[85]

The published version of Alexanderson's paper included a schematic diagram of the anticipated antenna and alternator arrangement at the central station being constructed on Long Island. The station was being designed with twelve multiple-tuned antennas with radial symmetry, and ten 200-kw alternators. Six of the antennas were intended for telegraphic traffic with Europe, with separate circuits to England, Germany, France, Sweden, Denmark, and Poland. Three antennas and two alternators were intended for South American traffic. The remaining three antennas and two alternators were to be used either for transpacific telegraphy or for telephony to Europe. Another illustration that accompanied the paper depicted the 200-kw radio alternators that were being completed at GE at the rate of two per month.[86] Intellectually it was a grand design, but it was destined not to be completed.

Traffic data for the last five months of 1920 showed a promising and steady trend for RCA radio circuits between the United States and England, Germany, France, and Norway. The total number of words transmitted increased from 231,000 in August to 362,000 in December, while the number of words received increased from 332,000 to 610,000 over the same period. An internal memorandum in January 1921 listed three factors essential for maximum efficiency at the high-power stations: maximum circuit capacity, a high traffic load factor, and an adequate tariff. The report stated that RCA's stations were presently operating at about an 80 percent load factor, since about one-fifth of the possible working time was not being used. It was noted that the load factor might be increased by adopting a lower weekend rate or by adding a low capacity but high-rate circuit with a country having poor cable facilities. The memorandum cautioned that RCA needed to engage in activities that would develop more traffic since they would soon have five transmitters in service.[87]

A Trip to Europe

Early in 1921, Alexanderson notified Francis Pratt, by now a GE vice-president, of an impending trip to London to participate in negotiations concerning the location of radio stations in Europe. Alexanderson mentioned that he would probably visit France, Germany, Belgium, Denmark, Norway, and Sweden while overseas, and would report on what he learned about both radio and power engineering.[88] Charles H. Taylor, the assistant chief engineer of RCA, provided Alexanderson with letters of introduction to W. H. Eccles, a British electrical engineer, and to the works manager of the British Marconi Company. Taylor asked Eccles to inform Alexanderson about research and development activities presently going on in the field of radio in England. There were so many difficulties in radio, Taylor stated, that RCA needed to be aware of all activities on both sides of the ocean in order to advance the art as quickly as possible. To the Marconi works manager, Charles Mitchell, Taylor wrote that Alexanderson would like to see how British Marconi was manufacturing radio equipment. Soon after his departure for England, Alexanderson received a telegram from Pratt asking that he obtain all possible information on high-power vacuum tubes and circuits and forward it to GE along with some sample tubes.[89]

Because Alexanderson's IRE presidential address was scheduled during the period that he would be out of the country, he sent a draft copy to Alfred Goldsmith and asked that Goldsmith deliver it in his stead.[90] George H. Clark of RCA introduced the talk given in absentia with an amusing commentary on the hectic life of RCA's chief engineer. According to Clark, Alexanderson had been writing instructions from himself in New York City to himself in Schenectady and then hurrying to Schenectady to answer himself in New York City. Clark also alluded to the dilemma faced by a chief engineer who was an inventor and who had to decide between his own ideas and those of others who might see different ways to achieve the same objective. He credited Alexanderson with a willingness to reach decisions on the basis of comparative performance of alternative devices or systems, for the good of the art. Clark concluded by calling Alexanderson "as delightfully human a man as I ever met."[91]

In his presidential address Alexanderson wrote, "Every true engineer is an inventor, though, unfortunately, every inventor is not an engineer." He noted that, because the IRE was an international organization, it was an ideal forum for the discussion of important issues in international communication, and he stated that most American experts were in favor of minimizing regulation of radio in order to provide the maximum opportunity for invention.[92]

After the London conference ended, Alexanderson reported to Francis Pratt that certain recommendations made by British Marconi had been contrary to the best interests of the United States and of RCA. Alexanderson stated that he was attempting to get all the high-power alternator stations into operation as soon as possible in order to strengthen the American bargaining position.[93] He informed Owen Young of his perception of the situation in Sweden, where he had discussed radio and railroad electrification with officials. Alexanderson mentioned that the Swedish government viewed the two technologies as being linked and as having important implications for future American trade competition with Germany. He added that it was his opinion that Sweden held the balance of power in northern Europe and the key to the Russian market. He wrote to David Sarnoff that RCA would be at a serious disadvantage in Sweden unless it arranged to deal with a local representative with the right connections.[94]

Another area that Alexanderson investigated while in Europe was work being done in England and Germany on electronic tubes. He reported to Albert Davis that German engineers were experimenting with grounded-grid circuits to achieve stable parallel operation of tubes, and that the British were seeking to reach the highest possible power-tube efficiency and were developing automatic protective devices for power tubes. He informed Francis Pratt that the German company AEG was making some 1,000-kw mercury-vapor rectifiers for use in a power substation. Alexanderson noted that as a result of impressions formed during the European tour and his own research with vacuum-tube devices, he had been thinking a great deal about the use of electronic tubes in railroad electrification.[95]

The Alternator versus the Vacuum Tube

In January 1921, during Alexanderson's absence, his assistant Samuel Nixdorff attended a conference with David Sarnoff and Charles Taylor in which Sarnoff outlined what RCA's developmental expenditures would be for the coming year. Nixdorff recorded that a total of $160,000 was being allocated, with $35,000 going to GE for the development of low-power equipment and the rest earmarked for the high-power receiving and transmitting system being installed by RCA and for Roy Weagant's work at RCA on static elimination. Nixdorff noted that Sarnoff, while expressing a belief that the art was shifting from alternators toward tubes, said that he was not yet willing to appropriate funds for development of high-power tubes. Sarnoff had mentioned that RCA had 21 radio alternators on order, 8 more than were needed, and would try to sell the surplus units in South America or Europe.[96]

Soon after his return to the United States, Alexanderson attended a meeting of RCA's Technical Committee where the issue of alternators versus tubes again came under discussion. According to the minutes of the meeting, the discussion was precipitated by an inquiry from the United Fruit Company about the price of a transmitting station. Alfred Goldsmith was recorded as being in favor of recommending an Alexanderson alternator station while Alexanderson favored developing a tube set. Sarnoff, the committee chairman, then decided to ask that comparative bids on tube and alternator sets be prepared at Schenectady.[97]

A few days later, Alexanderson explained his position on the matter in a memorandum to Edmund Edwards. Despite the simplicity and certain results that would result from the use of an alternator, he wrote, he preferred to develop a 50-kw vacuum-tube transmitter for the United Fruit Company. Alexanderson admitted that this was something of a commercial gamble but asserted that if GE and RCA mounted a concerted effort, the project should enable them to catch or surpass the competition. He requested a manufacturing order that would enable GE's Radio Engineering Department to develop high-power vacuum-tube transmitters.[98]

During 1921, Alexanderson's executive responsibility for the development of radio apparatus at GE was transferred to Edmund Edwards. Alexanderson stated, however, that he planned to maintain an active role in radio development as a consulting engineer within the company, unaffiliated with any department. He complained mildly about a policy statement by Edwards that henceforth all experimental design and manufacture of radio equipment at GE would be confined to the Radio Engineering Department. Alexanderson asked where this policy would leave the consulting engineer, whose advice would not be very useful unless his activities paralleled to some extent those of the designers. He added that he thought the consultant's most important function was to take the initiative in developments that might lead to new business. If the Radio Engineering Department made it impossible for the consulting engineer to give intelligent and continuous attention to radio problems, he argued, it would lose the benefit of ideas and inventions that might come from outside. As a case in point, he cited the United Fruit Company episode, in which other engineers had recommended the alternator and only his advice had resulted in the decision to make a tube set.[99]

RCA's Radio Central Station

The formal opening ceremonies of the Radio Central transmitting station that was to be the hub of the RCA world communication system were held at Rocky Point, Long Island, on November 5, 1921. President Warren G. Harding was among the dignitaries in attendance, and David Sarnoff read radiograms from several individuals who were unable to attend, including Marconi, Nally, and Alexanderson. The latter, who must have been quite disappointed to miss the event, had been sent to Europe to participate in negotiations on stations proposed for South America.[100]

Owen Young, RCA's chairman of the board, delivered the keynote address. He reviewed the circumstances that had led to the formation of RCA two years before, and stated that the company enjoyed certain geographic, commercial, and technical advantages that would make the United States the radio capital of the world. He explained that every country desired to have direct communication with the United States without using a cable that passed through other countries. Young mentioned that Poland had recently negotiated a contract for a high-power station that would enable it to "reach out directly to America." He reported that he had just returned from Europe, where an agreement had been reached with the principal radio companies of England, France, and Germany for cooperative development of radio in South America. As a consequence, he continued, only one station would be operated in Argentina instead of four competing stations. He concluded that there could be neither world public opinion nor a successful disarmament program without "cheap and adequate communication in the world."[101]

At the time of the formal opening of Radio Central, only 2 spokes of the planned 12-spoke antenna system had been completed. Each spoke consisted of six 410-foot-tall steel towers spaced 1,250 feet apart, with 150-foot crossarms and with 16 horizontal cables suspended from the crossarms. Two of the intended ten 200-kw Alexanderson alternators had been installed; because the shift to tube transmitters was beginning, they turned out to be the only alternator sets operated at Radio Central.[102] The Radio Central station was given the call letters WQK.

Rocky Point, where the Radio Central transmitter was located, appears to be a very undeveloped area in a photograph album belonging to Edward D. Sabin, an RCA construction engineer who spent several months there in 1922. One of his photographs shows a sleet-melting house located adjacent to down leads from the antenna. This was installed because of the likelihood that ice would form on the antenna in winter. Sabin's photographs also show that a community center, cottages, and a hotel were constructed for use by permanent RCA staff assigned to the station or by visitors. Sabin recorded that preliminary work began in 1922 on an RCA "Camp Colony" overlooking a beach near the transmitter but that it was never completed.[103]

A Speech on Radio Broadcasting

Although he was not actively involved in the rapid growth of the radio broadcasting industry during the early 1920s, Alexanderson observed the phenomenon with great interest. In a speech to Union College alumni in April 1922, he described radio as a technology with the "brake off." Although he had long been connected with the field, he continued, the developments that had occurred over the past year were difficult for him to comprehend. He noted that it seemed that everyone in the country was clamoring for radio receiving apparatus. Perhaps, he suggested facetiously, future electrical engineers "should study mob psychology more than Ohm's Law and . . . psycho-analysis is more necessary than Fourier's Series." He observed that broadcasting enabled the phonograph to "get in its deadly work at long distance" rather than being confined to the home where it should be if at all. He suggested that adults were really using radio broadcasting to amuse themselves, although they might try to give the impression that it was educational. Alexanderson defined a radio broadcasting system as a radio transmitter "entirely surrounded by listeners" and predicted that it would be impossible to continue to supply programming at no cost to the listener beyond the cost of receivers. He pointed out that whereas the public received radio broadcasting for free, they had to pay for the use of transoceanic radio service.[104]

The Beverage Antenna and Other Developments in Antenna Technology

Even as new stations using the Alexanderson alternators were still being built, research continued which resulted in a better understanding of propagation phenomena, and incremental improvements in the design of antennas, insulators, and other elements of the system. A significant improvement in longwave receiving antennas resulted from observations made by one of Alexanderson's assistants, Harold Beverage. He had come to work at GE in 1915 after graduating in electrical engineering from the University of Maine. He joined RCA in 1920 and eventually became a vice-president and research director before his retirement in 1958.[105]

In December 1919, Alexanderson informed Albert Davis of the results of an investigation of long-wire receiving antennas that Beverage had begun at a station in Bar Harbor, Maine. He had discovered that an antenna that terminated in an open circuit at the end farthest from the receiver was quite directive. Alexanderson had assigned Beverage to install and continue testing long-wire antennas at a site on Long Island, New York. Subsequently Beverage had found that grounding the remote end of the wire gave even more directivity, and that the directivity increased with the length. He installed an experimental antenna wire along the road from Riverhead to Southampton, and measured field strength and static at intervals over a distance of about 10 miles. His graph of signal-to-static ratio revealed a steady improvement as a function of antenna length.[106]

Chester Rice and Edward W. Kellogg, who were engaged in theoretical work on longwave antennas, learned of Beverage's observations from Alexanderson. Rice and Kellogg used an analysis based on transmission-line theory to determine that a long horizontal wire should act as a unidirectional antenna if terminated in its characteristic, or "surge," impedance. Beverage and Rice verified this theory using a long wire terminating in a resistance. This type of antenna became commonly known as the "Beverage antenna." A 9-mile-long Beverage antenna was installed at Riverhead and used for reception of incoming radio traffic beginning in the summer of 1921. Beverage, Rice, and Kellogg disclosed the properties of the new directive antenna in a 1923 AIEE paper.[107]

The results of experiments on antenna insulators were reported by William Brown, another of Alexanderson's assistants, in an IRE paper published in 1923. According to Brown, the insulators used previously at New Brunswick and Marion had been found inadequate once the big 200-kw Alexanderson alternators had been installed. Tests had been undertaken at Schenectady to evaluate various designs of insulators over a frequency range of 18 to 28 kHz, and it had been determined that the most satisfactory insulators were long cylinders of porcelain with electrostatic shields and rain shields at one or both ends. He mentioned that heat and flashover voltages had been measured, and simulated rain from sprinklers had been used. Brown stated that the most severe conditions encountered for the insulators in actual service were at the station in Hawaii, where salt spray from the ocean posed a serious problem. He described a standard insulator that had been developed: a 40-inch-long porcelain cylinder with a cast-aluminum corona shield at the lower end and a formed-aluminum combination corona and rain shield at the upper end. He noted that two such insulators were connected in series at the Radio Central station.[108] A discussant of Brown's paper pointed out that the problems that would be encountered in insulating the high-voltage power transmission lines of the future were likely to be similar in many respects.[109]

The large inductance coils used for antenna loading were another important component of the Alexanderson system which was improved and standardized. They were the subject of an IRE paper by William Brown and J. E. Love published in 1925. Brown and Love noted that wood frames had initially been used for the large coils but that this had caused numerous fires and had required a reduction of the permissible power and voltage. The wood frames had subsequently been replaced by porcelain supports that provided increased efficiency and four times the kilovolt-ampere rating of the older designs. Brown and Love included a formula, which they credited to Alexanderson, that could be used to calculate the power factor of the coils as a function of the DC resistance and the ratio of the diameter of the conductor to the diameter of the coil. They stated that the formula had been found correct to within a few percent for all conductors commonly employed in high-power tuning coils.[110]

In a 1926 IRE paper, Nils E. Lindenblad and William Brown discussed the electrical, mechanical, and economic factors important to the design of transmitting antennas of the type used in the Alexanderson system. They included a graph of cost as a function of antenna size to show that when initial cost and maintenance costs of both the generator and antenna were considered, an optimum antenna size could be determined. They explained that a scale model was often useful in determining weight stresses and wind-loading effects. Lindenblad and Brown emphasized the importance of soil resistance at antenna sites and noted that Alexanderson's multiple-tuning invention had been the single most important step toward reducing the effective resistance to ground currents. However, they stated, conditions frequently justified the addition of an artificial ground, in the form of buried wires or an aboveground "counterpoise." They reported that RCA had come to favor the use of buried wires because a buried wire was less subject to mechanical or electrical failure.[111]

Charles W. Hansell, another of Alexanderson's assistants in the early 1920s, later recalled an instance where soil resistance had been a serious difficulty to overcome. He stated that the soil resistance at the Rocky Point site chosen for the Radio Central transmitter had been "near the highest on earth," and that when this was learned, the engineers had considered stopping further construction and moving to a better location. Hansell remembered that, characteristically, Alexanderson had ordered that they go ahead with construction at Rocky Point and rely on him and his assistants to devise a technical solution. Hansell and P. S. Carter had assisted in the design of an artificial grounding system that had cost about $120,000, but Hansell noted that this had still seemed modest compared to the total expenditure of about $3 million for the Radio Central station.[112]

Lindenblad and Brown also noted that scale models were useful for estimating antenna capacitance that was difficult to calculate from theory. Charles Hansell alluded to the capacitance problem some years later in an anecdote about Alexanderson. According to Hansell, J. Leslie Finch had asked Alexanderson to make a guess as to the capacitance of a particular antenna. Alexanderson did so. After several days of laborious calculation, Finch had been astonished to find that he had arrived at the same value that Alexanderson had estimated. Hansell believed that Alexanderson had used a simplified calculation based on model tests.[113]

In 1919 a model antenna located in a field at Schenectady was used to simulate some of the large, multiple-tuned antennas that were to be constructed at sites around the world. In a January 1920 report to Albert Davis concerning experiments with the model antenna, Alexanderson commented that he favored locating stations where the land was cheap enough to permit expansion and the use of highly directive (and thus very long) antennas.[114]

The Installation of a Radio Alternator Station in Poland

An interesting example of technology transfer resulted from a decision in 1921 by the newly independent Poland to install a high-power Alexanderson radio transmitter. In this instance the nationalistic goal of gaining an international means of communication which did not depend on the facilities of neighboring states took precedence over what probably would have been a more prudent expenditure of scarce resources. As Jensen and Rosegger have pointed out, "technological and political ambition" may override considerations of economic feasibility in a new nation-state.[115] During the same period, Poland was attempting to establish an electrical supply industry while trying to unify a country composed of portions of territory formerly under German, Russian, and Austrian control. According to an article published in *Electrical World* in 1924, Poland was quite eager to attract American trade and investment, since Polish dislike of Germany and Russia was so intense that Poland would conduct only essential transactions with either country.[116]

A contract for building the Polish transoceanic station was signed by RCA president Edward Nally and a Polish official August 1, 1921. The transmitter was to use two 200-kw Alexanderson alternators. An RCA engineer, William G. Lush, was sent to Poland in February 1922 to serve as the resident engineer in charge of construction, and several other GE and RCA engineers, including J. Leslie Finch, were sent over to assist at various stages of the project. A Polish technician who had been to the United States and could speak English was given the responsibility for recruiting a local labor force. The transmitter site was located about 10 kilometers from Warsaw.[117]

The American engineers were surprised to discover that the Polish workmen required not only living quarters but individual cooking facilities rather than the community dining facilities that were used in the United States. Lush and his RCA colleagues later reported that the American supervisors had found it necessary to be "most particular in their dealing with everyone." They had "found at once that some of the rough and ready methods of action in common use on the projects in the United States would not apply and it was necessary to substitute carefully thought out ways, more in conformity with Polish ideas." The buildings for the station were brick with ornamental concrete and were described as being designed to reflect the "true architectural feeling of Poland." The project attracted great public interest in Poland, and various groups came to see the work in progress. The American engineers reported that Polish students were keenly interested and that the universities were attempting to include modern engineering in their curricula.[118]

Most of the electrical apparatus and the structural materials for the antenna were made in the United States and shipped to Poland through the port of Danzig and over the Polish state railroad. The transmitting antenna was similar to those used in the United States, with 10 towers approximately 400 feet in height and spaced about 1,250 feet apart in a straight line. Power for one of the alternators was provided by a 500-kw steam turbine, while the second was driven by a 750-hp diesel engine. The receiving station was located at a separate site near Grodzisk, about 19 miles from Warsaw. It employed a Beverage antenna oriented at right angles to the transmitting antenna and about 16.2 km in length. The system was designed so that the two alternators could be operated simultaneously at different frequencies or in parallel on the same frequency. In practice, one was generally used with the other on stand-by. The station began commercial operation on October 4, 1923, and was formally accepted by the Polish government in November 1923. Some of the RCA engineers remained for some time to train and advise the Polish staff.[119] In February 1924, it was announced that Alexanderson had been awarded the Order of Polonia Restituta by the Polish government in recognition of his creation of the system employed by the Polish station.[120]

The Polish transmitter, which had the call letters SPL, provided reliable 24-hour communication between Poland and the United States at its normal operating frequencies, 14.29 kHz and 16.42 kHz. It was operated by the Polish Ministry of Posts and Telecommunications until the fall of 1939, when it was captured by German forces. The station was then used by the Germans to communicate with submarines until early in 1945, when it was destroyed by the German army to prevent its capture by the Russians.[121]

The Installation of an Alexanderson Alternator Station in Sweden

Alexanderson, of course, was quite eager to have a station located in his native Sweden. Early in 1920 he received word from a contact in Europe that a Swedish radio company might be organized and that both Germany's Telefunken Company and British Marconi were already negotiating. Alexanderson was urged to "act immediately" if he wished to enter the competition. His informant pointed out that Sweden might be the key to getting a station located in Russia.[122] GE decided to bid on the Swedish contract, but in May 1920 Alexanderson learned that GE's bid was higher than those of its competitors, which now included a French radio company. However, he was told that the Swedes were very eager to have an American-equipped station since the station was to be used for communication with America.[123]

In November 1920, Alexanderson reported to David Sarnoff that he had escorted some representatives of the Swedish Telegraph Administration on a tour of the RCA station at Marion, Massachusetts, and the GE plant in Schenectady. Alexanderson stated that they had discussed a high-power station to be installed in Sweden, and that if GE were to remain in the running for the contract, it would have to make some guarantees and agree to take responsibility in case of patent litigation. He added that the Swedish officials had been quite impressed by the possibilities of vacuum-tube transmitters, which were being promoted by British Marconi, but that the GE-RCA engineers still regarded the alternators as preferable. However, he pointed out that GE and Western Electric were leaders in vacuum-tube development and would probably be able to furnish a tube transmitter by the time the Swedish station was completed. As a compromise, he recommended that they propose a two-alternator station and install one alternator quickly. He pointed out that this would enable them to decide later whether to install a second alternator or a vacuum-tube unit. He suggested a contract price of $515,215 for two 200-kw alternators, receiving apparatus, antenna insulators, and engineering services.[124]

The fact that Alexanderson was Swedish and had direct access to the engineering community in Sweden was probably more important to RCA's success in selling a transoceanic station than were political considerations. Alexanderson was able to take a personal hand in the negotiations, which culminated in the signing of a contract in 1922. Accompanied by his wife and his daughters Amelie and Edith, he went to Sweden in the late summer of 1922 after attending a radio conference in Brussels.[125] While in Sweden he signed the contract with the Swedish Telegraph Administration for the construction of a two-alternator transmitter and a receiving station by RCA. A published report of the contract stated that it was for $2 million and that RCA had won the contract in competition with British, French, and German firms.[126]

The transmitter site, where two 200-kw Alexanderson alternators were installed during 1924, was located at Grimeton, near Varberg on the west coast of Sweden. The alternators were driven by electric motors powered by two remote hydroelectric plants. The transmitting antenna consisted of six towers 400 feet in height, with 150-foot crossarms and a spacing of 1,250 feet between masts. An article on the construction of the antenna appeared in the Swedish technical journal *Teknisk Tidskrift* in December 1924. The receiving station was built at Kungsbacka, about 30 miles north of Grimeton.[127]

Sometime prior to the completion of the facility in Sweden, a Swedish engineer came to the United States to observe operations at Radio Central and Riverhead. According to Alexanderson, this engineer, who was to be chief engineer of the Swedish station, wrote down the dial settings on the receiver at Riverhead and later found it necessary to use the same settings at Kungsbacka to pick up the American station successfully. Commercial operation at Grimeton began in December 1924, and the official opening was in July 1925.[128]

Alexanderson was present when the opening ceremonies were held and was awarded the Order of the North Star in recognition of his achievements in radio.[129] The Swedish station remained in commercial use until 1946, with the call letters SAQ and a frequency of 17.2 kHz. In a radio talk given during World War II, Alexanderson remarked that the alternator at the Swedish station was known affectionately as the "old war horse, because it always gets through with the message." One of the two alternators was scrapped in 1960, but the other has been preserved and in 1986 was still operable.[130]

An Overview of the RCA Transoceanic System

A comprehensive overview of the Alexanderson alternator system was given in a paper presented at the annual AIEE convention in June 1923. In the paper, Alexanderson, Alexander E. Reoch, and Charles Taylor reviewed the history of developments in transoceanic radio since around 1920 and noted that after the war "practically every European country demanded direct radio communication with the United States." In order to meet this demand, they explained, RCA had expanded its Atlantic Coast facilities from its initial two transmitting stations at New Brunswick and Marion to a unified group of six alternator transmitters controlled from a single traffic office in New York City. They pointed out that the Radio Central station was the first station that RCA had planned and designed from the start as a modern plant rather than as an adaptation of an older facility. They listed four primary considerations that had led to the selection of the Long Island site for Radio Central: proximity to New York City, the availability of a large tract of land at moderate cost, an adequate source of electric power to run the alternators, and direct wire communication from the site to New York City. They mentioned that the principal liability of the Rocky Point site had been the quartz sand that gave a very high soil resistance.[131]

Alexanderson, Reoch, and Taylor wrote that the "method of centralization" had been carried to a logical conclusion by the elimination of separate receiving stations and the location of all East Coast transoceanic receiving apparatus at Riverhead, Long Island. They noted that new European stations could be received at minimum cost by simply installing standardized receiving apparatus on shelves at the Riverhead station, and they added that the Radio Central transmitting station had been designed for similar centralization with as many as twelve 200-kw alternators, although only two had actually been installed.[132] It is perhaps worthy of note that the RCA system was being designed and built at a time when the centralization of electric power systems was taking place and "superpower" systems covering large areas were being promoted.[133] The superpower proponents could only be envious of the rapid deployment of an international "superpower" communication grid.

In their paper, Alexanderson and his RCA colleagues devoted considerable attention to a discussion of the considerations that had led to the design of the multiple-tower transmitting antenna of the type used at Rocky Point, in Sweden, and in Poland. The use of towers with cross-arms instead of a spring-stay suspension between towers had reduced the detuning effects sometimes encountered under strong wind conditions, they explained, and improvements in the design of antenna insulators had raised the antenna working voltage from 60 kv to 150 kv. They went on to say that a series of measurements on buried wires had been carried out to determine the optimum dimensions of the buried-wire ground system; the ultimate outcome of this research was that the Rocky Point antenna "in effect, stands on a plate of copper 2000 feet wide and 3 miles long." The artificial ground actually consisted of 750 copper wires, each 2,000 feet long and spaced 10 feet apart along the 7,500-foot-long antenna array.[134]

The three engineers stressed that the transmitter and receiver should be viewed as part of a single system and subject to economic

Figure 6.1. Alexanderson adjusting the transoceanic radio receiver at the Riverhead station on Long Island, N.Y., in May 1922. Courtesy of the General Electric Company, Schenectady, N.Y.

laws, and that development should continue until the total cost reached "its ultimate minimum." They reported that in recent years a great change had taken place in the speed and method of sending and receiving messages. This had been achieved through the substitution of mechanized methods for manual keys, which had limited speeds to about 40 words per minute. For transmission by machine, they explained, messages were transferred to a punched paper tape, which in turn actuated an automatic key; for automatic reception at 100 words or more per minute, an ink recording device had been developed which was an adaptation of the siphon recorder used in cable telegraphy. Messages were transcribed from the automatic recorder by skilled typists.[135]

Alexanderson, Reoch, and Taylor concluded their paper with a discussion of the factors that limited the number of longwave channels that were feasible. Pointing out that the spectrum between 11,000 meters and 22,000 meters could only accommodate 70 channels of 1-percent bandwidth, they asserted that modern transmitters and receivers such as the RCA stations on Long Island were capable of the frequency stability and discrimination necessary for 1-percent channel separation but that older radio systems would have to be replaced in order to realize the full potential of longwave communication. Despite international agreements on wavelengths to be used, they stated, ether congestion was becoming a serious problem. Although further improvement in existing longwave technology was possible, "radically new methods" such as the use of directional radiation using shorter wavelengths ought to be considered. Nevertheless, they concluded, "guess work [has] been eliminated from the development of radio communication, and . . . sound foundations, both technically and financially, can be laid for all future expansions of our system."[136]

In November 1923, as the RCA international communication network neared completion, Alexanderson informed Francis Pratt of GE that with the completion of stations in South America and China, RCA would have a chain of high-power stations around the world. All the transmitting stations, except for three in Europe and the one in China, would employ the 200-kw alternator; the Chinese station would employ a Federal Telegraph Company arc transmitter. Alexanderson commented that he had advised RCA to purchase a high-power tube set from GE in order to prevent control of the new technology from going to AT&T. He suggested that it was the manufacture of radio broadcast receivers that was the "money making end" of radio and that those engineers who were attempting to create a business of manufacturing vacuum-tube transmitters would be better employed in an organization devoting all its efforts to the broadcasting industry.[137]

As the RCA transoceanic system neared completion, Alexanderson engaged in some reflections on the general significance of the system that he and his staff had wrought. One opportunity to look back was provided in April 1923, when he responded to an inquiry about the background of his inventions. He commented that he could not answer a question about the invention of the radio alternator in a simple way, since the alternator was not a distinct invention made on a specific date. Rather, the alternator had resulted from a process of continuous development during the period 1904–18, and much related work had been done in which the alternator had served as a tool rather than the object. Again Alexanderson stressed that the most significant feature of the entire development had been the introduction of the methods and viewpoints of electric-power engineering into radio. He wrote that he found the most interesting feature of the radio alternator to be not that it differed so much from power alternators but that a machine so similar to power generators could be designed to suit the requirements of radio communication.[138]

Alexanderson continued in a reflective mode in a short paper published in the *General Electric Review* in June 1924. He recalled an episode during the war when a marine guard had become alarmed by seeing a flickering light due to an electrical discharge emanating from a shack housing a tuning coil. The guard had fired into the shack and the insulator had exploded, knocking the station off the air. Alexanderson commented facetiously that it had not been firearms that had solved the insulator breakdown problem but rather "technical knowledge acquired by scientific investigation." He added that the commercialization of radio had "afforded an unusual opportunity for the application of scientific engineering methods." He characterized the transoceanic radio transmitter as a "power station" with an input of kilowatts and an output of words. "The problem of radio engineering," he asserted, "is to establish the relation between kilowatts input and words output."[139]

He included in his paper a qualitative discussion of a principle of what later became known as information theory. He pointed out that the maximum speed of radiotelegraphic signalling "is directly proportional to the ratio between the wave amplitude and the amplitude of the atmospheric disturbances." The total energy contained in a telegraphic dot needed to be greater than "the maximum energy of a single atmospheric impulse," he continued, and "if the wave amplitude is doubled the length of the dot may be shortened to one-half." As an indication of the capabilities of the RCA chain of stations, Alexanderson predicted that it would soon be possible to relay automatically a signal that would "circle the earth with the velocity of light." He explained that such a signal would travel from New York and be relayed by the stations in California, Hawaii, Shanghai, and Poland or Sweden until it came back to New York.[140]

In a letter written early in 1925, Alexanderson included some of his thoughts on the strategic and financial importance of the RCA transoceanic system to the United States. He suggested that the transoceanic cables had been a significant force for the development and maintenance of the British empire, and that the RCA system was in the process of making New York City the financial and commercial center of the world.[141]

Alexanderson pursued the same theme in a paper given at an AIEE meeting in Cleveland, Ohio, in March 1925. He stated that long-distance communication had "always been the force that developed and maintained great civilizations," and he cited as examples the Roman and British empires. Because of the RCA system, he continued, New York City was replacing London as the hub of the world. Commenting on the enjoyment that the building of this system had brought the active participants, he noted, "It is difficult to convey by words the appeal to the imagination which the development and operation of this world wide spider web of communication had to those engaged in it." He mentioned the obstacles that had been overcome before he and his colleagues had been able to receive signals from all over the world, "concentrating the operation of the whole system in one large room in New York." Finally, he wrote that the goals that had motivated the engineers responsible for the RCA system were essentially the same as those that had motivated those who built and operated a railroad system, namely "reliability, service and speed."[142]

Alexanderson in Corporate Context: Management Style, Career Developments, and High-Level Changes

Alexanderson faced a challenging managerial problem during his tenure as chief engineer of RCA. Normally he spent Monday, Tuesday, and Wednesday of each week with RCA in New York City and Thursday, Friday, and Saturday with GE in Schenectady. He also found it necessary to make frequent trips to Europe. The evidence concerning his managerial style and his relations with his assistants is fairly sketchy and ambiguous. A newspaper article by William J. Butler published in the summer of 1922 stated that Alexanderson was not regarded as an easy taskmaster and had, in fact, been called a "hard boss" by some employees. The article, entitled "Alexanderson the Great," commented on his virtual obsession with work and invention. Butler stated that Alexanderson "was born with a text book in his hand and ever since he has been everlastingly night and day at text book and mechanical invention."[143]

Charles Hansell later recalled that Alexanderson's ability to concentrate had been phenomenal. According to Hansell, Alexanderson sometimes would call a group of assistants in to discuss a difficult problem and, pausing to think before speaking, would "promptly [forget] we were there. Eventually he would be startled by our presence and want to know what we wanted."[144] William Brown characterized Alexanderson as "methodical and deliberate" and always willing to listen to the suggestions of assistants when there were differences of opinion. Brown called him a "fine leader with abundant energy."[145]

According to Howard I. Becker, who worked under Alexanderson in the early 1920s, Alexanderson had been "a very honest man about patents, never hesitating to give his assistants full credit, but on the other hand, when his assistants were working on his ideas, there was never any question whose idea it was." Becker recalled an occasion at the end of a group meeting when Alexanderson had said, "I feel just like a pliotron—somewhat unstable but highly efficient."[146]

Alexanderson commented on employee relations at GE in a December 1922 memorandum in which he wrote that it was his impression that the principal reason for GE's success was the absence of autocracy within the organization. He expressed his belief that it was inspirational to the GE workers to know that they were contributing to world leadership in power, lighting, and communications technology, and he concluded with the view that he personally had found workers and shop foremen to be just as susceptible to a feeling of personal pride in accomplishment as were engineers.[147]

Alexanderson's patent productivity continued at a surprisingly high level during his four-year tenure as chief engineer of RCA, a time when he had numerous duties that might have been expected to interfere with invention. During 1920–23 he applied for 29 patents that were eventually issued. The average, about 7 per year, was almost exactly the same as his average for the 46-year span 1903–48. Approximately half (14) of the 29 were related to the RCA transoceanic system, while the remainder were linked to GE and to power systems. A majority of the radio patents concerned methods of achieving optimum operation of several transmitters or receivers at one location. One covered the buried-wire grounding system of the type used at the Radio Central station.[148]

The contract that made Alexanderson an employee of both GE and RCA was renewed on March 31, 1923. Under the terms of the contract, Alexanderson was paid $15,000 per year by GE and $5,000 per year by RCA. He was to assign all radio inventions to RCA. The agreement was to run only through June 30, 1923. Possibly it was at this time that an arrangement was made for him to relinquish the title chief engineer in favor of that of chief consulting engineer of RCA, although that was not explicit in the contract. Beginning in 1924 and continuing until 1930, he listed his corporate positions as consulting engineer of GE and chief consulting engineer of RCA.[149]

SEVERAL CHANGES in top management took place more or less simultaneously at GE and at RCA. At GE, it was announced in May 1922 that Owen Young was to succeed Charles A. Coffin as chairman of the board, a position that Young already occupied at RCA. Also Edwin

Rice, who had long been supportive of Alexanderson's projects, retired as GE's president and was succeeded by Gerard Swope; and Francis Pratt was made a vice-president in charge of engineering at GE.[150] At RCA, Edward Nally in February 1922 informed Young of his desire to be relieved of the company's presidency. After a search of several months for a suitable successor, James Harbord was chosen; he retired from the army to take the post. An item in *Electrical World* which announced Harbord's selection included the comment that the position was deemed of vital importance by the United States government. Harbord, a major general, had achieved some eminence during World War I as chief of staff under Pershing and had been in charge of supplies.[151]

Young's choice of Harbord to succeed Nally was motivated to a considerable degree by a desire to divert criticism of RCA as a radio monopoly and to reinforce RCA's image as an "all-American company" motivated by patriotism. Young wrote to Nally in October 1922 that he considered it "particularly important that no one should be able to question his [Harbord's] Americanism."[152] Harbord had the advantage of coming highly recommended by members of the military establishment in Washington, although it seems somewhat surprising that a former naval officer such as William Bullard was not given the position, since the navy had been intimately involved in the formation of RCA. Harbord, who retained the RCA presidency until 1930 and was chairman of the board until 1947, seems to have introduced a rather military style of management at RCA. For example, he would sometimes issue a "General Order" for circulation among his subordinates.[153]

The Vacuum Tube Supplants the Radio Alternator

The chain of Alexanderson-alternator stations installed by RCA by 1924 represented a remarkable technological achievement, the significance of which has tended to be somewhat obscured by the incredible proliferation of radio broadcasting and the advent of a less-expensive shortwave technology for international radio. The Alexanderson system was an achievement in the tradition of power engineering at GE, as Alexanderson stressed frequently. Ironically, it was the high-power vacuum tube, a descendent of the de Forest audion that Alexanderson had brought to the GE Research Laboratory in 1913, that became the principal cause of the early obsolescence of the radio alternator. During 1922, Irving Langmuir announced that GE had developed a 20-kw pliotron vacuum tube, the most powerful yet made, and that it might soon displace the Alexanderson alternator. Langmuir stated that even more powerful tubes were feasible if needed.[154] A transmitter using six of the new tubes and developing about half of the power of a 200-kw Alexanderson alternator was tested successfully the same year. Within a short time, GE introduced a 100-kw tube that was about 5 feet long and 6 inches in diameter and cost about $1,000.[155]

The public was informed of the revolutionary nature of the power vacuum tube in October 1922, when a New York newspaper carried an article entitled "Vacuum Tubes vs. Alternator." The writer observed that the lower comparative cost of the vacuum tube seemed to mean that the radio alternator was destined for the scrap heap. Another newspaper item published a few weeks later concluded that the alternator system had become old in two years. The writer observed that an invention that had been regarded as having revolutionary importance in the radio field was already giving way to another invention.[156]

A second important factor in the rapidly declining importance of the Alexanderson system was the concurrent discovery that long-distance communication by means of much shorter wavelengths was possible much of the time. A shift to higher frequencies generated by vacuum tubes meant that the transmitting and receiving antennas could be much more compact and less expensive than the huge antenna structures at Rocky Point and the long antennas at Riverhead. The change to shorter waves also opened the possibility of far more channels, thus overcoming a fundamental limitation of the alternator system.[157]

Some attempts were made to adapt the alternator to generate higher frequencies. During 1925, Nils Lindenblad and William Brown presented an AIEE paper on some experiments with a frequency-multiplier circuit suggested by Alexanderson for use with the 200-kw alternator. The circuit employed an iron-core autotransformer and tuned circuits to accentuate the harmonics of the alternator output. Lindenblad and Brown reported that they had been able to convert a signal generated at 23.5 kHz to 94 kHz with a conversion efficiency of about 90 percent. They stated that a 20-kw frequency multiplier would be used at the station in Marion, Massachusetts, to achieve multiplication by a factor of five in a single stage. With the 200-kw alternator operated at 25 kHz, this would produce a 125-kHz carrier for use in marine communication. However, the shortwave vacuum-tube transmitters soon came to share the very facilities that had been built to house the Alexanderson system. By 1930 there were 19 shortwave transmitters with ratings of 20 kw to 40 kw in use at RCA's Rocky Point (Radio Central) station. Most of the other RCA transoceanic stations also had several shortwave units by that time.[158]

Actually, the alternator did demonstrate greater survivability than many anticipated in 1922. Its success in maintaining communication with remote sites and under conditions that interfered seriously with shortwave transmission kept it in use until well after World War II.

Alternators proved almost indispensable for communication with naval ships and submarines during World War II. An alternator in Hawaii was operated by the U.S. Navy until 1957, and another at Marion, Massachusetts, was used by the U.S. Air Force from 1949 until 1961 for reliable communication with bases in Greenland and Iceland.[159] RCA retained large tracts of land at Rocky Point and Riverhead until 1979, when 5,100 acres at Rocky Point and 2,000 acres at Riverhead were donated to New York State. The 400-foot antenna towers at Rocky Point were removed for safety reasons.[160] Very-low-frequency methods are still employed to communicate with nuclear submarines, since very-low-frequency radiation is the only form of radiation capable of penetrating seawater to considerable depths. Alternators are no longer used in such systems, although physically large antennas are still required.[161] Only the single 200-kw alternator in Alexanderson's native Sweden is still maintained in an operable condition and is a fitting monument to the architect of the system that once occupied the center stage of international radio communication.

Even before deployment of the radio alternators had been completed, Alexanderson's interests were already shifting to "new fields for radio signalling" and to new applications of high-power electronic tubes.[162] One of the first indications of a new interest that would occupy much of his time for the rest of the decade was a memorandum to GE's Patent Department written in January 1923. It concerned the transmission of pictures by radio. Alexanderson outlined how this might be done by adapting apparatus already used to send pictures by telegraph wire. He stated that the only novelty would consist of adapting the methods to the existing RCA system so that it would not interfere with routine traffic.[163]

Within a few weeks, Alexanderson was considering the design of a system to transmit moving pictures or live images by radio. In March 1923 he drafted another memorandum, in which he suggested that pictures could be produced by 30,000 individual impressions, each completed in a 20th of a second. This would require the transmission of 600,000 individual impressions per second. It could not be achieved using present radio systems, he commented, but it might be achieved if shorter wavelengths, of the order of 10 meters, were employed. He proposed a method that would employ 100 photoelectric cells, 100 oscillographs, and a rotating mirror. The following summer he wrote a patent memorandum entitled "Television by Radio."[164] He had developed a new obsession.

AN UNUSUAL intersection between Alexanderson's public and private life occurred when his only son, six-year-old Verner, was kidnapped while playing near their home on April 30, 1923. The following day, Alexanderson requested aid from the Radio Relay League. Stating that his son had last been seen in the company of a man of medium build, aged 25 to 30, he offered a reward for information leading to Verner's return. Through the pages of *Electrical World*, Alexanderson appealed to everyone in the electrical industry for help in the search.[165] Descriptions of Verner also were broadcast over the GE station WGY, and these proved instrumental in his recovery after a three-day search. A caretaker at some summer cottages on the Indian River heard the WGY announcements and recognized that the description fitted a young child he had seen at one of the cottages.[166] Although the caretaker immediately notified the authorities and Verner was reunited with his family, the kidnappers were not apprehended until years later. Harry C. Fairbanks was arrested for the kidnapping in July 1925 and given a 50-year sentence. A second suspect, S. G. Crandall, was arrested in 1927 in Washington, D.C. The reported motive for the kidnapping was that Crandall had wanted a child; it was charged that he and Fairbanks had conspired to kidnap one for the Crandall family.[167] Apparently they had been unaware at the time that they were abducting the child of a celebrated engineer. The incident served to generate publicity for radio and for RCA. Alexanderson provided RCA's Publicity Department with details of the role of WGY and its broadcasts in the recovery of his son, and RCA's president, James Harbord, encouraged RCA publicity on radio's contribution in the case.[168]

Chapter Seven

Television, Thyratron Applications, and Engineering Philosophy

During the 1920s, Alexanderson functioned in a dual role—research manager and engineer-inventor—in projects funded by General Electric, the Radio Corporation of America, and the U.S. Navy. As a research manager, he directed the activities of GE's Radio Consulting Department (formerly the Radio Engineering Department) after ending his four-year tenure as chief engineer at RCA in 1924. His patented inventions encompassed the fields of communication, electronics, and electric power, and specific inventions often combined elements from two or more fields. Among his principal areas of investigation were shortwave propagation, radio facsimile, mechanical-scan television, aircraft altimeters, and applications of the gaseous triode known as the thyratron. As something of an engineer-celebrity, he also was afforded frequent opportunities to express his views on factors that affected technological creativity and change.

Research on Shortwave Propagation

A revolution in long-distance radio communication took place in the 1920s after radio amateurs discovered that shortwave transmitters with relatively low power could, under certain conditions, be used to communicate over remarkably great distances. This discovery stimulated radio engineers at GE, RCA, and elsewhere to carry out systematic investigations of shortwave propagation phenomena. In the process they collected much data on the properties of ionospheric layers of the upper atmosphere, including diurnal and seasonal variation. They also studied the effects of polarization on radio communication and developed new types of directive antennas. These investigations revealed the economic advantages of using shortwave vacuum-tube transmitters instead of longwave equipment such as the Alexanderson radio alternator for transoceanic communication. RCA and other companies involved in international radio communication soon introduced the use of shortwave transmitters in commercial service.[1]

Beginning in 1924, Alexanderson and several of his engineer assistants initiated a systematic research effort to learn more about shortwave radio propagation and directive antennas. In the process, they discovered unanticipated polarization effects that led to some modifications in propagation theory and to the development of new techniques for long-distance radio communication. In particular, they found that radio waves with horizontal polarization were much more useful in long-distance communication than had previously been thought. Although the polarization of electromagnetic waves had been demonstrated by Heinrich Hertz in the 1880s, longwave radio systems had come generally to rely on vertical polarization owing to the rapid dissipation of waves with horizontal polarization that traveled along the earth's surface.

In October 1924, Alexanderson began a series of shortwave experiments using an elevated loop antenna. His assistant Samuel Nixdorff recorded in his laboratory notebook that the goal of the research was to obtain unidirectional radiation by creating an unbalanced set of currents in the loop.[2] In January 1925, Alexanderson addressed a memorandum on radiation with horizontal polarization to GE's Patent Department. This led to a patent application filed in May 1925 covering the use of a loop antenna to produce horizontally polarized waves.[3]

In March 1925, Alexanderson reported on his shortwave research in a technical paper presented at a meeting of the American Institute of Electrical Engineers. Contrary to earlier experience, he stated, he and his colleagues were finding that extraordinary distances could be achieved by short waves transmitted at low power levels. He interpreted this to mean that high-angle shortwave propagation was "a new phenomenon or law of nature" that could not as yet be explained adequately, for the observed effects were not in agreement with the "old theory of the Heaviside layer as a conducting and reflecting surface." He suggested that a theory proposed by Joseph Larmor, which postulated change in the refractive index of the upper atmosphere due to the presence of electrons, was more promising. Alexanderson pointed out that shortwave techniques not only opened "new paths for wave propagation but give us an almost inexhaustible scale of wavelengths, provided that we utilize it to full advantage."[4]

Nixdorff mentioned the propagation and antenna experiments frequently in his notebooks during 1925. In early February, he recorded a radiation experiment using a shortwave antenna that could be switched from vertical to horizontal polarization. In March, he mentioned that Gregory Breit of the Carnegie Institution had been to see Alexanderson and that they had discussed the theory of the Heaviside layer. Later the same month, Nixdorff listed those who were engaged in various experiments at GE. He stated that two engineers were investigating the Heaviside layer and that Chester Rice was at work on directive radiation. Early in April, Nixdorff wrote that he had taken a member of GE's Patent Department out to the test site to explain the design of a doublet-loop antenna and the difference between vertical and horizontal polarization in wave propagation.[5]

On April 17, 1925, Nixdorff summarized a conference with Chester Rice and Alexanderson during which future research was discussed. Rice had recommended that the engineers continue to investigate shortwave propagation in the range of 30 meters to 300 meters wavelength. Nixdorff recorded that Alexanderson and Rice had decided on a series of experiments using loop antennas with receiving stations located at distances between 10 and 3,000 miles away in various directions.[6]

Alexanderson planned to report some of the newly discovered wave propagation phenomena at an AIEE meeting in June 1925, but his plans were altered because of his trip to Sweden. He decided instead to disclose the latest findings at a meeting of the Swedish Academy of Engineering Science. Abstracts of his talk appeared in American newspapers shortly afterward. One report, in the *New York Herald,* credited Alexanderson with the "rediscovery" of horizontally polarized radio waves at his laboratory in Schenectady. The writer speculated that the discovery might be the most important contribution to radio that year. According to the newspaper account, the discovery had resulted from the investigators having made a wrong connection during an antenna experiment and having found reflected waves to be horizontally polarized.[7]

Alexanderson gave further details on the shortwave research in an article published in *Wireless World* in September 1925. He pointed out that two forms of polarized propagation were well known in optics but that only vertically polarized waves had been used in radio until recently. He attributed the neglect to the fact that receiving instruments had given no indication of the presence of waves with horizontal polarization near the earth's surface. He explained that short waves did not follow the surface in the manner of the long waves that had been used commercially, and he predicted that the new form of propagation would become very useful in radio communication, although the high-power longwave system was still more reliable for continuous service.[8]

Alexanderson suggested a potential commercial application of short waves with horizontal polarization in a memorandum written in October 1925. In it, he stated that comparative tests of shortwave systems with vertical and horizontal polarization had been made between Schenectady and the West Coast during the past two months, and that these experiments had demonstrated "conclusively that the horizontal polarization transmission is superior." The tests had been done at a wavelength of 80 meters. Alexanderson added that RCA had decided to adopt short waves for its transpacific chain and had asked GE for recommendations on antenna design. Commenting that it was fortunate that research carried out in the past year placed GE in a position to respond to RCA's request, he concluded by saying that he and his colleagues planned to recommend the use of horizontal polarization of the type they had tested at 80 meters.[9]

Alexanderson included a summary of the results of the shortwave propagation experiments in his annual report on the activities of GE's Radio Consulting Department for 1925. He mentioned that several new shortwave properties had been discovered having to do with polarization, twisting of waves, high-angle radiation, and skip distance. He stated that he and his colleagues had learned that a wavelength could be selected that would make transoceanic shortwave communication possible at all times during daylight hours. He pointed out that this discovery was especially important in such new radio fields as facsimile and television.[10]

In early February 1926, Alexanderson proposed a new interpretation of longwave propagation in a brief communication submitted to the IRE and later published. His communication took the form of a comment on a paper by Greenleaf Pickard in which Pickard had stated that low-frequency waves had been found to be vertically polarized at all distances from the source. Alexanderson cited some irregularities in longwave measurements which he interpreted as indicating the presence of a horizontal component. He added that recent tests had revealed that radio waves were propagated with both vertical and horizontal components that traveled at different velocities. Plane polarized waves were produced when the two components happened to be in phase, he explained, and circularly polarized waves resulted from a 90-degree phase difference. He concluded that horizontal polarization was not confined to short waves.[11]

In a lecture given for the Radio Division of the National Research Council in Washington, D.C., in April 1926, Alexanderson discussed what he called the "corkscrew wave" theory. He stated that radiation data gathered at various distances from transmitters had revealed a "continuously twisting plane of polarization" between the two extremes of plane and circular polarization. Direct measurement of the horizontally polarized component of short waves could be made at moderate distances above the earth, he explained, but the evidence for a horizontal component of long waves was indirect because direct measurements could only be made at great heights. He suggested that tests be made using airplanes to verify the theory and to learn more about the exact mechanism of wave propagation.[12]

Alexanderson speculated about what might cause the corkscrew effect. He stated that it could not be explained fully as a Faraday effect, due to the earth's magnetic field; or as a Kerr effect, due to an electrostatic gradient around the earth. However, he commented that it might be analogous to the Kerr effect, even though the latter had never been observed in gases. The Kerr effect, he pointed out, could be interpreted as a wave consisting of two components with different velocities. He mentioned that a mechanical model had been constructed which gave a visual display of how different wave velocities in two planes could produce "the composite effect of a spiral wave." He suggested that the velocity difference between the two components in actual radio waves might be the result of greater absorption of energy from the horizontal component by the earth.[13]

An article about the ongoing propagation experiments at GE was published in *Scientific American* in April 1926. The author, Orrin E. Dunlap, Jr., stated that Alexanderson was "devoting much of his time to shortwave experiments." Dunlap mentioned that GE had seven transmitters that could be operated at different wavelengths with a variety of antennas.[14] Later in the year, Dunlap contributed another article to *Scientific American* on the GE radio research facilities. In the paper, entitled "Acres of Radio," he reported that GE had a "radio reservation" of 1,154 acres in Schenectady where the engineers were investigating wave-propagation phenomena over a range of wavelengths from 5 to 3,000 meters. He stated that the power level of the seven transmitters ranged from 50 kw for WGY to 600 watts for the experimental station 2XAW.[15]

Alexanderson delivered a paper on the theoretical and experimental aspects of radio-wave polarization at an AIEE meeting at Niagara Falls, New York, in May 1926. He pointed out that polarization effects in radio had seemed to be of only academic interest until quite recently, since the waves actually used for communication were thought always to be vertically polarized. There had been some unexplained anomalies, he continued, but they had been ignored until new concepts had emerged from shortwave experiments. He discussed some difficulties that had been encountered with radio direction finders in airplanes and used the polarization theory to explain the probable cause. He pointed out that the theory also could be used to explain the phenomenon of fading in radio.[16]

In June 1926, an article on polarization based on an interview with Alexanderson appeared in the journal *QST.* The interviewer, Robert S. Kruse, reported that Alexanderson had turned his own car into a "radio exploration car" and had taken Kruse out to demonstrate the effects of a twisting plane of polarization at a wavelength of 50 meters.[17] In an unpublished paper written at about that time, Alexanderson expressed a high regard for the contributions of amateur radio operators to the research effort on shortwave propagation and polarization at GE. Reports of reception had been received from as far away from Schenectady as Australia. He wrote that the GE researchers' goal was to encourage both the professional and amateur radio experts of the world to participate in the research. He stated that although the physicists had taken away the ether from the radio experimenters, the latter might discover a replacement by using the world as a radio laboratory.[18] It was at about this time in 1926 that a newspaper article on Alexanderson compared him to Charles Steinmetz. The writer stated that since the death of Steinmetz, no one had an undisputed claim to having the preeminent mind in the electrical field. He concluded that Alexanderson probably came closer to the title of preeminent electrical expert than anyone else.[19]

Radio Facsimile and Mechanical-Scan Television

The researches on shortwave propagation and the transmission of photographs and documents by radio and television that were undertaken by Alexanderson and his team of assistants in the 1920s were closely linked, and tended to exemplify his theory of the dynamic interaction of fundamental research, invention, and engineering development. On several occasions, Alexanderson used the metaphor of the stepping-stone to explain his belief that the introduction of a complex technology such as television could best be approached in a series of well-defined stages. In a memorandum written in April 1926, he pointed out that he regarded facsimile radiotelegraphy as a necessary stepping-stone to television, and that the funds being expended on facsimile development had as their ultimate objective a television system. He stated that he was already building a strong patent position in television.[20]

In June 1926, Alexanderson reported that he and his colleagues were engaged in a program of development which included several steps, with each having a potential commercial application. The first step would be to improve the speed and quality of radio facsimile transmission, the second step would be to transmit motion-picture film by radio from Europe to the United States so that it could be shown the same day, and the final step would be to devise a system of direct vision by radio suitable for theater and home reception.[21]

According to Alexanderson, the initial stimulus for beginning a research project on radio facsimile had been provided by Owen Young, GE's chairman of the board, in a banquet speech delivered in 1923. Young had called attention to the need for a method that would make possible the transoceanic radio transmission of documents without the need to send the text word by word in telegraphic code.[22] Soon after Young's speech, Alexanderson and his engineering staff began work on a facsimile system that could be adapted to the RCA transoceanic radio network. In October 1923, Alexanderson and an assistant, Richard H. Ranger, applied for patent coverage of a method for facsimile radio transmission. Early in November 1923, it was reported that Alexanderson and his colleagues had transmitted a picture from the United States to Poland and back by radio.[23]

By the summer of 1924, Alexanderson's interest had progressed from the transmission of still pictures to television. By then he had perceived the advantage of using shortwave transmission to accommodate the greater bandwidth needed by a television system. In July 1924, he drafted a memorandum to GE's Patent Department entitled "Television by Radio," in which he forecast that advances in shortwave techniques eventually would make television by means of radio waves feasible. He predicted that television would be achieved first by scanning different points in sequence but that it might in the future become possible to project an entire picture simultaneously. For example, he wrote, the transmission of 10,000 signals separated by 100 Hz would require a bandwidth of 1 megahertz (MHz). If a carrier frequency of 10 MHz were used, he noted, a picture could be transmitted in the band between 10 MHz and 11 MHz.[24]

In August 1924, Alexanderson reported to Francis Pratt, GE's vice-president for engineering, that the Radio Development Committee had met and decided to recommend continued investigation into the feasibility of television. Alexanderson stated that he believed that three areas should be considered: optics and photoelectric phenomena, wave propagation, and the process of picture reproduction. He added that television was apt to emerge as a natural byproduct of further developments in shortwave techniques. He stressed that the entire chain of elements of a television system should be examined closely to determine whether there were any missing links. If this were done systematically, he pointed out, the GE organization would be in a good position to establish a dominant patent coverage in the new art. He mentioned that a television system might be employed in conjunction with radio broadcast stations to show current events accompanied by sound in moving-picture houses.[25]

The rotary scanning disk became a key element of the television system developed by Alexanderson in the 1920s. The use of a high-speed perforated disk in television had an obvious appeal to an engineer-inventor who had spent two decades on the development of rotary-disk radio alternators. Some years later, he responded to an inquiry about the similarity between the two disk applications by agreeing that the design of the scanning disks had been almost the same as that of the alternator rotor. He stated that he had worked out the design of the scanning disk, had entrusted its fabrication to the best mechanic he could find, and had followed the fabrication closely and made several changes in the process.[26] The idea of a scanning disk was not original with Alexanderson but had been proposed by Paul Nipkow in the 1880s. Nipkow had not, however, put the invention into practice successfully. A scanning-disk television was developed independently by John L. Baird in England and demonstrated publicly early in 1926.[27]

In early December 1924, Alexanderson reported an optical experiment by an assistant, Charles Hoxie, using a rotating steel disk with a small hole near the rim, and expressed his belief that a disk might prove to be more practical than a rotating-lens scanning system of the type being used in television experiments by Charles F. Jenkins.[28] Two days later, Alexanderson wrote to E. W. Allen of GE's Patent Department that Hoxie's experiment had convinced him that a photoelectric cell was sensitive enough to be used with a scanning disk to convert varying shades of light into electrical signals. Alexanderson concluded that it was still uncertain whether light of sufficient intensity to reproduce pictures on a screen could be controlled. Alexanderson also informed David Sarnoff of RCA that GE was actively engaged in television development and that he was quite enthusiastic about its possibilities.[29]

A television demonstration was conducted for E. W. Allen and other GE officials in January 1925. Alexanderson wrote to a witness of the demonstration that it had served to show in operation the elements he and his colleagues considered necessary for the development of a practical television system. He explained that the image had first been passed through a pattern of holes in a rotating scanning disk to a photoelectric cell. The varying current from the cell had then been converted by a toothed-wheel chopper into a 1-kHz signal that in turn modulated the radio signal produced by station WGY, and the WGY signal had been picked up by a suitable receiver and used to control the mirror of an oscillograph. Finally, the optical signal from the oscillograph had been sent through a rotating Benford crystal and projected on a screen.[30]

Alexanderson's assistant Samuel Nixdorff became an active participant in the television experiments at GE and recorded their progress in his laboratory notebooks. In January 1925, Nixdorff mentioned that Alexanderson had started him working on television experiments.[31] During his trip to Europe in the summer of 1925 to attend the official opening ceremonies for the new Swedish radio alternator station, Alexanderson stopped off in Germany and learned of the Karolus system of light control.[32] When August Karolus visited Schenectady in October, Nixdorff reported that he had gone to the illuminating laboratory with Alexanderson and a German who had invented a television system. The next several pages of Nixdorff's notebook described an effort to employ Karolus's methods at GE. When the response obtained by a two-stage amplifier proved weak, Nixdorff built a third stage and a lightproof container for the Karolus cell. In December 1925, Nixdorff wrote that Alexanderson had been in the laboratory all afternoon and had asked Albert Hull of the GE Research Laboratory to help in getting the photoelectric cells to function properly. A few days later, Nixdorff reported that, by using five stages of amplification and direct light, Alexanderson and his colleagues had managed to swing the oscillograph beam. He mentioned that Alexanderson had outlined a program of continued research on the television system.[33]

The Karolus innovation consisted of a Kerr-effect cell that employed a solution such as nitrobenzol between two electrodes. When a voltage in the range of 500 to 1,000 volts was applied to the electrodes, the polarization of light passing through the solution could be varied. With a proper setting, the cell functioned as a light valve with negligible inertia. Germany's Telefunken Company later used a version of the Karolus system to transmit photographs by wire.[34]

The effort to develop a commercial radio facsimile system continued in 1926. Alexanderson attended a facsimile conference at GE in January 1926 along with Walter R. G. Baker, F. A. Benford, Lawrence Hawkins of the GE Research Laboratory, and others. They decided that the next demonstration of facsimile transmission would be between Schenectady and New York City. During the meeting, they visited the radio laboratory to see the operation of a speed regulator designed to synchronize the transmitter and receiver. During the demonstration, a typewritten letter was projected onto a photoelectric cell that was used to modulate a radio alternator.[35] In March 1926, Alexanderson provided GE's Patent Department with sample copies of facsimile pictures sent by radio. He reported that he and his assistants had solved the problem of transmitting pictures directly from a typed sheet. The sheet of paper was mounted on a rotating drum and a lens was used to focus a light spot on the paper, he explained. During 1926 he filed applications for three patents covering apparatus for transmission of pictures by radio.[36]

Alexanderson gave an invited lecture on radio facsimile at the annual summer camp for GE engineers at Association Island in Lake Ontario in 1926. The island had been owned by GE since 1912 and was the site of various camp meetings during the 1920s. Camp Engineering combined athletic competition, rituals, and reports on research in progress at GE. Alexanderson was asked to serve as "medicine man" of his tribe at the 1926 gathering.[37] In his talk, he stated that radio picture transmission across the Atlantic had been introduced recently and required about 20 minutes for each picture. He demonstrated a projector that included a light source, a lens, and a rotating drum. As the drum revolved, a spot of light could be observed moving across a screen. Alexanderson commented that the mechanism used the basic principle employed in all remote transmission of pictures, whether by wire or by radio. He mentioned that several means of controlling the light spot's intensity were under investigation, including a mechanical device similar to an oscillograph, a sensitive arc light, and the Kerr cell.[38]

Alexanderson concluded his presentation with some speculations on the future of television. He pointed out that transmission of about 160,000 "brush strokes per second" would be needed but that the use of short waves made such a rate entirely feasible: a picture could be transmitted in 0.125 second if a wavelength of 12 meters instead of 12,000 meters were used. He predicted that moving pictures soon would be received from Europe by radio.[39]

In December 1926, Alexanderson spoke on the current status of radio photography and the prospects for television at an AIEE meeting in Saint Louis. He began by describing a scene from George Bernard Shaw's play "Back to Methuselah," which had depicted the communication of life-sized images of people over distances of hundreds of miles in the year 2170. Alexanderson commented that great writers or statesmen frequently are the first to envision what is to come. Later, "the inventors and engineers take hold of the ideas and dress them up in practical form." He reviewed some of the problems that had been encountered and overcome during the development of commercial telephotography and mentioned that the effort had been an international one, with important work being done in France, England, and Germany as well as in the United States. He reported that photographs had been transmitted by the GE station WGY and pointed out that existing radio broadcasting stations could be used to provide such a service if the public became interested. He noted, however, that at broadcast frequencies good-quality photographs could be sent only at a slow rate. He stated that he could foresee the use of short waves as a "wonderful medium of communication to transmit pictures, facsimile[s] of letters or printed pages, and moving picture films, and ultimately to see by radio."[40]

In the remainder of his paper, Alexanderson addressed the challenging problems that would need to be overcome in the development of television. He noted that GE's research on radio photography had shown that at least 300,000 picture units per second would need to be scanned to produce satisfactory television pictures. Such a rate posed major problems for the design of a projector of life-size images of the type used in Shaw's play, he explained: the most brilliant light source available would not produce adequate illumination if it remained in one spot for only one-300,000th of a second. As a possible solution to this problem, he mentioned, experiments were under way at GE on a projector that used seven light spots from separate sources in an arrangement that gave 49 times as much illumination as a single-spot scanner.[41]

Alexanderson identified the dependence on moving mechanical components as the root cause of many of the difficulties in television. He noted that the system could be simplified greatly if a ray of light could be made to scan without the use of mechanical motion, but he observed that although a way to accomplish this might be discovered, "we are not willing to wait for a discovery that may never come." He mentioned that a cathode-ray oscillograph had been proposed for television but that it would not be suitable for projecting a picture on a large screen. He did not know how long it might take to introduce television, he concluded, but research had shown already "that it may be accomplished with means that are in our possession at the present day."[42]

In January 1927, Alexanderson gave a lecture-demonstration about television before an audience of more than 500 at an IRE meeting in New York City. During the lecture, pictures broadcast over station WGY in Schenectady were received. A newspaper account of the demonstration stated that no one who had been present would disagree with Alexanderson's prediction that every home would have television within ten years.[43] Nixdorff mentioned in his laboratory diary early in February 1927 that Alexanderson had given a talk on television and radio photography at the radio laboratory. The room had been crowded, and there had been a lot of interest in the subject.[44]

Alexanderson's lectures on television stimulated a flurry of newspaper articles about the Schenectady inventor and his predictions.[45] In a report on Alexanderson's AIEE talk in Saint Louis, the *New York Times* characterized him as "a typical industrial research scientist of the modern school."[46] A newspaper in Oklahoma predicted that television would soon become as familiar as the telephone, and called Alexanderson the inventor of television. Such innovators as Alexanderson, the article continued, "affect our lives beyond our comprehension. They and not the Valentinos are the real men of our time."[47] Another newspaper commented in January 1927 that Alexanderson's predictions had helped send the imagination of the average man "soaring to realms hitherto undreamed of."[48]

In May 1927, the *New York Herald* included a feature story on Alexanderson in a series devoted to prominent men which emphasized turning points or crises in their careers. In the interview, Alexanderson emphasized that he did not claim to be the inventor of television; rather, television was the conception of many minds. He stated that he tended to be skeptical of accounts of inventors whose great inventions made them suddenly wealthy. He had not had that kind of turning point, and inventors generally started inventing when young and kept right on. He indicated that a "sort of turning point" had occurred when he began to work in the field of radio after Fessenden had come to GE with an order for a machine no one had. He concluded with the observation that there were not many turning points in the professional lives of men like himself who "go straight ahead and one thing leads to the next."[49] He expressed a similar view in a talk to members of Sigma Xi in April 1927 which included a television demonstration. He commented that the "march of progress is necessarily slow; it is an evolution and not a revolution."[50]

Alexanderson proposed some possible military applications of radio and radio photography in a memorandum to David Sarnoff of RCA in January 1927. Alexanderson suggested that unmanned aircraft controlled by radio and loaded with bombs could be equipped with a "radio photograph" transmitter with its lens directed toward the ground; the resulting radio maps would be sent back and used to direct the craft toward a target.[51] Evidently this idea was not pursued until about three years later, when a renewed proposal to develop military uses of television attracted attention from a number of high-ranking military officers.

The multifaceted nature of the telecommunication research of GE's Radio Consulting Department was clearly evident in Alexanderson's outline of the activities planned for 1927. Photographs would be broadcast over station WGY (an inexpensive recorder was being developed to be used for reception); facsimile letter transmission by radio would be initiated between Schenectady, New York City, Pittsburgh, and San Francisco; and research would be done on wave propagation, time standards, photoelectric cells, amplifiers, optics, photography, and elimination of static and fading. Alexanderson added that television not only involved the problems encountered in radio photography but also raised the difficulties to a higher level and would necessitate research on the Kerr effect and radio by multiple channels. He pointed out that many branches of science and research were involved and that an important function of his department was the coordination of the efforts of research groups at GE, RCA, Westinghouse, and the National Broadcasting Company (NBC) toward a common end.[52]

Alexanderson suggested that the Radio Consulting Department could best serve this coordinating function by concentrating on developments well in advance of commercial activities in order that a comprehensive program could be formulated before other corporate departments became interested. In its role of exploring new territory and educating men for new technical fields, the Radio Consulting Department was the key to countering the tendency for a large organization to become too conservative. There were cases that called for a gambler's risk and transcended the mission of existing branch organizations, Alexanderson concluded. He asserted that he considered it his duty to alert the central management to such cases and to recommend an appropriate course of action.[53]

Alexanderson's conception of the proper role of the internal consulting engineer such as himself showed the formative influence of Steinmetz's prewar Consulting Engineering Department but had been elaborated to cover the case of the joint interests of "associated companies." In a later memorandum concerning the radio research of the associated companies, Alexanderson stated that the merit of a certain line of research could best be judged by the ultimate results. He commented that once development had reached the practical level, corporate executives who were responsible for financial transactions were in a better position to decide on funding than were individuals or committees responsible for the technical work.[54]

In October 1927, Alexanderson reported that a television receiver designed for home use had been tested successfully in his own home. He stated that it had been possible to recognize individuals who posed for the transmitting unit in the GE radio laboratory. He explained that he and his colleagues had used rotating disks but had added a simplified method of synchronizing the transmitting and receiving disks devised by his assistant Ray D. Kell. Alexanderson commented that the simplicity of the receiver should be an important factor in popularizing television.[55] A few days later, he wrote to E. W. Allen of GE's Patent Department that the experiment had not been the first transmission of television by radio but that it had been the first to demonstrate the practicability of home television. The next step, Alexanderson stated, would be to demonstrate home entertainment by means of voice and pictures transmitted over the local radio stations.[56]

The home television receiver was demonstrated at a meeting of the Fortnightly Club in Schenectady on January 13, 1928, and regularly scheduled broadcasts began on April 22, 1928. Alexanderson drafted a statement inviting the public to participate in the experimental broadcasts. He stated that the pictures would be broadcast at 37.8 meters and the voice on WGY.[57] A press release in 1928 stated that several homes in Schenectady had receiving equipment. The statement continued that it was possible to see lips move, eyes blink, and cigarette smoke

Figure 7.1. Alexanderson and his family watching a mechanical-scan home television receiver, c. 1928. Alexanderson holds a motor control necessary to synchronize the picture. Courtesy of the General Electric Company, Schenectady, N.Y.

rise in pictures transmitted over a GE shortwave station.[58] GE was almost deluged by letters from people wanting to assist in the experiments. Many were sent a copy of a form letter signed by Alexanderson stressing that television was still in the experimental stage.

An article about the Alexanderson home television system appeared in *Scientific American* in 1928. The author, A. P. Peck, reported that television had arrived at the "headphone stage." He observed that the apparatus seemed quite simple and that the problem of synchronization of disks, which had frustrated many, had been solved by Alexanderson "with his usual direct method." Peck explained that, in fact, no transmitted synchronizing signal was used. Instead, the viewer employed a hand control that could vary slightly the speed of the disk-drive motor until a clear and stable picture resulted. Peck wrote that the receiver included a vacuum-tube amplifier that fed a special neon tube invented by D. McFarlan Moore which could respond to variations at the rate of 1 million per second. Peck explained that the light produced by the neon tube illuminated a 24-inch scanning disk. Peck also described the transmitter, where the performer sat before a similar scanning disk, which was illuminated by an arc light. The light of varying intensity that was reflected from the subject's face was picked up by photoelectric cells. The output of the photoelectric cells was applied to the modulator of the shortwave radio transmitter.[59]

In 1928 Alexanderson again was asked to prepare a presentation for those who would attend Camp Engineering at Association Island that summer. He gave a lecture entitled "Progress in the Art of Television," in which he commented that the prospect of television had fired the imaginations of writers and engineers all over the world. He reported that the GE experimenters now could reproduce the image of a human face by means of a 24-line scanning system. The use of shortwave transmitters would make possible the use of finer detail in the pictures, and a 48-line picture had been produced by means of the Karolus system of light control. He noted that greater detail might also be obtained through the use of several light beams and several radio channels.[60]

A televised play was broadcast by GE on September 11, 1928. The play, entitled "The Queen's Messenger," was a 40-minute performance that used two shortwave radio channels for the pictures and a radio broadcast channel for the audio. In an article about the broadcast in *Radio News*, Robert Hertzberg predicted that the television "camera" would soon be seen as frequently at banquets and lecture platforms as the ubiquitous radio microphone. He wrote that the show had only been feasible since Alexanderson had reduced the complexity and size of the television apparatus so it could easily be moved. Hertzberg mentioned that WGY had recently gone to Albany to televise Governor Smith's acceptance of the Democratic party's nomination for president.[61]

Hertzberg explained that three cameras had been used in the broadcast of "The Queen's Messenger." One had shown the face of each of the two actors, and a third had been used to show props and various special effects. The program director could select which picture to broadcast and fade the images in or out. Hertzberg described the transmitting apparatus, which was housed in a 20-inch-long wooden box mounted on a camera tripod. The key element was the 12-inch-diameter scanning disk driven by a small electric motor. The receiver projected the picture on a 12-by-12-inch screen. Hertzberg mentioned that Alexanderson had been observing pictures televised from Schenectady at his summer place at Lake George, about 50 miles away. Alexanderson had found that the reception was hampered by a "mirage effect" that he believed might be due to multiple reflections from the Heaviside layer or layers. Alexanderson had stressed that it would be some time before television became generally available to the public for entertainment.[62]

Alexanderson had become interested in the possibility of color television after a visit to George Eastman's laboratory in Rochester, New York, in August 1928.[63] Afterwards, he informed GE's Patent Department that he had observed a demonstration of color photography at the Eastman laboratory and had been quite impressed by the quality and detail of the photographs. This had stimulated him to think that the introduction of color would be the next step in the development of television. He suggested that the use of color should give a higher-grade picture even with a 24-line scan, and he described a method that might be used for color television. It would employ Kerr cells in three separate radio channels, with red, yellow, and blue screens. He explained that each of the Kerr cells would be controlled by a radio channel and that each in turn would control a photoelectric cell and color screen.[64] A few days later, he reported that an experimental test had confirmed the principle of three-channel color television. As a result of his interest in color photography, he decided to purchase an Eastman color-film motion-picture camera and projector for the radio laboratory.[65]

Alexanderson sought the help of his friend Irving Langmuir of the GE Research Laboratory in developing an improved source of light for television pictures. In August 1928, Alexanderson reported that his discussions with Langmuir had led to a test of a new arc lamp that could be modulated electrically as a source of light. He stated that the lamp had produced a brighter picture than any tried previously.[66]

Alexanderson recognized the potential advantages of the cathode-ray tube as an electronic alternative to mechanical scanning in television. In a memorandum to GE's Patent Department in October 1928, he described a television receiver that would employ either electromagnetic or electrostatic deflection of a beam of electrons to produce a picture. He noted that the fluorescent material used to produce the image placed a practical limit on the size of the picture, a limit he hoped could be overcome by projecting the image through a lens.[67] A few days later, he wrote that numerous difficulties would have to be overcome before the cathode-ray tube could serve successfully in a television receiver, but that he did not doubt that the adoption of the electronic picture tube would "be a step toward the television receiver of the future."[68]

In a progress report written in November 1928, Alexanderson described the experimental radio facilities in Schenectady as the best in the world, but noted that he would like to establish a receiving station in California and enlarge the receiving station in Schenectady. He explained that these new facilities would be a permanent asset and would be used in future television and radio relay experiments. In another memorandum concerning recent television demonstrations at GE, he stated that television had become a valuable tool for the investigation of propagation phenomena.[69]

In his annual report covering the activities of his department during 1929, Alexanderson wrote that office correspondence, drawings, and calculations had been received on a daily basis from California by radio facsimile. This system had replaced airmail and also had served as a valuable aid in propagation research. In regard to television, he commented that the researchers' goal was to introduce it initially in theaters, where the apparatus could be operated by experts. He reported that a television projector had been developed which could produce four-foot-square pictures, and he mentioned that methods being developed to relay radio broadcasts might be used eventually in television.[70]

Difficulties were encountered in communicating research and inventions in the field of radio and television from GE and Westinghouse to RCA, which had been given responsibility for commercial operation. Alexanderson addressed this problem in a letter to Gerard Swope, president of GE, in March 1929. Alexanderson credited GE's generous support with having made it possible to undertake a comprehensive program of radio research. He added that he had taken the initiative in this area and wanted RCA to derive the maximum benefits. Observing that personal contact between technical experts in the various laboratories was the best way to coordinate a research effort in which several independent organizations cooperated, he argued that a committee that kept formal minutes did not provide a very favorable atmosphere, and proposed holding periodic radio research luncheons at the radio laboratories in Schenectady, Pittsburgh, and Long Island to facilitate the voluntary exchange of information. The success of such meetings would depend on mutual confidence and professional ethics, he concluded, but he believed they would be productive.[71]

The problem of intercompany communication was also a concern of RCA's top management, as evidenced by a memorandum from James Harbord, president of RCA, to RCA's vice-president for engineering in October 1929. Harbord requested that meetings of the Technical Committee be resumed under the auspices of RCA. He pointed out that RCA had contributed to the research expenses of Alexanderson, who had done much for the field of telecommunication. Harbord asked that monthly meetings of the Technical Committee be held so that Alexanderson and others from outside RCA could discuss technical problems and developments with RCA representatives. As Harbord had asked, meetings of the Technical Committee resumed in November 1929.[72]

At the December 1929 meeting of the Technical Committee, Alexanderson reported that GE had developed an improved radio facsimile transmitter. He advised that RCA use it instead of undertaking a similar development on its own. He also stated opposition to the idea that the work of the associated manufacturing company should cease as soon as a commercial service was established. He contended that RCA could not afford to have GE and Westinghouse drop out, and that their usefulness would continue if their research received support from RCA.[73]

At the next meeting of the committee in January 1930, Alexanderson continued to encourage RCA to begin long-distance facsimile service. He pointed out that GE had been authorized to demonstrate a practical facsimile service across the continent two years before and could now transmit documents from California to Schenectady at the rate of 8 square inches per minute. He expressed the desire to turn further facsimile work over to RCA so that the Schenectady station would be available for the experiments needed to produce a satisfactory television system. He exhibited copies of facsimiles received from Oakland, concluding by saying that the GE system was capable of producing clear pictures of commercial quality.[74]

The goals of the GE radio-television research effort directed by Alexanderson were indicated in his proposed budget for 1930. Explaining that the budget covered only research, as opposed to development, he defined the research he directed as work toward the objective of producing "knowledge and invention rather than apparatus for practical use." He noted that the principal research tools were the stations located in Schenectady and in California. He estimated a total expenditure in the range of $200,000 to $250,000, which would include $60,000 to $80,000 devoted to television research. He wrote that one goal of planned research was to determine the relative merits of cathode-ray tubes and rotating disks for use in television receivers. He added that his research team also planned to investigate the relative merits of frequency and amplitude modulation and to determine what wavelengths and atmospheric conditions were best for long-range television, and he mentioned that he wanted to determine whether television could be broadcast using ultrashort waves. He also commented that he planned to maintain close contact with Charles Taylor of RCA, and that he considered personal contact between responsible individuals to be far superior to committees.[75]

In response to a 1939 inquiry concerning the cost of the television research he had directed in the 1920s, Alexanderson estimated that between 1926 and 1930 his group had spent $750,000 on visual radio research, which included radio, facsimile, wave propagation experiments, and television. He explained that although it might be claimed that the entire amount had been in preparation for television, it would be more accurate to conclude—since the wave propagation studies had wider significance—that about $500,000 had been spent directly or indirectly on the development of television.[76]

The Development of a Radio Altimeter

A visit to the GE radio research facilities by a group of army officers from Wright Field in the fall of 1927 provided the initial stimulus for Alexanderson to begin an effort to produce a radio altimeter for aircraft. After the visit, he reported that he had discussed the problem of altitude "sounding" from aircraft with the officers and that he already had a solution in mind, which would involve the use of a small oscillator and an antenna. He explained that the altitude measurement would be based on a shift in the frequency of the oscillator which was expected to occur as the plane approached the ground.[77]

With Samuel Nixdorff's assistance, Alexanderson began experiments on the radio altimeter in 1927 and continued them into the following year. In January 1928 Nixdorff informed GE's Patent Department that he and Alexanderson had tested an altimeter that employed two 75-foot-long antennas attached to an aircraft's wings. A 1-kHz beat tone had been obtained at an altitude of 600 feet but had disappeared at 400 feet; no results had been obtained for altitudes above 600 feet.[78] Later in the year, Alexanderson reported that a radio echo effect had been demonstrated using a wavelength of 100 meters during several flights and that the effect established the basis for the design of a direct-reading instrument for altitude. He wrote that a cyclic half-wave variation had been obtained and that this would indicate altitude changes with great accuracy. He filed applications for patents on a radio altimeter in November 1928 and April 1929.[79]

Alexanderson presented a paper on the radio altimeter project at a meeting of the National Academy of Sciences held in Schenectady in November 1928. Summaries of his presentation were published in *Science* and in *Airway Age*. He mentioned that the theory that had been the basis of the initial experiments "had to be thrown overboard as soon as we left the ground." Once the theory of interference between the direct and reflected wave was introduced, he continued, he had found it simple to interpret the observed phenomena. He asserted that he had been able to determine altitude by means of an earphone while wearing a blindfold.[80] During his lecture, Alexanderson exhibited charts from a graphic recorder used on test flights which indicated the feasibility

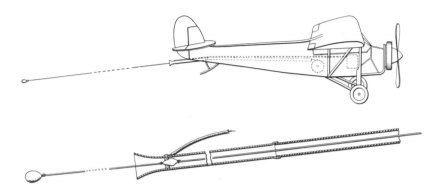

Figure 7.2. Rectractable antenna used with radio altimeter tested in 1929. The antenna consisted of a single wire attached to a weight that trailed behind the aircraft during flight and was retracted before landing. The system warned the pilot when the ground was within 600 feet. Courtesy of the General Electric Company, Schenectady, N.Y.

of accurate altitude measurements to 4,000 feet above ground level. He outlined an alternative method that would actuate green or red lights automatically at certain altitudes, and finished by saying that experienced pilots had told him that the radio altimeter should make it possible to land safely with virtually zero visibility.[81]

The radio altimeter investigation caused Alexanderson to recognize that there might be attractive opportunities for GE in the manufacture of aviation products. In December 1928, he wrote a memorandum calling attention to the potential market for navigational aids for airplanes. He recommended that GE acquire a plane to use as a laboratory for the development of instruments that would use radio echo and beacon methods to ensure safe flight above clouds and through fog. He also suggested that GE consider becoming a manufacturer of aircraft engines or even airplanes and concluded by predicting that the aviation industry and aerial transportation would enjoy a great future.[82]

F. G. Patterson, a GE engineer, was assigned the task of developing the radio altitude meter into a practical instrument and working out optimum methods for its use. In 1936, Alexanderson gave the Institute of the Aeronautical Sciences an account of improvements made since

1928. He explained that since he was not a pilot he had not felt it appropriate for him to promote the method after announcing the results of the early tests, for he believed that "the demand for such service should naturally come from the aviators."[83]

The altimeter, as improved by Patterson, was housed in a compact package weighing 6.5 pounds. It utilized a single vacuum-tube receiver with a single wire antenna that was trailed from the plane during flight. The standing wave pattern was displayed on a dial that alternated between maximum and minimum readings at 200-foot intervals. Patterson had added auditory signals that warned the pilot when the plane descended to 1,000- and 600-foot altitudes. In practice the altimeter could warn the pilot of approaching hills when visibility was poor. It was used in conjunction with a conventional aneroid altimeter housed in the same package as the radio altimeter.[84]

A paper on altimeters published in 1944 included a discussion of the Alexanderson altimeter as developed in the late 1920s. The author expressed puzzlement that the device appeared not to have been used commercially. He speculated that it might have required too much attention from the pilot.[85]

Power Transmission and Naval Gun Control: Applications of the Thyratron

Although the telecommunication research that he orchestrated was far more newsworthy, Alexanderson continued to take an active interest in the development and applications of electronic tubes in the fields of electrical power transmission and control during the 1920s. The wisdom of maintaining a diversity of technological programs was confirmed when Alexanderson and his colleagues at GE were prevented from continuing their radio-television investigations in the 1930s owing to a drastic change in the relationship between GE and RCA. The "thyratron," a gas triode tube developed at GE, became the focus of attention as an attractive alternative to the vacuum tube in electrical power applications. The thyratron provides an instructive example of the close interaction between scientists and engineers at GE. The device served as a vehicle for scientific research by Irving Langmuir and others at the GE Research Laboratory into the fundamental aspects of gaseous conduction. At the same time, Alexanderson and several assistants worked to develop engineering applications that could take full advantage of the unusual properties of the new tubes.

During the fall of 1921, Alexanderson and some of his assistants at GE worked on various applications of vacuum tubes in railroad electrification. In September 1921, he reported the successful operation of a DC motor by means of a vacuum-tube switching circuit connected

Figure 7.3. Alexanderson and Samuel P. Nixdorff in an airplane testing an experimental radio altimeter in June 1929. Courtesy of the General Electric Company, Schenectady, N.Y.

to a 20-kv transmission line. The following month, he informed Albert Davis in GE's Patent Department that the three most promising applications were a 1,500-volt rectifier substation, an electronic transformer designed to operate 1,500 volt DC motors from a 20-kv trolley line, and an electronically controlled motor that would operate directly from a 20-kv line.[86]

In January 1922, Alexanderson reported the results of an experiment using a technique he called "inverted rectification." He stated that the experiment was part of an investigation into power applications of vacuum tubes: an electronic rectifier had been used to produce DC at 10 kv, which in turn had been transformed back to AC by means of triode tubes and stepped down by a transformer to drive a variable-speed motor.[87] A few days later, Samuel Nixdorff recorded in his laboratory notebook that at Alexanderson's request a demonstration of inverted rectification had been given for Edwin Rice and Albert Davis. A later entry in the notebook mentions that Alexanderson and Irving Langmuir had discussed various methods of converting high-voltage DC to AC with Nixdorff and Alexanderson's other assistants Albert Mittag and H. D. Brown.[88]

Gas tubes began to emerge as an attractive alternative to high-vacuum tubes for use in AC-DC transformation systems during 1922. In February, Alexanderson reported to Albert Davis that he had concluded that it was feasible to use three-element gas tubes to obtain rectification with high efficiency. Later in the year, Alexanderson informed GE's Patent Department of a successful demonstration that had used a gas triode to convert AC to DC. He expressed his belief that this was the starting point of an important new application that should be protected by patents as soon as possible. He explained that there were four basic ideas in electronic power conversion, two of which were already covered by patents. The first idea was a method he had invented in 1911, which involved the initiation of an arc in a gas tube on each half cycle; the second was covered by a Langmuir patent on tube design; the third, which had grown out of work carried out by Nixdorff in 1921–22, was a method of using tubes in power circuits; and the fourth was the method just demonstrated using the gas triode.[89]

In his annual report on activities during 1922 in his department at GE, Alexanderson claimed results of far-reaching importance in the field of power electronics. He stated that he had foreseen the possible development of efficient DC transformers and had initiated a systematic investigation approximately a year earlier. High-vacuum triodes had proved to be too inefficient, he continued, but a very satisfactory alternative had been found in a different direction than he and his colleagues had anticipated. He remarked, parenthetically, that this often seemed to be the case in research work. He observed that they had not really discovered a new phenomenon, since he and Langmuir had 10-year-old patents: the novelty consisted of a better understanding of how the properties of gas tubes could be utilized effectively by means of suitable circuits.[90]

By early 1923, the GE researchers had adopted the name *thyratron* for the newly developed gas triode. In a memorandum written in April 1923, Alexanderson gave the scientists at the GE Research Laboratory suggestions related to their thyratron studies. He stated that the fundamental characteristics of thyratrons should be determined so that engineering development could be carried out in conjunction with improvements in the tubes. Among these characteristics were the magnitude of the "hold-back voltage" when the grid was negative, and the minimum value of the plate voltage required to initiate current flow when the grid was positive.[91] Later in the year, he reported that work on power electronics in the Radio Consulting Department was being concentrated on DC transformation that might be used eventually in DC power transmission for railroad electrification. He mentioned that the department had done tests comparing thyratrons with other electronic tubes such as pliotrons and magnetrons.[92]

During 1923, Irving Langmuir undertook an investigation of the space charge and the potential distribution around electrodes located in ionized gases. The study culminated in his "ion-sheath" theory, which was applicable to the design of thyratrons. Langmuir announced the theory in a paper published in *Science* in October 1923. According to the theory, a negative electrode in a gas-discharge tube attracted positive ions and formed a sheath that contained only positive ions and neutral atoms. Langmuir pointed out that the theory could explain why the electrode exerted little effect on the discharge once it had started. He included theoretical equations describing tubes with several different electrode configurations, derivations of the equations and experimental data that had confirmed the theory.[93]

In November 1924, Alexanderson reported that thyratrons had been used in a successful demonstration of DC transmission at a power level of 25 kw. He termed the demonstration an important step in the developmental program. He suggested that the ultimate development of such a system would be a superpower transmission system at possibly 100 kv, and observed that this would make it feasible to supply power to remote communities that could not be reached economically using AC transmission.[94]

Several additional applications of the thyratron were tested during 1925. In the annual report of the Radio Consulting Department in January 1926, Alexanderson mentioned that he and his assistants had constructed a thyratron speed-control circuit for DC motors and also had devised a thyratron circuit to replace the conventional exciter on an AC generator.[95] A thyratron was even used in radio facsimile experiments as part of a circuit between a photoelectric cell and the modulator of a radio alternator.[96]

DURING 1922, Alexanderson became interested in a military application of power electronics which was to occupy his attention intermittently for the rest of his career. The problem that attracted his attention was the use of selsyns for the remote control of naval guns. His work on electric gun-control systems during the next two decades helped to maintain his close ties with the military research establishment. The selsyn, or self-synchronizer, was a system that used devices similar in construction to AC induction motors to give remote indication of angular position without the need for mechanical interconnection. Edward Hewlett of GE had designed a selsyn system to provide centralized control of the Panama Canal locks in 1914.[97] However, the selsyns used in Panama only indicated canal lock gate position at a control center, since they lacked sufficient torque to move large weights. It was this deficiency that Alexanderson hoped to remedy through the use of auxiliary electronic amplifiers.

In late May 1922, Nixdorff recorded in his notebook that Alexanderson had become interested in the use of selsyns for gun control and had proposed an arrangement that would use pliotron tubes in order to obtain more power from a selsyn control system.[98] A few days later, Alexanderson sent Albert Davis of GE's Patent Department a memorandum entitled "Vacuum Tubes for Gun Fire Control." He reported that he had just demonstrated the use of vacuum tubes for the control of selsyn motors. He explained that this was a solution to the problem of amplifying the output of a master selsyn so that a number of remote selsyn motors could be operated from a single master. The tests had been so successful that he and his assistants would now be able to write the specifications for a system that could be used in the control of guns on battleships, he added. He mentioned that the demonstration had been witnessed by Edward Hewlett and several assistants.[99]

In June 1922, Alexanderson wrote to Hewlett concerning the possible use of selsyns in aircraft to make radio control from the ground possible. Alexanderson expressed confidence that there would be many future applications of such a system. He commented that using an on-off control system would probably be better than using continuous intensity variation of radio signals. He welcomed Hewlett's proposal to experiment with the selsyn control of elevators, noting that this would be an application of electronic tubes similar to a gun control system and that he would like to see GE install an experimental elevator so that the technique could be evaluated carefully. He went on to say that he intended to develop a variety of applications for the new system which combined selsyns and vacuum tubes to produce synchronized motion.[100]

In 1925, Alexanderson devised an interesting and entertaining application of synchronized motion planned as part of the summer's Camp Engineering program at Association Island. His trusted assistant Samuel Nixdorff recorded in a laboratory notebook entry in March that Alexanderson had outlined some proposed "horseplay experiments" for the upcoming Association Island meeting. Nixdorff wrote that a designated "convict" was to be seated in a chair on a turntable that could be operated by remote control. A "judge" was to be able to cause sudden and unexpected movement of the turntable. Nixdorff also described a proposed arrangement that would make it possible to control a water fountain and electric lights by means of a microphone and an oscillator hidden in a baby carriage. In early May, he wrote that he had demonstrated to Alexanderson and Langmuir the control of a 1-hp motor with the baby-carriage microphone; he had been able to operate the motor—which was 200 feet from the carriage—by standing 5 feet from the carriage and shouting. He mentioned that he and his colleagues had operated the "convict's turntable" but that it had tended to hunt and that a corrective measure would be needed.[101] In a memorandum written in 1933, Nixdorff observed that a gun-control system GE developed for the navy had the same general features as the playful remote-controlled turntable of 1925 except for an added antihunt circuit. He estimated that during the three years 1924–26, approximately $40,000 had been spent on work applicable to gun-control systems.[102]

Alexanderson called attention to possible industrial applications of the thyratron in a memorandum outlining proposed research for 1927. He pointed out that the gas triode provided a new tool for the delicate control of power flow and opened the prospect of electronic control of heavy machinery, and he explained that thyratrons might be used to control large variable-speed machines in factories.[103] During 1927 an experimental thyratron control for an electric locomotive and an electronic voltage regulator for a power plant in Alabama were designed.[104]

An analysis of Alexanderson's patents reveals a pronounced peak in his inventive activity in the application of power electronics in 1925 and 1926. During these two years, applications were filed for 17 patents in this field which were eventually issued. Four of these were joint applications with his assistant Albert Mittag, and one was a joint application with Samuel Nixdorff. There followed an abrupt decline: Alexanderson filed only 2 power-electronics-related applications in 1927 and none in the next three years. It seems that his early enthusiasm for thyratrons was diminished by unexpected difficulties, although a diversion to television development and the adverse budgetary impact of the depression that began in 1929 probably contributed to the decline.[105]

Just as the efforts by Alexanderson and his assistants to harness the thyratron to power machinery were abating, Albert Hull of the GE Research Laboratory made a breakthrough that led to dramatic improvements in gas-tube design. Before Hull's investigations in 1927, the use of thermionic emission in thyratrons had proved to be impractical because of the rapid disintegration of the cathode due to ion bombardments. While attempting to devise a circuit breaker that functioned as a result of disintegration of a tungsten filament, Hull discovered that the energy of the ions had to exceed a critical value. He determined that the critical point, which he called the "disintegration voltage," was about 22 volts for gas tubes using mercury vapor. His discovery opened the way for the design of gas tubes with thermionic-cathode emission and external protective circuits to guard against exceeding the disintegration voltage.[106]

In 1928, Hull presented an AIEE paper on the new thermionic-emission gas tubes. He explained how the problem of rapid cathode disintegration could be avoided in tubes using any inert gas over a wide range of gas pressures and levels of electron emission, and he stated that he had already designed gas-tube cathodes that enabled tube efficiencies of up to 98 percent with life expectancies of several years. He pointed out that a principal advantage of the hot-cathode thyratron was that it was so sensitive that the application of 0.1 microwatt for 10 microseconds could control 1,000 watts.[107]

Hull reviewed the characteristics and a variety of potential applications of the hot-cathode thyratron in a paper published in the *General Electric Review* in 1929. He asserted that the improved thyratrons could be used in many applications in which mechanical switches or rheostats had formerly been employed. He stated that the thyratron control was so sensitive that it could be used to sort fruit on the basis of color and size. Thyratron circuits might be used to turn off lights automatically, to detect smoke, to hold furnace temperatures constant, and to respond to slight motions, he added. In a test of a thyratron circuit at the GE Research Laboratory, a physical displacement of only 0.00001 inch had been sufficient to start or interrupt a current of several amperes. Hull mentioned that tubes were already available with a capacity of 100 amperes at 10,000 volts, and asserted that he knew of no reason why they could not be designed with 10 or even 100 times that capacity.[108]

Alexanderson's Philosophy of Engineering and His Cyclical Theory of Development

In an AIEE paper presented in March 1925, Alexanderson suggested that a "breathing spell" was an essential stage in engineering development. As an example, he mentioned the stage reached in the propulsion of ships by reciprocating steam engines before the advent of improved methods. He commented that both transoceanic and broadcast radio seemed to have reached a stage where "the technique has, after strenuous effort, caught up with the commercial requirements and is now enjoying a breathing spell." This was beneficial to the engineering art, he argued, since engineers needed time to engage in the fundamental research that would pave the way for the next major advance. As a more current example, he cited recent experiments with shortwave radio which had revealed previously unknown phenomena. He pointed out that the segment of the frequency spectrum below a wavelength of 100 meters provided space for 1,000 times as many messages as were handled currently by existing longwave stations. There was no immediate need for such a great capacity, he conceded, but "we have a new tool to do things with and we must use our imagination as to what to do with it."[109]

Alexanderson elaborated on his philosophy of engineering research and invention after a week in residence at Yale in March 1926. He wrote to Charles F. Scott, a former Westinghouse engineer who was now a professor at Yale, that he had found it inspiring to see how engineering and research were being taught at Yale and that at Scott's suggestion, he had decided to write a few paragraphs outlining his personal views. Alexanderson suggested that modern society was characterized to a great extent by physical research, engineering research, and industry. The primary contribution of physical research was the discovery of new phenomena of nature, and engineering research adapted these to useful purposes. The products of engineering research were called inventions, and these were exploited by industry. Alexanderson observed that there seemed to be a tide in this process of research and invention just as there was in human affairs in general. He claimed that the electrical industry had already gone through its first complete cycle of adapting discoveries made during the nineteenth century and that past inventions and discoveries had largely been capitalized.[110]

In his letter to Scott, Alexanderson again stressed the need for periodic breathing spells so that the sources of financial support would not dry up and the time needed for fundamental research would be available. He pointed out that it was during a breathing spell that the engineer came to realize the limitations of his tools and design principles and turned to physical science for inspiration for new inventions. He predicted that radio would prove to be the forerunner of a new cycle in the electrical industry and noted that radio had not become a large-scale public utility until certain missing links had been found through physical research. In conclusion, he observed that an awareness was developing that more might be done with electric power if a few missing links could be provided by inventions that engineers were ready to make, once the necessary knowledge and materials became available.[111]

In December 1926, Alexanderson discussed engineering creativity and the changing context of engineering in a letter to a writer, Edgar H. Felix. Alexanderson expressed agreement with Felix that engineers were little understood by the public, and went on to say that the engineer who had creative ambition experienced a struggle for success similar to that of any pioneer. Alexanderson contended that the disappointments and joys of the engineer were different in kind from those of individuals whose success was commonly measured by wealth and power. The lone inventor was becoming an exceptional case, he added, and the majority of engineers-inventors were now employed in industry.[112]

He went on to say that the engineer-inventor who worked for a corporation had probably settled in his mind that he would neither be rich nor poor. He suggested that such a person faced a choice between the complacent life of corporate inventor, with its economic security, and a career as an executive. However, he stated, for a few the creative instinct became "an all dominating passion" like that which motivated writers and artists. He wrote that the reticence of many engineers, which might be attributed to modesty, was more apt to be a "suppressed pride, a consciousness that his heartaches and success will not be understood, also a kind of family pride and loyalty to his profession, in which his friends and associates are also his rivals."[113]

In a second letter to Felix, written in January 1927, Alexanderson commented on the qualities needed by an engineer responsible for leading a concerted effort to introduce a radical innovation. He wrote that the problem should first be stated clearly in order to be certain that a solution was possible without violating the laws of nature. "Unless we can see in our mind's eye an ultimate good worthy of our efforts," he continued, the project should not be undertaken. Alexanderson pointed out that the actual process of engineering development required time, patience, and the cooperation of numerous specialists. He explained that the leader had to educate all those who had to cooperate to reach the common objective, and that the process required coordination and simplification of the devices that were produced. The normal life of the engineer, he concluded, constituted a "fair balance between difficulties and successes"; extreme fluctuations were more characteristic of fiction than of reality.[114]

Alexanderson discussed the dynamic interaction of science, engineering, and invention in a talk prepared in September 1928. He observed that industry had learned that research was a necessity and that corporate executives and stockholders no longer were aghast when an annual appropriation to support a search for new things was proposed. He insisted that scientists and inventors served as the backbone of research and should work in concert to furnish engineers with ideas that the latter could convert into finished products. The average engineer, he suggested, did not investigate the subvisible field of matter or take on the inventor's role. Instead, the normal task of the engineer was to collect and adapt facts and incorporate them into an artifact of value to the world. Alexanderson went on to urge that corporations begin to lay the foundation for change a decade or so ahead. He finished with the observation that isolated groups of scientists or of engineers tended to work along well-established lines, whereas close cooperation between the two groups served as an inspiration for the ideas that would keep an industry up-to-date.[115]

In a talk delivered in February 1930 to an organization of Swedish-born engineers employed in the United States, Alexanderson asserted that engineers had certain common traits and constituted a "brother-hood bound together by common ideals," no matter what their nationality or language. He stated that the engineering fraternity possessed bonds of friendship and understanding that transcended the bonds of nationality. He suggested that the whole social structure was dependent on engineering achievements, and predicted that material prosperity would be destroyed if the creative imagination of engineers ceased to function. In another talk delivered later the same year, Alexanderson contended that engineers "would have no interest in research and invention unless we thought it would result in benefit and pleasure to others." He observed that unlike the actor, who received immediate rewards, the research engineer generally had to wait for a decade.[116]

Alexanderson's Review of Developments in the 1920s

The end of the decade of the 1920s was the occasion for Alexanderson to write a thoughtful essay entitled "A Review of Electrical Developments from 1920 to 1930." In looking at the past, he identified a cyclic pattern in the history of electrical engineering, with mileposts "alternately labelled 'Direct Current' and 'Alternating Current.'" In effect, he reiterated his "breathing spell" theory: a major innovation was followed by a several-year period of large-scale commercial development, with a portion of the resultant profits going to support new research in science and engineering. He argued that it was during this period that the stage was set for the beginning of a new cycle. The decade of the 1920s had been the culmination of a cycle, he wrote, with very little change in the technology of electric power but with a great increase in the volume of business and in the size of units. Estimating that total revenue for electric service had increased from $880 million in 1920 to $1.8 billion in 1930, he observed that "the profits reaped from this great commercial expansion have enabled the electrical industry to support scientific and engineering research on a large scale."[117]

Alexanderson asserted that the most outstanding new development in the electrical industry during the first half of the decade had been radio, including transoceanic service and broadcasting. He suggested that an important indirect effect of radio had been to alter the viewpoint of electrical engineers. Instead of regarding electricity as a sort of "juice" in wires, they had learned to consider it in terms of electromagnetic fields and electrons or ions in rarefied gases. Alexanderson characterized radio as a "great social force" that had "done more than the printing press, the railroad and the automobile in bringing the entertainment and thought of the great cities to every town and farm house." He credited the reduction in the cost of electric power generation and transmission which had occurred during the decade with making power available over large geographic areas at reduced rates and with conserving natural resources.[118]

Alexanderson identified two innovations, the electronic power converter and the power condenser, as having been the most significant advances in the technique of power transmission over the past five years. Observing that the power condenser was not really new but had not been used on a large scale until recently, he stated that its rediscovery had been the direct result of a new concept of transmission borrowed from radio, which amounted to "an adoption of Maxwell's theory by the practical engineer." He explained how power condensers and the transmission concepts drawn from radio or telephone engineering could be applied to the design of AC power transmission lines up to 1,000 miles in length. As a practical example, he reviewed the case of a 500-mile power line from Canada to New York City in which power condensers had been inserted at 100-mile intervals. He noted that the technique was analogous to the loading coils used in wire telephony, and he calculated that an investment of $3 million in power condensers had increased the capacity of the existing lines by 50 percent, with a saving of $9 million in capital costs. He also mentioned that an electronic voltage regulator had been used at the generating station. Alexanderson pointed out that instability would have made a 500-mile power line impractical if the old method of design, without the power-tuning condensers and electronic regulator, had been used. He concluded that the new method opened the prospect of large-scale "super-power transmission," and that energy could be transmitted for 500 miles more cheaply in the form of electricity than in the form of coal.[119]

Alexanderson included in his review an analysis of the economics of rural electrification and railroad electrification, and characterized the farm and the railroad as "the two great new consumers of the super-power system." He estimated a potential increase in revenue of $1.3 billion per year from these two sectors, "an amount equal to the total revenue from electric service in 1925." He pointed out that farms' and railroads' peak demands for power would come at different times of the year, so that the combination would tend to improve the load factor of the supply system. His estimate of the revenue from railroad electrification was $600 million per year, assuming a rate of 1.5 cents per kilowatt-hour. His calculation for rural electrification was based on an expected average annual revenue of $100 from each of the 7 million farms in the United States and a rate of 7 cents per kilowatt-hour.[120]

In the concluding paragraph of his essay, Alexanderson alluded to the policy debate over whether the needed national superpower system of electric supply ought to be controlled by the government. He used his cyclical theory that technology advanced "by periodic leaps" as an argument in favor of private ownership, asserting that the activities leading to major innovations were "hindered when regulated by cumbersome governmental processes." Only the private owner, he contended, could "take the gambler's chance," supporting research that had no immediate commercial value and then risking financial resources "in picking the winner" when a new cycle began.[121] What had begun as a retrospective account of electrical developments of the decade of the 1920s ended with a catalog of technical and economic opportunities for the industry and a defense of private enterprise.

Alexanderson's review of the 1920s reflected the beginnings of a change in the external context of his inventive activities as a decade of economic depression began. The new cycle of commercial expansion which he anticipated failed to materialize, at least in the short run. The new decade would see a continuation of his inner drive to innovate but some changes in the focus of his inventive effort, and in his patrons. Power electronics would remain a principal interest, but his telecommunication research, intended for commercialization by RCA, would soon be terminated. The severance of his RCA connection and a sagging industrial market for innovation combined to shift his attention toward military technology, a shift that was facilitated by a network of friendships with naval officers in key positions.

During the six-year period 1924–1929, Alexanderson applied for 52 patents that would eventually be issued. Of this total, 28 (53.8 percent) were in the field of power and power electronics, while 22 (42.3 percent) were in the field of telecommunications. Two patents from this period fell outside the two major categories. One was an ignition system for internal combustion engines, and the other was an invention intended to measure the relative speed of two moving vehicles.[122] By the end of 1929, the 51-year-old Alexanderson had filed applications for a grand total of 241 United States patents that had been or would be awarded.

DURING THE 1920s, Alexanderson received further recognition for his inventive achievements. In 1926, he was awarded an honorary doctorate by Union College in Schenectady,[123] and in 1928 the American

Society of Swedish Engineers awarded him its John Ericsson prize. In his acceptance talk, Alexanderson discussed the influence of Charles Steinmetz on American electrical engineering and on his own work. Alexanderson expressed his belief that he had been the first engineer to apply Steinmetz's methods to the field of radio, and he concluded that Steinmetz's method of dealing with electrical phenomena had helped make the United States a world leader in radio as well as in electrical engineering.[124]

By the late 1920s, Alexanderson was finding time to engage in his favorite recreation, sailing at Lake George. He purchased a boat, called the *Nordic,* that had been built in Denmark in 1927. In August 1928 he responded to an inquiry about the boat that it was now at Lake George and was not for sale. He commented that the *Nordic* was the "most able and comfortable one man boat" he had ever seen, and that drawings of it had appeared in *Yachting* in July 1927.[125] In a light-hearted vein, he wrote to Leo H. Baekeland, a chemist and the inventor of Bakelite, concerning the recipe for "viking ale" that he took along when sailing on the *Nordic.* Alexanderson explained that he had on one occasion experienced a thirst that could only be satisfied by a glass of ale, and that he had mixed some strong muscatel with a near ale to produce a good imitation of a heavy ale.[126] In an earlier instance of good fortune with beverages, he informed the Taylor Wine Company that he had bought a five-gallon keg of grape juice from them which had become excellent port wine after a year. He asked for a price quotation on another keg of the same preparation.[127]

Alexanderson's daughter Edith later recalled an episode of her father's absent-mindedness which took place at around this time. They had been scheduled to join the rest of the family at Lake George on a Fourth of July outing in the early 1930s. According to Edith, her father had gone as usual to the GE plant and later had called her to report that it seemed to be some kind of holiday, since only he and a watchman were there. As a result, they had been quite late in getting to the lake. Edith also remembered that Alexanderson had understood perfectly when she had lost her car in a parking lot.[128]

Chapter Eight

Electronic Engineering: The Invention of the Amplidyne

The year 1930 was a year of transition for Alexanderson, for General Electric, and for the Radio Corporation of America. In January, RCA's president James Harbord issued a "general order" announcing the elevation of David Sarnoff to the company's presidency.[1] Later that year, Sarnoff's plan to have RCA, rather than the associated companies GE and Westinghouse, manufacture radio apparatus was approved by RCA's board of directors. The terms of the agreement gave RCA exclusive rights to manufacture under the radio patents held by the associated companies for 10 years, with GE and Westinghouse to receive compensation in the form of stock. However, the federal government initiated an antitrust action that challenged the agreement for unlawfully restricting competition in the field. After a period of protracted negotiations among the parties, a consent decree was issued in November 1932 which gave GE the right to resume the manufacture of radio tubes and receivers 30 months after the date of the decree.[2] From Alexanderson's perspective the 1930 agreement was a disaster, since it had the effect of eliminating RCA's support of the radio-television research and development at GE. His only recourse in the short run was to seek military support for some continuation of telecommunication research and to shift his focus toward industrial applications of electronics which were unrestricted by the agreement. He became a persistent critic of the agreement because it limited the scope of research at GE, and he used every opportunity to lobby within the company for a less restrictive arrangement.

A Demonstration of Theater Television

A well-publicized demonstration of a large-screen television system in a theater in Schenectady in May 1930 served Alexanderson as a vehicle for contacts with potential military patrons of his research. During the April 1930 meeting of RCA's Technical Committee, he reported that television apparatus had been installed in Proctor's Theater in Schenectady, and that he expected it to provide entertainment of commercial quality. He proposed holding a private demonstration of the system for the benefit of RCA officials. Alexanderson expressed doubt that a demonstration would lead to an immediate demand for home receivers, since the cost per unit would be approximately $1,000. However, he added, television might have military applications. The day after the Technical Committee meeting, he wrote to Sarnoff to suggest that a television show for the public be given at the theater in Schenectady and later repeated in New York City, where it could be staged properly. He enclosed the draft of a talk he had prepared that gave his views on the future of television.[3]

Alexanderson's speech was entitled "Television and Its Uses in Peace and War." In it, he mentioned a talk he had given three years earlier, just before television "broke out like measles," and commented that the engineer and inventor seemed "peculiarly blind" on the destiny of major innovations. He cited his own inability to foresee the "great social significance" of radio broadcasting before the 1920s and expressed uncertainty as to whether television would repeat the history of radiotelephony. He reported that television pictures had already been transmitted from Schenectady to Australia and back, although with substantial distortion.[4]

Turning to television's military uses, Alexanderson predicted that the airplane might be "many times more deadly" if combined with television and that the development of television would be pursued for that reason, whether or not television was found suitable for entertainment. He described a hypothetical naval battle in which television cameras on remote-controlled scout planes were used to locate enemy ships. Future wars, he anticipated, would not be fought by millions of men in trenches; instead, war would be "a battle of brains between experts[,] conducted with the chivalry of the knights of the air, which was the only redeeming feature of the World War." Finally, he asserted that "weapons have always been a dominant factor determining the social and economic order of the world."[5]

Sarnoff responded to Alexanderson's letter and its enclosure by stating that he had read the paper on the prospective military application of television and had already discussed the matter with Capt. Stanford Hooper and other naval officers. They had been quite interested, Sarnoff stated, and he was also informing members of the War Department. He expressed agreement with Alexanderson that there should be a public showing of television in New York City.[6]

During April 1930, Alexanderson was invited to Washington, D.C., to discuss military applications of television with both army and navy officers. Major Mitchell was among the officers he met. Mitchell later came to Schenectady to see a television demonstration. In early May 1930, Alexanderson reported to a GE executive that Major Mitchell was convinced that the tactics of war would be changed if the performance of television demonstrated at GE could be duplicated in airplanes. Mitchell had offered to send an airplane to Schenectady for tests when needed.[7] Shortly afterward, Alexanderson requested that GE rent an airplane to use in television relay experiments. He pointed out that the experiments would be of interest to the military and might also lead to news broadcasts by television. As a further indication of the navy's interest in his research, Alexanderson was invited to get a firsthand look at operations on the aircraft carrier *Saratoga* during a passage through the Panama Canal to the Pacific in June 1930.[8]

Alexanderson and his assistants gave a demonstration for the press of their latest television apparatus at Proctor's Theater on May 22, 1930. Before the demonstration, an internal GE memorandum stated that the show would establish GE's priority and might be an advantage in later years in the same way that KDKA, in Pittsburgh, Pennsylvania, had received credit as the first radio broadcast station. An advertisement of the show appearing in a Schenectady newspaper called it "a Supreme Event of This Century," which would inaugurate a "New Era of Entertainment."[9] Another newspaper item credited Alexanderson with being the "father of television," who was predicting "future marvels as [his] child increases in size."[10]

An account of the Proctor's Theater show and its significance was written for *Radio News* by Edgar Felix. He reported that pictures transmitted from Alexanderson's laboratory had been projected on a 6-foot-square screen at the theater. The show had featured a performance by two vaudeville actors, of whom one was onstage at the theater and the other was seen over television. Also, an orchestra at the theater had been directed by a conductor at the television studio. Felix reported that the picture's brilliance was about half that of a motion picture film and its large size made it easy to detect defects due to the 48-line scan. The scan was produced by a perforated disk rotating at 16 revolutions per second. According to Felix, the most significant feature of the system was the Karolus light valve, which had enabled the use of a powerful arc lamp in the projector instead of a neon tube. Felix likened the importance of the new light valve to that of the grid in a vacuum tube. He quoted Alexanderson as saying that the main bottleneck was in the narrow television bandwidth—23 kHz—that had been used and pointing out that a bandwidth of the order of 400 kHz would be needed for good fidelity at a scanning rate of 200 lines per frame.[11]

The End of RCA Support and a New Research Agenda

Soon after the television demonstration at Proctor's Theater, Alexanderson boarded the *Saratoga* at Old Point Comfort, Virginia, to accompany a battle fleet on its passage to the Pacific Ocean. During the

Figure 8.1. Mechanical-scan television projection system used at Proctor's Theater in Schenectady, N.Y., in 1930. Alexanderson's assistant, Ray D. Kell, operated the projector. Courtesy of the General Electric Company, Schenectady, N.Y.

cruise, he discussed and compiled a virtual shopping list of needed innovations as perceived by naval officers. Shortwave detection of the enemy, radio control of aerial torpedoes, a television system for scout planes, and research on ultrashort waves were among the desiderata listed in a memorandum he wrote when he returned to Schenectady.[12] Unhappily, during the *Saratoga* trip Alexanderson received the news— transmitted by radio—that his father had died in Sweden at the age of 89.[13]

A letter from GE to the secretary of the navy seeking support for continued telecommunication research was one apparent result of Alexanderson's *Saratoga* cruise. In a memorandum to GE's Ray Stearns, Alexanderson suggested that the letter stress that the creation of RCA had resulted from a cooperative effort between GE and the navy during the war. He noted that renewed cooperation might be expected to have important consequences in the future, and that the potential benefit of continued radio-related research at GE to the government made it very desirable for this research to be continued despite the transfer of commercial radio interests to RCA. He concluded by urging that the letter state that GE's radio research organization was now available to work on government problems.[14]

At the August 1930 meeting of RCA's Technical Committee, Alexanderson reported that his organization's research that was of interest to RCA had been reduced to a minimum owing to a lack of appropriations and that his staff was being shifted to other lines of work.[15] At the November 1930 meeting, he reported that his department at GE was still doing some work on wave propagation and on ultrashort waves. He commented that this research had potential military applications with a range of up to 140 miles from a ground transmitter to airplanes. He also mentioned that some work was being done on radio aids to aerial navigation.[16] In his budget request for 1931, he indicated that research on government and aviation apparatus would constitute one of the major areas to be pursued in his department. He included in his list of projects a radio-echo altimeter, direction finding, fog navigation, radio facsimile for the military, and television.[17]

Alexanderson tried without success to interest Sarnoff in supporting further GE research on television. In a letter addressed to Sarnoff in October 1930, he described recent work done jointly with August Karolus which had produced some impressive pictures and could be adapted to color television. Alexanderson estimated the cost of continuing the work at $5,000 per month, or $60,000 per year. Sarnoff replied that he could not authorize the expenditure. In response, Alexanderson wrote that although Sarnoff's letter contained bad news, he realized that he was not alone in encountering financial difficulties in the present crisis. He expressed the hope that he might continue to have contacts with the engineers at RCA if only for the "love of the art." Sarnoff assured Alexanderson that he had many friends at RCA who would want him to continue to attend meetings of the Technical Committee, and that he appreciated Alexanderson's attitude.[18]

Despite the loss of RCA's financial support of television research at GE, Alexanderson's interest in and enthusiasm for commercial television continued. In November 1930, he wrote a memorandum listing some requirements for a home television receiver. He included reliability, synchronization, an adequate light source, and a picture dimension of at least 5 by 6 inches. Moreover, the receiver should be as simple to operate as a radio receiver and no larger than a radio console, and the picture detail should be sufficient to permit the viewer to watch a football game.[19]

The following month, Alexanderson addressed a memorandum to Laurence Hawkins of the GE Research Laboratory on the subject of new devices and systems such as the cathode-ray tube and color television. Alexanderson wrote that his conception of a research program was that it should be such that "we will be allowed to go wherever a natural evolution takes us."[20] During 1931, he applied for two patents, which were later issued, covering improvements in a disk-scan television. One of his principal assistants in the television experiments, Ray Kell, left GE to work for RCA. Kell published a paper in 1933 describing an experimental disk-scan television transmitter located in the Empire State Building in New York City. By that time, a cathode-ray tube instead of a synchronous disk was being used in the receiver.[21]

In December 1931, Alexanderson reported to GE's Patent Department on a demonstration of the use of a beam of light as a carrier of television. He expressed his belief that the logical progression to shorter waves in communication meant that light eventually would be used, and he noted that he had discussed with Irving Langmuir whether it might be possible to modulate a light source at frequencies of up to 1 mHz. He added that Langmuir had known of no fundamental reason that it could not be done. In the demonstration reported, Alexanderson explained, an arc light had been modulated at up to 25 kHz and the beam had been picked up at a distance of 130 feet. It had been used to control a television projector.[22] In 1933 he applied for a patent on a color television receiver using cathode-ray tubes in conjunction with red and green filters. As an alternative to the filters, he suggested the possibility of a cathode-ray tube producing different kinds of fluorescence.[23] This was his last patent application in the field of television during the 1930s.

The changes in the focus of Alexanderson's research after the termination of RCA support were discussed in his annual report on the activities of his department, now named the Consulting Engineering Department, for 1930, in which he stated that several projects sponsored by RCA, including facsimile and television, had been terminated during the first half of the year. He went on to say that the work had been reorganized during the second half of the year to concentrate on power problems, especially in areas where GE's radio experience was an asset. He noted that a revised program of radio research had been continued on government projects, along with some fundamental research on ultrashort waves. Among the specific projects he mentioned was a model transmission line that would be used to investigate the feasibility of wave transmission on a long power line. He explained that a key problem in such a system was to devise artificial means to control the phase relationship between the sending and receiving ends of the line. He wrote that calculations indicated that a line operated at 400 kv could transmit 600,000 kw for 1,000 miles. A second project mentioned in his report was a gun-control system that used thyratrons and selsyns. Alexanderson requested a total of $150,000 from GE to support the research of his department in 1931.[24]

In a November 1930 memorandum to Chester Rice, Alexanderson assessed the potential applications of radio principles and electronics in power transmission and control. He stated that the use of wave transmission on power lines, along with phase control of generators, would permit the exchange of power between large systems over lines that were not robust enough to maintain synchronization. He also suggested that a refined control system might permit centralized control of all the power systems in the country by means of a master signal distributed by telephone lines. As a second application, he listed the electrical control of guns or industrial machines. He noted that a suitable system using thyratrons and selsyns had already been demonstrated. He commented that the problem of hunting had been solved by means of a device based on the same principle used in the squirrel-cage winding in a synchronous electric machine. Another application of potential importance Alexanderson discussed was railroad electrification using AC distribution with DC motor propulsion. He noted that tests had been done preliminary to working out the design of an electric locomotive for the Pennsylvania Railroad or the New Haven Railroad.[25]

The Development of a Torque Amplifier and a Gun-Control System

In February 1931, Alexanderson reported on the development of a "synchronous torque amplifier" with thyratron control in a memorandum to GE's Patent Department. Alexanderson listed several potential applications, including gun control, steel-rolling mills, turbine governors, and ship propulsion. In a second memorandum, submitted two weeks later, he suggested that the torque amplifier might be used in conjunction with a gyroscope for ship stabilization.[26] In June 1931, he reported that the navy had requested that GE build a torque amplifier suitable for gun control. The navy, Alexanderson explained, had furnished a gear mechanism and flywheel that simulated the inertia of an actual gun. During tests with this device, slack in the gearing had resulted in self-oscillation of the control system, although the system had worked perfectly in earlier tests using a model. To overcome the oscillation, Alexanderson proposed the use of two selsyns, with one attached to the gun and the other geared to the drive motor. He concluded by asserting that this modification of the control system would cause it to imitate the motion of a skilled operator who corrected instinctively for slack in the gears.[27]

Alexanderson mentioned the use of the torque amplifier for gun control in a talk delivered to an audience of GE engineers and executives in the summer of 1931. After explaining that the problem was to maintain a large gun in synchronism with a sighting telescope, a task that presently was accomplished by a "highly trained man," he added, "We propose to replace the man by an automatic control using the thyratron as a synchronous torque amplifier between the telescope and the gun." He suggested that gun control would be only one of many applications of thyratron motor control in which quick response and accuracy were required.[28]

In a September 1931 memorandum to GE's Patent Department, Alexanderson disclosed a modification of the torque amplifier for applications where a drive motor was too large to be controlled directly by thyratrons. For such cases he proposed inserting a "booster generator" between the main generator and the motor, with the booster field being controlled by a thyratron torque amplifier. He asserted that a thyratron amplifier with a capacity of 2 or 3 kilowatts should be able to control a 100-kw motor without hunting if the booster were used. As possible industrial applications of the booster control system he listed paper mill drives and boring mills.[29]

In the same memorandum, Alexanderson pointed out that the torque amplifier control system might be adapted to the manufacture of turbine wheels, the shape of which would correspond to a template incorporated in the control. The possible use of the torque amplifier in the automatic control of machine tools attracted the attention of others at GE. In November 1931, Alexanderson mentioned that F. H. Penny and J. W. Harper of the Industrial Engineering Department had become interested in using the "thyratron follow-up device" to cut metal to correspond to a template.[30] In a talk delivered the same month, Alexanderson predicted that the automatic control of industrial processes would eventually be possible on such a scale that it baffled one's imagination.[31]

In November 1931, Alexanderson reported a successful test of a torque amplifier in the control of a 5-inch gun. He stated that the problem of self-oscillation due to gear slack had finally been solved through the use of antihunt coils. The coils, he explained, had acted as a filter to suppress the oscillation, which had a frequency of 10 cycles per second. He mentioned that the modification had been tried first in the laboratory and then on a gun. The following month, in a memorandum on the use of the torque amplifier in the control of antiaircraft guns, he stated that the amplifier would enable its user to combine the science of gunnery and "the personal skill of a duck hunter."[32]

By early 1932, the issue of military secrecy had arisen in connection with GE's development of gun-control systems. Alexanderson reported to an officer that patent applications covering gun control were to be kept in a special, confidential file, with the contents not to be revealed to GE's foreign associates so long as it was desired to maintain secrecy. In an internal memorandum later in the year, he stated that reports on the gun-control project were kept in a secret file in GE's Patent Department, with duplicate copies kept in a locked file in his office.[33] One result of the navy's desire for secrecy was that there was a long delay between the time of application and the date of issue of Alexanderson's patents on gun-control systems. He applied for his first gun-control patent, entitled "System for Reproducing Position," on May 19, 1932, but it was not issued until April 24, 1951.[34] In his budget request for 1933, Alexanderson mentioned that his department had been given an appropriation for secret research on a government project. In January 1933, he wrote that he had organized the secret project by assigning several groups to work on separate phases simultaneously.[35]

In May 1933, Alexanderson informed the Patent Department that the GE gun-control system had recently been tested on the navy cruiser *Portland*. He mentioned that thyratron control devices of the same type used for gun control might be adapted to remote control of target planes from the ground or from another plane. He explained that the use of such a system would require the transmission of angles by radio and selsyns. The same month, he reported to the GE Engineering Council, a group composed of leading GE engineers, and scientists from the GE Research Laboratory, that naval officers believed that the GE gun-control system ultimately would be used on all naval guns. He again emphasized that similar methods could be used for the automatic control of machine tools. Expanding GE's machine-tool-control business was, he asserted, a "matter of straight engineering development."[36]

The army also became interested in possible GE contributions to the control of antiaircraft guns. In May 1933, Alexanderson reported that he had spent several days at Fort Monroe discussing technical problems. Among the topics that had been addressed were triangulation by radio direction finders, visual indication of acoustic direction finders to aid spotters, and automatic transmission of angles.[37]

In June 1933, Alexanderson wrote a memorandum to GE's Patent Department concerning a secret patent application on the use of the torque amplifier for gun control. Alexanderson reported a design change in the torque amplifier which increased accuracy at all speeds without causing hunting. He stated that the change had resulted from experience with the 5-inch antiaircraft guns that were being installed on ships.[38] In August 1933 he requested that his assistant Martin A. Edwards be admitted to the GE plant at times other than regular working hours, since Edwards was engaged in "urgent work for the Navy."

Edwards had joined Alexanderson's Consulting Engineering Department in 1929 and had participated in television experiments for the next two years. Alexanderson later described Edwards as possessing an unusual combination of inventive, experimental, and practical engineering skills.[39]

Another of Alexanderson's assistants, Samuel Nixdorff, wrote a memorandum late in 1933 reviewing the research and the expenditures on the torque amplifier. He pointed out that GE work that was relevant to gun control dated back to 1924, although a separate work order for the torque amplifier had not been established until 1931. Since 1924, he recorded, a total of $82,800 had been spent on research applicable to gun control. The total included $19,600 for the period 1931–1933, and $23,000 had been spent in 1930 on the successful effort to overcome the hunting problem of the control system.[40] Applications of the torque amplifier constituted one of the four principal research categories listed in Alexanderson's budget request for 1934. He mentioned that he and his colleagues were preparing to apply the system on larger navy guns that used the Waterbury gear, and he predicted extensive use of the torque amplifier by industry.[41]

The introduction of the Waterbury gear between the electric control system and the gun created a new difficulty. In an April 1935 memorandum, Alexanderson reported that the use of the gear made it necessary to modify the torque amplifier to correct for speed. He described a method to do this by inserting an intermediate torque amplifier between selsyns so that the angular displacement of the selsyn stator could be made proportional to speed. This solution to the velocity-lag effect was covered by another of Alexanderson's gun-control patent applications; it was filed in September 1935 but the patent was not issued until 1947.[42]

Electronic Power Conversion and the Development of a Thyratron Motor

During the 1930s, Alexanderson played a significant role in the application of electronics in the fields of electric power transmission and variable-speed motors. The Camp Engineering meeting held at Association Island in the summer of 1931 gave him an opportunity to review developments in these fields with a sympathetic audience, "with the object that we can formulate a joint program of action." He began his talk by recalling that he and Irving Langmuir had presented their ideas on the subject at a Camp Engineering meeting held six years before. He admitted that they could not yet claim many practical achievements or show profits, and asked rhetorically whether they might not be "chasing a will-o-the-wisp." He suggested that every important development had passed through a similar stage of uncertainty, and that recent progress in electrical physics supported the conclusion that "we stand at the threshold of a major development in the electrical arts." He asserted that it was important to GE whether they would be "leaders or trailers" in the field of power electronics.[43]

Alexanderson went on to say that electronic tubes could be used to overcome certain limitations of both AC and DC power systems. He observed that the principal limitation of DC was the difficulty of transforming it from one voltage to another, and that tubes made this possible. He listed as the two most significant limitations of AC power the necessity for synchronization and the lack of a satisfactory variable-speed motor. He asserted that the synchronization problem could be solved by means of a thyratron stabilizer and that this had been demonstrated using a model artificial transmission line that was equivalent to an actual line more than 1,000 miles in length. He sought to counter the impression that AC power systems had reached a final stage of development, and stressed that his conclusions were "just the opposite": electronic tubes would make possible continued improvement in the power capacity and length of AC lines, and the development of new types of variable-speed AC motors. He did suggest that high-voltage DC cable transmission might eventually come into use once suitable tubes had been developed.[44]

Alexanderson singled out railroad electrification as another area where electronic methods would be beneficial. Pointing out that there had been two opposing tendencies, with locomotive design favoring the use of DC propulsion and the distribution system favoring AC, he observed that the two systems of electrification had enjoyed almost equal growth and that new techniques were needed to improve both. He stated that thyratron power conversion apparatus would enable a locomotive using a squirrel-cage induction motor to be operated with either an AC or a DC trolley; such a system had already been tested in the laboratory. Finally, he ended his talk by asserting that he and others interested in railroad electrification were participating in a process in which "a multiplicity of vague ideas" were "being sifted down to simple and concrete forms on which all can agree." He predicted that historians some day would realize that "the early disagreements were not due to the perversity of human nature" but reflected a "natural evolution of thought" in which both scientists and engineers had participated.[45]

In 1931 Alexanderson and his colleagues resumed the efforts they had initiated in the mid-1920s to develop variable-speed electric motors with thyratron commutation. In a memorandum to GE's Patent Department in February 1931, he reported preliminary tests of a variable-

speed motor using thyratron commutation and stated that preparations were being made for more comprehensive tests.[46] The same
month, he was quoted in a newspaper article as forecasting that thyratrons were likely to have a greater industrial impact than radio. The
writer called him the "genius of the GE Company."[47] Later in the year,
Alexanderson reviewed the history of the work on thyratron commutators in another memorandum to the Patent Department. He pointed
out that he had discussed thyratron commutation at Camp Engineering in 1925, and that some experimental work had been done at that
time. However, the available thyratrons had proved unsatisfactory. As
a result, the work had been discontinued until 1931.[48]

In November 1931, Alexanderson wrote to Charles E. Eveleth, a GE
vice-president, that GE's engineers and scientists were the only ones to
date who understood thyratron techniques, and that they were in the
process of building a strong patent position. He predicted that they
soon would have competitors and stressed that fundamental work
should be completed quickly.[49]

In his departmental budget request for 1932, Alexanderson listed
the thyratron commutator and the thyratron power system stabilizer as
two of the principal areas in which continued research was planned.
He stated that the stabilizer had been demonstrated in the laboratory
and that tests on an actual system were being prepared.[50] In December
1931, he reported the successful demonstration of a thyratron stabilizer on a 50,000-kw generator. He mentioned that power shocks of up
to 7,000 kw had been handled by a single pair of tubes. This experimental stabilizer was described in a patent application filed in December 1931; the patent was issued jointly to Alexanderson, Philip L. Alger,
and Samuel Nixdorff in July 1933.[51]

Albert Hull of the GE Research Laboratory provided data on the
characteristics of various thyratrons produced by GE in a paper published in 1932. He noted that thyratrons had a major advantage over
high-vacuum tubes in that the current density was 1,000 to 10,000
times as great. The average current rating for thermionic-cathode thyratrons discussed in his paper ranged from 2.5 amperes (for the FG
57) to 100 amperes (for the FG 53). Hull also described a multianode,
mercury-pool-cathode tube that was rated at 5,000 amperes, with an
overload capacity of 14,400 amperes. He asserted that thyratrons offered a spectacular range of power conversion possibilities and could
be used in inverters, rectifiers, and frequency converters. He mentioned a 400-hp variable-speed motor with thyratron commutation that
was under development at GE.[52]

The 400-hp motor mentioned by Hull provides an illuminating case
study of the interaction of three interested parties: scientists and en-

gineers at GE, and an outside client that planned to use the motor in a power plant. The American Gas and Electric Company (AG&E), the firm for which the 400-hp thyratron motor was intended, became a pioneer among private utilities in the use of power electronics. It had already installed thyratron voltage regulators at a substation in Ohio (in March 1930) and at a plant in Virginia (in April 1931). Philip Sporn, an engineer with AG&E, commented on the company's interest in power electronics and its support of GE's development of the thyratron motor during an AIEE meeting in 1935.[53] He stated that the power industry had been handicapped by the lack of a satisfactory alternative to the DC shunt motor in applications that required variable speed. Engineers at his company had followed developments in electronics with intense interest, he continued, and acting on the belief that "the best way to begin is to commence," AG&E had placed an order for a 400-hp thyratron motor to be used in driving an induced-draft fan in a power plant. Sporn acknowledged that because of the high cost of tubes the new motor was an expensive alternative, but he predicted a reduction in cost once the demand began to increase.[54]

In May 1933, Alexanderson informed the GE Engineering Council that the 400-hp motor had provided important data on practical limitations of power tubes. He argued that the intangible advantages arising from the introduction of new engineering methods would outweigh the high cost of tube replacement, and he requested authorization to begin development of a 3,000-hp thyratron motor suitable for ship propulsion.[55]

In a July 1933 memorandum on the current status of thyratron research, Alexanderson stated that there were two classes of problems: tube problems and power problems. He asserted that he and his colleagues had now solved the "arc-back" difficulty with tubes and were now working on power problems associated with motor operation. In his departmental budget request for 1934, he mentioned that 10-hp and 400-hp motors had been tested, and that a 3,000-hp motor was being built. He noted that the large motor could be adapted to demonstrate DC power transmission at 30 kv and had potential for ship propulsion and railroad locomotion.[56]

In November 1933, Alexanderson reported on an experiment intended to determine the stability of the 400-hp motor when it was connected suddenly to a 4,000-volt power line. He stated that the stability had been found experimentally to depend on the relationship between the line reactance, the motor transient reactance, the armature reaction, and the ampere-turns of the field winding. He went on to say that a design rule had been formulated: the line reactance should be at least as great as the motor transient reactance, and the ampere-turns of the field should be twice the ampere-turns of the armature reaction. He asserted that designing for stability was the responsibility of the motor designer just as tube life was the responsibility of the tube designer. Early the following month, he reported a test during which the 400-hp thyratron motor had been run with one bad thyratron. The motor had operated despite the defective tube, and he concluded that some past difficulties had not necessarily been the fault of the tubes but instead had been due to the fact that "we did not know how to use them." Copies of this memorandum were sent to Irving Langmuir, Albert Hull, William Coolidge, and William White, all affiliated with the GE Research Laboratory.[57]

Clodiun H. Willis, an electrical engineering professor at Princeton, was a parttime participant in GE's research on thyratron applications during the 1930s. In 1932 he published a paper on thyratron commutation, and in 1933 he published a paper on the thyratron motor. In the 1933 paper, Willis discussed some advantages of the synchronous motor with thyratron commutation—for instance, it could be operated with either a DC or an AC power supply. He pointed out that in the case of a DC source, there was no need for the brush shifting used with conventional motors, and better insulation of the armature was obtained. Among the advantages of AC operation, he mentioned the adjustable-speed feature and a high power factor. He explained that the speed could be changed by changing the voltage by means of a tapped transformer or by changing the phase of the voltage applied to the thyratron grids.[58]

The members of the GE Engineering Council held a roundtable discussion of the development and application of power tubes in March 1934, with 17 people in attendance. Roy C. Muir, a GE vice-president, commented that in view of the substantial investment, both in money and in the number of highly talented employees, there was a need to review the program at least every six months. Chester Rice reported that there was an unliquidated balance of $2 million and a current annual rate of expenditure of $300,000 on the power tube program. Charles W. Stone saw large power systems, in which there was a constant risk of power blackouts, as offering the prospect of a $150 million business in power electronics. William White identified the central problem as the high initial cost of tubes, along with the expense of tube replacement and maintenance. He pointed out that in the radio industry a tube life of a few thousand hours was acceptable but that in the power field a much longer tube life was needed. White added that the tube specialists at GE believed that the substitution of metal envelopes for glass could lower the manufacturing cost of tubes from the current figure—$2 to $3 per ampere—to as little as 10 cents per ampere. Albert Hull urged the development of a metal tube with conservative ratings be undertaken. Alexanderson commented that the power electronics field seemed to have emerged from a situation where there were as many systems as inventors, and that those in this field all were working with a unity of purpose.[59]

Of his requested total departmental budget of $130,000 for 1935, Alexanderson proposed spending $65,000 on thyratron applications. This included $20,000 on thyratron motors, $20,000 on a frequency changer, and $25,000 on thyratron control systems. To justify the continued concentration on thyratron applications, he stated that even though certain principles had been established and there was some commercial use, the developmental period was not over and the specific solutions that had been worked out should not be regarded as final. He stated that ultimate success of industrial applications of thyratrons depended on whether further improvements were made which would increase reliability and gain the confidence of industry. He asserted that GE should continue to push hard in order to maintain its leadership in thyratron applications.[60]

Alexanderson and his long-time assistant Albert Mittag prepared a paper on the thyratron motor for presentation at an AIEE meeting held in New York City in January 1935. The manuscript was submitted in September 1934 after having been approved by GE's Patent Department.[61] The authors described the new motor as providing power control with a "delicacy, quickness, and accuracy which is superior to anything heretofore available." They presented test data for and a physical description of the 400-hp motor that had been made for AG&E, and they mentioned that the auxiliary apparatus included a current-limiting reactor designed to limit the short-circuit current to five times the normal current. They asserted that this design feature had effectively eliminated arc-backs as a problem in normal operation and noted that the remedy had been to alter the circuit rather than the tube design. Their data indicated that the speed of a fan driven by the motor could be varied smoothly over a range of 0 to 750 rpm, with almost linear behavior from 400 to 750 rpm. The power delivered varied from 60 hp at 400 rpm to 400 hp at 750 rpm, and the system exhibited good efficiency and power factor characteristics. They stressed the stability and reliability of the motor: it would resume normal operation after momentary disturbances in the power supply, and would work even if one or more thyratron tubes were defective.[62]

During the discussion of the paper, Philip Sporn, whose company had ordered the motor, stated that he had been impressed favorably by its performance in tests although it had not yet been put into regular service. He described the motor as being "simple and rugged" and stated that it could start with some tubes out and could run on one phase although it was designed for three-phase operation.[63] Sporn and G. G. Langdon, both with AG&E, published a short paper on the 400-hp motor as part of a series on their company's use of power electronics techniques in the *Electrical World* in 1935. They revealed that the motor used 18 GE thyratrons of the FG 118 type, which had a rating of 12.5 amperes and 10 kv. They mentioned a planned modification in the tube cubicle which would make it possible to replace a tube without having to stop the motor, and noted that the removal of up to 4 of the 18 tubes resulted only in a speed reduction. They identified the life expectancy of the tubes as the principal unknown factor associated with the motor. Sporn and Langdon concluded that utility engineers needed to make a contribution toward advancing the state of the new art of power electronics.[64]

Alexanderson reported in February 1936 that the 400-hp motor had been shipped to AG&E but was not yet in service. The motor was installed at a power plant in Logan, West Virginia, and was placed in regular operation on May 29, 1936. Albert H. Beiler reported on several months of experience with the motor in an AIEE paper in 1938. Noting that the designers of new apparatus often concentrated on theoretical aspects and might overlook details that were important to the user, he stated the AG&E engineers had worked closely with GE during the final stages of development. He mentioned that the cooperation had led to several changes, including a dust-tight container for the distributor, a safety lock on the tube cubicle door, and a tube-failure alarm system. Eighteen glow lamps provided a remote indication of tube performance, he explained, and a bell would ring to alert the plant operator of abnormal tube firing.[65]

Beiler included in his paper a table of tube failures during the period from May 29, 1936, to July 31, 1937. The first of 20 failures had occurred after 191 hours and the last after 8,272 hours. He reported that the motor had been in operation 96.5 percent of the time, which he felt to be satisfactory for a new control system. He asserted that the plant operators were very enthusiastic about the motor's performance and called it convincing evidence of its practical success. He concluded that the ultimate success of the thyratron motor depended on whether the tubes could be made durable enough that maintenance costs would not be excessive.[66] During the discussion of Beiler's paper, William White of GE attributed the abnormally high rate of tube failures to mechanical vibration, and claimed that a minor change in some welded connections had seemingly eliminated the problem.[67]

In October 1935, Alexanderson wrote to William D. Coolidge of the GE Research Laboratory to discuss aspects of the design of thyratron tubes and circuits that used them. Alexanderson expressed his conviction that arc-backs were inevitable in mercury-arc devices of all types. He went on to say that whether a high-voltage surge occurred or not depended on the mercury temperature and the characteristics of the external circuit. The designer, he commented, was caught "between the devil and the deep blue sea, or in more classical terms between Scylla and Charybdis." He argued that the current surge that followed an arc-back was more important information for the designer than normal volt-ampere characteristics, and recommended that the stress due to an arc-back surge be determined for each device. Also, he suggested that the high and low danger points of temperature be measured so that circuits could be designed with an adequate safety margin.[68]

In January 1936, Alexanderson reported that some of the developmental costs of electronic power converters might be liquidated in a contract being negotiated with the navy. He mentioned that the contract involved the aircraft carrier *Langley* and would include a substantial sum for the development of both tubes and circuits; if the experiment were successful, it might lead to universal adoption of electronic power converters by the navy. His memorandum indicated that a total of $363,000 in developmental expenses of power converters had not yet been liquidated.[69]

In a report prepared in February 1936, Alexanderson wrote that GE had made a proposal to convert the *Langley* to propulsion by a thyratron motor. He stated that the proposal was strongly supported by Admiral Robinson and the navy's Bureau of Engineering. Alexanderson pointed out that work on projects for the navy might be adaptable to the commercial sector. For example, the power conversion method being developed for ship propulsion might be used in long-distance power transmission, and a DC system that had been demonstrated for submarines was applicable to electric locomotives.[70]

As in the case of AG&E and the thyratron motor, the navy's supplier-customer interaction with GE exerted a detectable influence on details of design and stimulated patentable inventions. In May 1935, Alexanderson informed GE's Patent Department of a significant improvement in thyratron-tube design which had resulted from the navy's dissatisfaction with existing tubes. According to Alexanderson, he had been told by Albert Taylor of the Naval Research Laboratory that the thyratron's five-minute warm-up time was unacceptable for use in gun-control systems. He had then discussed the problem with Coolidge, Hull, and White, who had felt that considerable improvement was possible. Only a few days later, Hull had produced a thyratron that could be operated at full power within 30 seconds.[71]

Alexanderson's work on the use of thyratron motors for ship propulsion received the strong support of an old friend, Adm. Samuel Robinson, who was chief of the navy's Bureau of Engineering from 1931 to 1939 and was appointed as the first chief of the Bureau of Ships in 1939.[72] In December 1938, Robinson wrote to Alexanderson concerning the urgent need for thyratron motors suitable for use with diesel-electric propulsion. Robinson noted that speed control would be difficult for the large number of engines that would be required for a battleship unless the thyratron control was available.[73] In his annual report for 1939, Alexanderson mentioned that a new system of electric ship propulsion was being developed with the cooperation of Admiral Robinson. In a memorandum to the Patent Department in April 1940, Alexanderson stated that Robinson was enthusiastic about the use of thyratron motors in ships and that his bureau was designing a battleship to employ a 300,000-hp electric propulsion system. Alexanderson reported that GE engineers also were working on the designs. In October 1940, he reported that thyratron motors were under construction to be used in a 60,000-hp diesel-electric destroyer.[74]

The Development of Direct-Current and Nonsynchronous Power Transmission

In addition to working on the torque amplifier, the thyratron motor, and gun-control systems, Alexanderson worked diligently on the development of DC power transmission and other methods for nonsynchronous power transmission during the 1930s. In September 1933 he wrote to Roy Muir that the power companies were facing serious problems, for their power distribution systems had become so large that they threatened to become unmanageable: the utility executives felt "as if they are setting on top of a volcano." More flexible tie lines were needed so that power could be transferred without the need for rigid synchronization. Alexanderson cited the Boulder Dam generating plant and the plan to transmit its power to Los Angeles. He stated that alternating current might be used, with a nonsynchronous thyratron link to make the power flow nonreversible so that a line failure would not destabilize the Los Angeles system. He mentioned direct current as another possibility, concluding that high-voltage DC transmission was a topic that deserved GE's attention. The following month he informed Muir of a discussion with Langmuir, Hull, and Chester Rice concerning a proposed demonstration of DC power transmission that would use thyratron converters. Alexanderson observed that success or failure could not be determined by circuit diagrams but only by testing all the elements of an actual system.[75]

In December 1933, Alexanderson prepared a summary of an internal GE conference on high-voltage direct current which credited Langmuir with having made a very significant statement about thyratrons. According to Alexanderson, Langmuir had discovered that the mercury-vapor thyratron could be designed for much higher voltages than had yet been reached. Alexanderson interpreted this to mean that the scientific basis existed to design thyratron tubes with a rating of 100 kv. If 100-kv tubes with current ratings equal to that of an FG 43 (a commercial thyratron) were available, he concluded, the design of power transmission systems that could operate at several hundred kilovolts with a capacity of several hundred kilowatts could be undertaken with confidence.[76]

Langmuir had mentioned that in a thyratron designed for 100 amperes and 100 kv, bombardment of the anode during deionization might cause difficulty. In the course of reflecting on this, Alexanderson conceived a new method of current extinction for power converters, an idea he reported immediately to GE's Patent Department. Early in January 1934, he reported that some experimental tests had confirmed the validity of his theory of the extinction inverter.[77]

In August 1934, Alexanderson reported a successful test of a DC power conversion system and discussed the potential advantages of DC transmission systems as compared to AC systems. He informed Roy Muir that a 3,000-kw rectifier and a 3,000-kw inverter had been tested, with each unit employing six thyratron tubes. Power from a 15-kv source had been rectified to produce direct current and then retranslated to alternating current by the inverter. Alexanderson asserted that these experiments confirmed that DC was feasible for large-scale power transmission without synchronization. He went on to say that the cost of power generation was comparatively low and that the high cost of electricity to the consumer was due mainly to distribution costs. He pointed out that a DC line would withstand three times as much voltage as an AC line could, and would be nine times as effective, since it could transport three times as much power over three times the distance. He described Thury's DC system as a classic development in electrical engineering, although it lacked a suitable large-scale DC-AC power translation device.[78]

The Thury system had been introduced in Europe in the 1890s as a constant-current system that used rotary motor–generator units as power converters.[79] H. M. Hobart, a GE engineer, prepared an analysis of it, entitled "Thury System as a Public Works Project," in September 1934. In his report, Hobart suggested that the exploitation of this system could provide employment and enhance the value of Boulder Dam and the Tennessee Valley Authority projects by creating a national power system. Alexanderson commented on Hobart's report and advised that it was overly ambitious to undertake a superpower system linking all the power systems in the country. He recommended instead that a less ambitious DC power project be tried in a few places so that technical problems could be solved and the economic value demonstrated. He urged that several alternatives be tested in order to determine which was the most promising.[80]

In his departmental budget request for 1935, which totaled $130,000, Alexanderson included $55,000 to be used for DC power transmission research. He stated in the accompanying memorandum that the DC system involved such radical changes that several evolutionary steps would probably be necessary. He mentioned two applications: exchange of power between nonsynchronous systems, and underground power cables. He observed that mercury-pool-cathode tubes might replace hot-cathode tubes but stressed that such a change would not be merely a substitution but would require developmental research and invention.[81]

Alexanderson continued to regard wave transmission (which treated power as electromagnetic waves guided by transmission lines) as an attractive alternative to both conventional AC power transmission and DC transmission. In a December 1934 memorandum to Roy Muir, he suggested that wave transmission might constitute one of the first large-scale applications of thyratrons. Alexanderson used the line from Boulder Dam to Los Angeles as an example. He stated that the distance—270 miles—was near the limit for synchronous AC transmission, but that the limit could be extended to 500 miles or more by means of wave transmission. He recalled that Steinmetz had proposed the use of wave transmission and had pointed out that a line of this type changed from constant voltage to constant current. Alexanderson noted, however, that Steinmetz had not had available the power electronics apparatus that now could be used. Alexanderson added that he did not yet wish to take a position on the issue of wave transmission versus high-voltage DC. He stressed that GE's situation would be strong with either system, since both would depend on thyratrons. He concluded with the recommendation that a large-scale demonstration of wave transmission should be conducted.[82]

Clodiun Willis, B. D. Bedford, and F. R. Elder reported on GE's ongoing research on DC power transmission at an AIEE meeting in New York City in January 1935. They described an experiment that had involved converting AC to DC, transmitting constant-current DC, and then reconverting to constant-voltage AC. The system had exhibited exceptional stability and reliability, they stated, and a temporary short circuit had damaged neither the line nor the conversion apparatus. They explained that their inability to correct problems with constant-voltage thyratron inverters had led them to try the constant-current inverter. They had used an artificial transmission line and a monocyclic network, with the latter serving to convert from constant voltage to constant current.[83]

Albert Hull of the GE Research Laboratory participated in the discussion and characterized the circuit as being unique in its good treatment of the thyratron tubes. He noted that the tubes were never idle or overloaded and were not "punished" by excessive current if they failed. Philip Sporn called the GE experiment a brilliant piece of research but doubted that it heralded a new era of power transmission. He predicted that in another 10 years, most power still would be transmitted by alternating current. Alexanderson was present at the discussion and commented briefly on the cases in which DC transmission would be advantageous.[84]

In his budget request for 1936, Alexanderson stated that the electronic tube research in his department could be classified as either hot-cathode applications or pool-cathode applications. He requested $77,000 of a total budget of $136,000 to support hot-cathode research that included the thyratron motor ($17,000), constant-current DC transmission ($40,000), and remote control systems ($20,000). He asked for $52,000 to support work on pool-cathode applications, including a frequency changer, railroad electrification, and constant-voltage DC transmission. Only $7,000 was requested to support work in the field of radio.[85] In a report prepared a year later, Alexanderson revealed that the Consulting Engineering Laboratory had actually spent a total of $143,732 over the previous 12 months.[86]

In a February 1936 memorandum to Roy Muir, Alexanderson called attention to the plans announced by the federal government to construct hydroelectric power plants with an aggregate capacity of 6 million horsepower. He pointed out that the government power projects might offer an opportunity for GE to supply some nonsynchronous power generators, and reported that a member of the Water Power Commission in Washington, D.C., had told him of related plans to develop a power distribution system with lines of up to 500 miles in length. Alexanderson asserted that asynchronous transmission would be more economical than a conventional synchronous AC system and that GE could accept orders for 50,000-kw asynchronous generating units. He noted that a typical line would require from two to four such units. Only a few days earlier, he had informed Muir that developmental work on a DC transmission line between Schenectady and Mechanicville, New York, was almost completed. He mentioned that it had been designed to use hot-cathode thyratrons, although they could be replaced by pool-cathode tubes.[87]

Alexanderson discussed trends in the power applications of electronic tubes in a paper published in *Electronics* in June 1936. He suggested that the power applications of electronics would be as significant as the radio applications in the long run, although they might not come about as quickly. He commented on the diversity of form of the mercury-vapor devices that were dominant in power applications, ranging from the hot-cathode thyratron with a single anode to the steel-tank, mercury-pool-cathode devices with multiple anodes. He noted that the diversity had caused some confusion in nomenclature and suggested that this might be alleviated by the adoption of the generic term *mutator,* which would be applied to electronic conversion units composed of "a group of tubes or tanks with the associated circuits." Gaseous devices used for rectification, power inversion, or frequency changers all would be covered by the umbrella term. Presumably the new term originated at GE, although Alexanderson did not indicate who had proposed it. Despite its "good linguistic as well as practical background" and its obvious suggestiveness as a biological metaphor, the word did not come into general usage.[88] Alexanderson again mentioned the term in a paper published in 1938, but he acknowledged that "the art is in such a rapid growth that the nomenclature cannot keep up with it."[89]

In his 1936 paper, Alexanderson stated that high-voltage DC transmission had achieved renewed attention for two reasons: the advent of the new electronic converters, and the government's interest in developing large power-generating facilities at locations remote from prospective consumers. He stated that the newly developed electronic converters meant that either DC transmission or asynchronous AC transmission might be used to extend power transmission beyond the "well recognized limits" of conventional AC. He asserted that there was no longer any practical distance limitation for power transmission for either alternating or direct current. Declining to predict which of various alternative systems might be adopted, he concluded that it was "the competition of ideas that is necessary for progress."[90]

In October 1936, Alexanderson reported to Roy Muir on the discussions at a power conference where various methods of transmission had been compared. Alexanderson cited a recent power blackout that had occurred in New York City as evidence of the need for introduction of nonsynchronous methods of interconnection. He urged that GE accept the challenge and determine the merits of various alternatives with complete impartiality.[91] Alexanderson had in fact made a start on such a study. In November 1936, he had informed Muir of the conclusions he had reached from an analysis of the comparative economics of DC power transmission. Alexanderson admitted that because it was difficult to find a case where the use of DC would be more economical than that of AC, DC would only be adopted because of superior performance. He suggested that a substantial portion of its cost could be justified by its assurance of continuity of service. He stated that a principal objective of GE's developmental work would be to demonstrate that a DC line could be operated at twice the voltage of an AC line, with greater reliability. Finally, he expressed his belief that important applications of DC transmission would be discovered but that GE would need to carry development beyond the current state of the art to do this.[92]

In his 1937 budget request, Alexanderson asked for $41,200 to support his department's research on DC transmission. His total budget request, $165,000, showed an increase of 15 percent over the actual 1936 expenditure, $143,732. Of the total, $106,276 was allocated to shop expenditures, with the remainder going to salaries and other costs. In addition to DC transmission research, the budget memorandum mentioned power-tube applications ($64,200) and control applications ($59,600).[93]

In April 1937, Alexanderson gave a lecture on power electronics for an advanced engineering course taught at GE. In his notes, which were distributed to the class, he included a general discussion of thyratron motors, the torque amplifier, and DC power transmission. He stated that thyratron DC motors were satisfactory for variable-speed applications that required power ranging from a few horsepower to about 200 hp. He mentioned that this type of motor had been installed on a number of ships and also was being used for the automatic control of machine tools. For higher power levels, extending to around 20,000 hp, he added, the thyratron synchronous AC motor could be used. He pointed out that the thyratron control added the advantage of speed control to the inherent lightness and efficiency of the synchronous motor. Citing as an example the 400-hp motor produced for AG&E, he explained that a larger version could be used for ship propulsion. In a brief comment on GE's applications of the torque amplifier, Alexanderson reported that the torque amplifier was being used by the navy, but he did not mention gun control, at least in his written notes. He did state that a large boring mill at the GE factory in Schenectady was being equipped with an automatic control system of the type used by the navy.[94]

On the subject of DC power transmission, he reported that a 5,000-kw system was already being operated between Mechanicville and Schenectady, and that tests were in progress to determine how to design larger installations. He mentioned that studies of the economics of DC transmission had indicated that underground transmission lines that would not be affected by sleet or lightning would be one case in which the use of direct current would be technically superior to overhead AC transmission, although not less expensive. Alexanderson ended his lecture by asserting that all available evidence indicated that the technical problems of DC transmission could be solved, but that at present the question of whether the power industry would find "this radical innovation" worthwhile could not be answered.[95]

Alexanderson gave a wide-ranging talk on "thyratrons and their uses" at a meeting of the United States Patent Office Society, and his talk was published in the February 1938 issue of *Electronics*. He began by suggesting that inventions could be broadly categorized into one of two groups, with the use of electronics in the power industry being in the second group. In the case of the first category, there existed "an insistent public demand for a certain service," and this demand stimulated scientists and inventors to seek a solution. The automobile and aircraft industries were examples of this class. In the case of the second category, a discovery in science led to the gradual development of a new technology. Alexanderson asserted that for the second category, "the fundamental invention and many of its refinements are usually established before the public is conscious that it has an important new tool at its disposal." He added that it often required longer than the life of a patent before inventions of this type were exploited commercially.[96]

He then reviewed the history of mercury-vapor, or mercury-pool, tubes and their applications, stressing the importance of the contributions of Irving Langmuir, whose basic patent had already expired, and Albert Hull, who had developed the hot-cathode thyratron. He reviewed the various thyratron applications devised by GE, including thyratron motors, inverters, torque amplifiers, and frequency changers, and he mentioned that he and his colleagues now were able to achieve position control with an accuracy of 0.001 inch or a few arc minutes in angular position by means of the torque amplifier. He ended his paper with a few remarks about the use of electronics in long-distance nonsynchronous power transmission, noting that there was a "temptation to let our imaginations run away" but then engineers were "confronted with the hard facts that even at present the limit of power transmission is not technical but economic." However, he wrote, "for those of us who are interested in electronic engineering . . . this subject has fascinating possibilities even if the most ambitious dreams do not come true."[97]

Reports of a plan to transmit power from the Bonneville hydroelectric plant—which was nearing completion on the Columbia River—to Idaho attracted Alexanderson's attention. In March 1938 he wrote to James D. Ross, head of the Bonneville project, to request information and mentioned the recent progress at GE in the use of electronics in power transmission. Ross that he was quite interested to learn what was being done at GE and that he anticipated seeing Alexanderson to discuss it in the near future.[98]

The same month, Alexanderson reported to GE's Patent Department a laboratory demonstration of the latest DC power transmission techniques, conducted for the benefit of several GE executives. He stated that the significance of the demonstration could best be understood in the context of the contemplated 500-mile-long DC line from Bonneville Dam to Idaho.[99] In the annual report on the activities of his department, now known as the Consulting Engineering Laboratory, for 1938, Alexanderson included DC transmission as one of six areas of research. He stated that a new mercury-pool-cathode tube had been developed which had a rating of 75 kv and 600 amperes and could be used in a DC transmission system with a capacity of 300,000 kw at 500 kv. The following year he reported that the DC line from Mechanicville to Schenectady was in its third year of operation but would be replaced by an improved system. By 1940, the Mechanicville DC line was described by Alexanderson as being in regular use by the New York Power and Light Company.[100]

The Development of the Amplidyne

In September 1937, Alexanderson reported on initial tests of a "metadyne amplifier" in a memorandum for GE's Patent Department. An alternative to the thyratron torque amplifier, this machine soon became known as the amplidyne. Alexanderson wrote that the metadyne generator had been tested as an amplifier and found to respond quickly to frequencies up to 60 Hz. The tests had been conducted by his assistants Martin Edwards and Kenneth K. Bowman as part of an investigation of the possibility of using the machine as an amplifier for various purposes. As one application, Alexanderson mentioned that it might be used as an exciter for synchronous motors.[101]

The following month, Alexanderson provided additional information on the genesis of the research at GE on the "dynamo amplifier," another term sometimes used before the adoption of the term *amplidyne*. Some years earlier, he stated, his group had experimented with a torque amplifier in which the final stage was a dynamo, and had improved the speed of response by installing a "booster" between the generator and the motor. These tests had convinced him that a special generator that would respond more quickly than an ordinary salient-pole machine was needed. He and his assistants now were in the process of trying this idea, he noted, and he expressed his belief that the dynamo amplifier should take the form of a machine with a distributed compensator winding. He explained that he and his colleagues probably would design a machine with one set of short-circuited brushes so that a single dynamo would behave as a two-stage amplifier; however, they were establishing the principles of a single stage before trying a machine with one armature and two sets of brushes.[102]

He reported that he had undertaken an analysis of how to make a dynamo with a single set of brushes as efficient an amplifier as possible. It was important, he observed, that as little energy as possible be stored in the magnetic field. Thus, a magnetic structure similar to that of an AC commutator motor gave better results than a salient-pole design. He suggested that a simple and effective dynamo amplifier could be constructed by adopting the structure of a well-designed single-phase series motor. He explained that it would only be necessary to rewind the field for use as the input circuit. The armature in series with the field winding would be used as the output. He concluded by asserting that further improvement could be gained by connecting the field winding in series with the compensator winding to counter the tendency to hunt.[103]

In a follow-up memorandum written in early November 1937, Alexanderson stated that experimental tests had confirmed his expectations and had improved the performance greatly: he had demonstrated the dynamo amplifier to representatives of the navy, and a sample gun had been operated with an error of only 5 arc seconds at the gun. He added that the simplification achieved by the use of the compensator windings had made an intermediate stage unnecessary. In his budget request for 1938, Alexanderson noted that many problems could be solved either by electronic methods or by the dynamo amplifier. As an example, he cited the use of a thyratron torque amplifier in a large boring mill and predicted that similar installations in the future were likely to use the dynamo amplifier.[104]

In a February 1938 memorandum, Alexanderson informed GE's Patent Department of further progress with the dynamo amplifier. He reported that a full test had been run which had yielded an amplification factor of 10,000 from the equivalent of a two-stage amplifier in one machine. Later in the year, he wrote that the dynamo amplifier was analogous in many respects to the electronic amplifier. He noted that both exhibited a basic amplification ratio that might be increased by regenerative feedback. Since regeneration tended to decrease the speed of response, he continued, he and his colleagues were doing an analysis of the dynamo amplifier to establish its basic amplification ratio and determine how much improvement by regenerative feedback could be used in practice. He stated that the practical amplification could be expressed as a product of the basic amplification ratio, a factor for series regeneration, and a factor for shunt regeneration.[105]

In his annual report on the activities of the Consulting Engineering Laboratory during 1938, Alexanderson stated that a boring mill at the GE works in Schenectady was being equipped with a dynamo amplifier that seemed better adapted to factory conditions than the torque amplifier that had been installed earlier. He mentioned that the dynamo amplifier was sometimes referred to as a metadyne or as an amplidyne, and he noted that many applications for it were being found in industry (e.g., in steel and paper mills) and that it was an ideal exciter for synchronous machines. Explaining that the dynamo amplifier was a two-stage amplifier in one machine, with the first stage located between the control field and a set of short-circuited brushes, he reported that an overall amplification of the order of 10,000 to 50,000 was possible.[106]

In a May 1939 memorandum to C. E. Tullar of GE's Patent Department, Alexanderson wrote an illuminating review of the historical roots of the amplidyne and summed up what he and his assistants had contributed to the invention. He noted that the GE amplidyne system used a machine that was called a metadyne in recent references but even earlier had been associated with the names of Winter-Eichberg and Rosenberg.[107] The current machine was characterized by having a set of short-circuited magnetizing brushes and had not been changed very much in its physical structure from earlier forms. Alexanderson stressed that the uniqueness of the GE amplifier system depended on the use of the machine rather than on the machine's physical structure. He explained that the developmental ideas of the GE team could be traced to their early experience with radio amplifiers and that they had recognized a close analogy between the dynamo amplifier and the vacuum-tube amplifier. He pointed out that in the process of developing the dynamo amplifier, his group had translated several ideas from vacuum-tube amplifiers into dynamo design terms. For example, he noted, they had adjusted feedback coupling to obtain a high amplification without instability and also had used the principle of geometric progression, which he had introduced for tuned radio amplifiers, to obtain both high gain and quick response.[108]

After this background discussion, Alexanderson turned his attention to the circumstances that had led the GE engineers to begin work on the dynamo amplifier. He noted that they already had developed a thyratron amplifier for gun control and had conducted tests comparing it with the metadyne developed by Joseph M. Pestarini. They had built a metadyne based on Pestarini's specifications to use in the tests. Before the tests, Alexanderson commented, he had only a vague idea of what Pestarini had proposed; he had therefore participated in the tests, along with Martin Edwards and Kenneth Bowman, in order to gain familiarity with the machine.[109]

According to Alexanderson, he and his GE colleagues had found the metadyne to be unsatisfactory because of its low amplification and its tendency to hunt. He explained that these deficiencies had led them to approach the problem in a different way, one that was analogous to the design of vacuum-tube amplifiers. He commented that he had provided the historical details of the case to help the patent specialists understand the relationship of GE's work to that of Pestarini. Alexanderson pointed out that GE already owned the rights to Pestarini's patent claims under a license agreement, but expressed his doubts that the patents could be construed broadly, since they covered inventive claims GE had not been able to use. In conclusion, he asserted that the dynamo amplifier was the beginning of an important new development on which GE was entitled to broad claims.[110]

Joseph M. Pestarini (1886–1957), whose metadyne invention was given a rather negative evaluation in Alexanderson's memorandum, was a European engineer-inventor with good credentials. Born in Athens, Greece, Pestarini received an electrical engineering degree from a French technical school in 1910 and was employed by the French Thomson-Houston Company from 1911 to 1938, rising to the position of chief engineer. During the 1930s he had an international consulting practice, and his clients included electrical manufacturers in England, Germany, Italy, and the United States. In particular, he was a consultant to GE during the period 1934–1939.[111]

In 1930 Pestarini published a series of articles on the theory of the metadyne in a French electrical journal, and during the 1930s he was awarded several U.S. patents, some of which were assigned to GE.[112] After World War II, he lectured on the metadyne to graduate students at the Massachusetts Institute of Technology and Columbia University; he was a visiting professor at the University of Minnesota in 1956, shortly before his death. His book *Metadyne Statics,* published in 1952, was on the general theory of the metadyne and was reminiscent of the abstract and highly analytical style of Charles Steinmetz in an earlier time.[113] In his book, Pestarini stated that the metadyne had been used for gun control by the Italian navy, and that, with the consent of the Italian government, he had suggested its use to the navies of both Britain and the United States in 1936.[114] Pestarini's initiative may have stimulated the U.S. Navy, as a GE client, to request comparative tests of the metadyne and the torque amplifier, although this has not been documented.

In a March 1940 memorandum, Alexanderson again addressed the issue of Pestarini's patents and the GE amplidyne. Alexanderson stated that an investigation by GE's Patent Department had determined that Pestarini did not have broad patents that covered the devices being constructed at GE. However, Alexanderson recommended that GE make some settlement to avoid controversy and hard feelings. He commented that he thought GE owed Pestarini possibly more than the patent situation alone would justify since it was because of their association with him, in a consulting engineering capacity, that they had embarked on a line of work that had resulted in important new business for the company.[115]

Engineers from GE presented three papers dealing with the theory and applications of the amplidyne at the AIEE meeting held in New York City in January 1940. The first of the three papers was by Alexanderson, Martin Edwards, and Kenneth Bowman, and contained a general discussion of the link between electronic amplifiers and the dynamo amplifier. The authors stressed that they had been guided in their approach to the amplidyne by the same principles used in the design of radio amplifiers. They pointed out that although there was a superficial similarity between the amplidyne and other machines such as the Rosenberg generator and the Pestarini metadyne, the amplidyne was quite different in its function. They explained how the amplidyne could be regarded as a cascaded two-stage amplifier, with each stage having a gain of about 100. In cases where an even higher amplification was needed, an additional stage—in the form of an electronic-tube amplifier—could be added, so that a few microwatts could be used for "quick and accurate control of power flow of many horsepower."[116]

A second paper, by Alec Fisher, included design considerations for practical amplidynes and performance characteristics of a 4-kw unit.[117] The third paper, written by David R. Shoults, Martin Edwards, and Frederick E. Crever, all GE employees, listed a number of industrial applications that already had been made, including a 50-kw amplidyne used as an exciter in a steel mill.[118] None of the three papers made mention of military applications.

In the discussion of the papers by Alexanderson and his colleagues, Philip Alger pointed out a promising trend toward the synthesis of knowledge from very different branches of engineering in innovations such as the amplidyne. In the past, he suggested, experts in electronics and those engaged in the design of industrial power equipment had seemed to live in separate worlds. Alger credited the engineers who had developed the amplidyne with having demonstrated that the concepts of electronics could be embodied in power equipment and having shown that there were remarkable similarities between the two fields of engineering. He characterized the amplidyne papers as being of "historic importance" in that, for the first time, the talents of electronic and industrial motor specialists had been joined in the creation of a power-control system of great potential value. He predicted a continued fruitful interaction, with the power engineers perceiving that electric-motor design had more than a "prosaic character."[119]

The interdisciplinary quality of the amplidyne innovation also was the theme of an editorial by R. H. Rogers in the issue of the *General Electric Review* that reprinted the three amplidyne papers. In his editorial, entitled "From Cobblestones to Amplidyne Generators," Rogers used the metaphor of two valleys, separated by a mountain range, in which the two cultures of electronic and power engineering had developed more or less independently. Each group had moved down its respective valley, and eventually "they [came] together and we have the amplidyne generator, partaking of the characteristics of both cultures."[120]

Radio Observations during a Solar Eclipse

Although his radio research was curtailed sharply by RCA's termination of its support in 1930, Alexanderson did not abandon the field entirely. Curiosity about the effect of a solar eclipse on radio propagation led him into a brief excursion into radio science in August 1932. He was joined in the experiment by the GE scientist Irving Langmuir, who was to receive a Nobel Prize in December of the same year. The *Albany News* of August 17, 1932, reported that Alexanderson and some associates planned to be in Maine on August 31 to conduct radio tests during the eclipse.[121] Alexanderson notified the eminent British radio physicist Edward V. Appleton that a signal would be transmitted from Schenectady at a frequency of 8,655 kHz during the experiment.[122]

An item in the *New York Times* of August 31, 1932, included a photograph of Alexanderson in a portable radio laboratory to be used during the eclipse experiment. The article stated that radio facsimile reception of a transmission from Schenectady was being used in order to produce a permanent record. It stated that Irving Langmuir of the GE "House of Magic" was a participant, and that he and Alexanderson hoped to learn more about the nature of the Kennelly-Heaviside layer.[123] A report carried by a Schenectady newspaper the day after the eclipse revealed that the GE team had discovered that a "radio electronic eclipse" (i.e., a blackout of terrestrial radio signals from the GE transmitter) preceded the optical eclipse by about two hours. The radio signals had returned to full strength about a half hour before the optical eclipse became total, and a sudden increase in radio signal intensity had occurred a minute before totality.[124]

Alexanderson contributed a paper on the radio eclipse experiment to the October 1932 issue of *Radio Engineering*. He stated that the initial stimulus for the investigation had been a suggestion by Langmuir that the GE observers collect data related to the theory that there was a substantial emission of charged corpuscles from the sun which traveled at a rate of 1,000 miles per second. Alexanderson explained that the frequency of 8,655 kHz had been selected so that the observation point would be located close to the boundary of the skip distance (where signals reflected from the ionosphere returned to earth), where the fringe effects of fading would be enhanced. He noted that such areas of fringe reception tended to create a "dance of ghosts" during television experiments. According to Alexanderson, the most unexpected result of the experiment was the discovery that normally strong radio signals vanished during the two-hour period preceding the visual eclipse. He interpreted the effect as confirming astronomers' calculations that this should be the time of a corpuscular eclipse. He speculated that the corpuscular bombardment from the sun might produce another reflecting layer—known as the "Appleton layer"—that seemed to be needed, in addition to the Kennelly-Heaviside layer, to explain radio propagation.[125]

Several months later, in April 1934, Alexanderson gave a presentation on the radio eclipse observations at a professional meeting in Washington, D.C. Soon after the meeting, he submitted a paper to the *Proceedings of the Institute of Radio Engineers* for publication. Subsequently, its editor, Alfred Goldsmith, wrote to Alexanderson that some unnamed referees of the manuscript had objected that the abrupt disappearance of the signal two hours before the optical eclipse was not conclusive proof of a corpuscular eclipse. They had pointed out that the transmitter frequency was so close to the critical frequency for penetration of the F layer of the ionosphere that a small fluctuation in ionization might have caused the effect reported. Goldsmith asked whether, in view of the referees' comments, Alexanderson wished to publish the paper in its present form. Alexanderson replied that he agreed that the recorded effects were not conclusive proof. However, he told Goldsmith confidentially, he had discussed the data with Albert Taylor of the Naval Research Laboratory, who had regarded the interpretation as reasonable and advised that the paper be sent to the IRE.[126] The paper was published in the May 1935 issue.

In his paper, Alexanderson listed several earlier publications by other researchers that dealt with radio phenomena during the eclipse. He noted that none of them had revealed any indication of a corpuscular eclipse, but he stressed that the other experimenters had used lower frequencies and shorter distances and had had different objectives in mind. He described the GE apparatus and stated that the actual recordings had been made by Earl L. Phillipi and L. J. Hartly, both of whom had extensive experience with radio facsimile and television investigations at GE. Alexanderson asserted that their experience with facsimile and television propagation gave the GE team a good background for analysis of the experiment's results.[127]

He acknowledged that he and his colleagues had not had a well-developed theory before the experiment; rather, they had hoped to discover something "accidentally." They would not have been surprised by a negative result, he commented, since the corpuscular eclipse was anticipated to be east of their transmission path from Schenectady. He explained the process by which he had deduced that a corpuscular eclipse had occurred before the optical eclipse. In support of his interpretation, he cited earlier observations by Albert Taylor, who had discovered that reflections sometimes seemed to come from an area located about 1,000 km beyond the point of reception. Alexanderson mentioned that the GE observers had monitored radiotelegraph transmissions from Germany during the eclipse period which seemed to provide further support for the corpuscular-eclipse interpretation. He speculated that the earth's magnetic north pole might be involved, since it was well known that the aurora borealis caused disruptions in shortwave communication with Europe.[128]

The Development of an All-Wave Radio Receiver

During the 1930s Alexanderson's Consulting Engineering Department continued to engage in a modest research effort in those areas of telecommunication that were regarded as outside the restrictions of the RCA-GE agreement. In August 1931, Alexanderson applied for a patent on a high-frequency railroad communication system that used the rails as a transmission line.[129] In his budget request for 1932, he mentioned that a railroad communication system had been developed and demonstrated. He explained that it would allow communication between trains on the same railroad and might be used for automatic signalling. He requested continued support for this project in his 1933 budget request, stating that an installation of the system on either the New York Central or the Pennsylvania Railroad was planned.[130] Some work was done in 1931 on a radio warning system intended to alert pilots of aircraft in danger of colliding. Alexanderson received a patent on this system and a joint patent with John H. Hammond, Jr., that covered a radio landing-aid system for planes.[131]

Alexanderson welcomed news of the consent decree of November 1932 which opened the way for GE to resume the manufacture of radio receivers, even though there was to be a delay of thirty months before they could be marketed. He wrote a memorandum in November 1932 that outlined "activities dependent upon new radio contract." He listed several developments that might be undertaken if GE resumed the manufacture and sale of radio apparatus. These included a printed book that would read aloud and a library of moving pictures that would use a GE radio for sound reproduction. He also suggested the possibility of broadcasting over beams of light; this would enable radio to assume a role like that of a national newspaper.[132]

Alexanderson expanded on the speaking-book idea in a memorandum to GE's Patent Department. He suggested that the GE engineers could take advantage of their experience with facsimile in designing an optical scanner of printed pages. He explained that a variation in the width of inked lines could be used to produce sound. In another memorandum to the Patent Department in November 1932, Alexanderson described what he believed would be an ideal broadcast receiver for GE to develop. He recommended that the simplicity and efficiency of the regenerative receiver be combined with the selectivity of the superheterodyne receiver. This could be accomplished with only two tubes and would make possible the production of an inexpensive, high-quality receiver.[133]

The implications of the resumption of radio receiver manufacture by GE was the principal theme of a talk Alexanderson delivered in February 1933 to a meeting of radio sales managers. He told them that GE intended to get back into the radio business but would have to overcome the handicap of having been left out of the design and manufacture of receivers for several years. Remarking that cooperative activity among those responsible for design, research, manufacture, and sales would be required in order to regain a position of leadership, he observed that it was the aggregate momentum of all these sectors that constituted the great strength of the company. He characterized GE as an organic entity that had "life, character and permanence like a beehive, quite apart from the individual bees."[134]

He admitted that GE had fallen behind in radio design and manufacture but contended that this specialized know-how could be obtained quickly when one knew where to go and had a flexible manufacturing plant like GE's. He noted that there might even be some advantage in making a fresh start without being encumbered with a manufacturing setup that already had gone through several evolutions. Speaking of television, a field in which GE's earlier work had become obsolete, he expressed confidence that a company with employees such as Langmuir and Coolidge could soon become a leader in an art that had turned to the cathode-ray tube. Alexanderson concluded his talk by stressing the importance of communication between those in research and those in sales. He credited the research group with exploring physical possibilities and the sales group with determining public reactions. He deemed it necessary that GE give adequate attention to both of these functions.[135]

In September 1933, at a meeting of radio dealers, Alexanderson spoke on the significance of a new all-wave radio receiver being developed by GE. During his talk, he reminisced about the past history of radio and called attention to the interaction of science and engineering that had been stimulated by radio. He recalled the time when spark sets and ship communication had been dominant; this had been followed by a period of longwave radio, during which GE had entered the field because longwave techniques had resembled electrical power engineering so closely. He told his audience that he could see clear indications that radical improvements in radio and television were imminent.[136]

Alexanderson stated that the new GE receiver was not just a new model; rather, it was an early sign of a turning point. He asserted that because the standardization and mass production of broadcast receivers had reached a point of stagnation, something new was needed, and he expressed the hope that the all-wave receiver would revive the spirit of adventure that had once been present in radio but had been missing in recent years. He urged that the enjoyment of the "sport of fishing in the ether" be encouraged, and he commented that dealing with uncertainties and acquiring skill in coping with the vagaries of nature was for many a pleasurable experience. Claiming that science and art had been widely separated before the advent of radio, he credited radio with having caused the whole world to become interested in science. He also observed that engineers at GE benefited greatly from being constantly in touch with scientists.[137]

In January 1934, Alexanderson expressed his delight with the GE all-wave receiver in a memorandum to GE's radio sales manager. Alexanderson reported that he had already received signals originating in London, Germany, and South America. He was confident, he continued, that the public would feel the same enthusiasm and that the new receiver would "revive the interest in radio."[138] In his budget request for 1935, Alexanderson mentioned that his department was continuing its work on a radio receiver attachment that would read books aloud. The research effort on radio and related topics was allocated $10,000 out of a total departmental budget request of $130,000 for 1935.[139]

AN AMUSING EXAMPLE of Alexanderson's imaginative response to opportunity occurred in 1933, in conjunction with the approaching end of Prohibition. He drafted a letter to GE's Publicity Department suggesting that GE capitalize on the coming legalization of beer drinking. He pointed out that draft beer should be cooled to exactly 8 degrees centigrade to maximize the flavor. Therefore, he suggested that GE

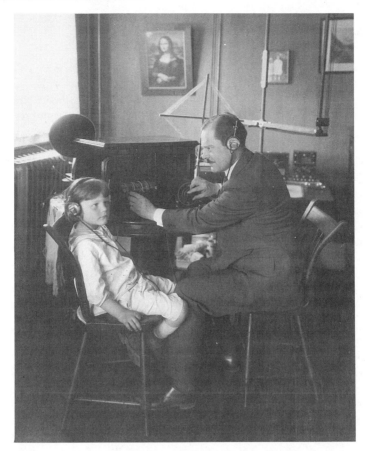

Figure 8.2. Alexanderson and his son Verner listen to a short-wave radio receiver. Courtesy of Verner Alexanderson.

place advertisements advising the public to buy 5-gallon kegs and install them in GE refrigerators, from whence beer could be drawn at the optimum temperature at any time. He painted a word picture of a foaming stein, with the flavor and temperature of the beer ensured by the refrigerator; the advertising would show the stein—and refrigerator—in the social atmosphere of a happy home. He advised that the advertising stress the 8 degrees centigrade rather than 46 degrees Fahrenheit to suggest subtly a European milieu, with the dignified participation of the family. GE, Alexanderson concluded, ought to "take advantage of the present enthusiasm" for beer and be the first to present the idea of refrigerated beer in an attractive advertisement.[140]

Alexanderson's Opposition to Restrictions
on Radio Research at GE

In a December 1937 memorandum to Roy Muir, Alexanderson urged that GE take the steps necessary to expand its activities in radio and related fields. He was especially critical of the policies that had prevented GE from competing fully with RCA, and recommended that a new agreement be negotiated which would give complete equality. He asserted that by restraining research, invention, and progress, the existing agreement had had a devastating effect. GE's record in innovation had been outstanding, he continued, and he expressed his belief that the company had the right and the duty to exploit radio and electronics. He deemed it most unfortunate that GE had given the most valuable rights in these fields to RCA. He called the arbitrary division of electronics into radio and power applications "contrary to nature and experience."[141]

Continuing his critique of the RCA agreement, he wrote that "inventions spring from freedom of thought, opportunity and personal contacts." He contended that "the perpetuation of a creative organization depends on holding out opportunity and careers to young men who desire recognition[,] and the paralyzing effect of the radio agreement is heavily felt by the man who should be building the future." GE also ought to welcome RCA's entry into the field of power electronics, he commented, for this would contribute to the solution of the technical problems and would also develop markets that would benefit GE. He asserted that engineers who had worked on radio electronics would come up with ideas that might not occur to those whose work had been on power applications, and vice versa, and he predicted that contacts between engineers in the power and radio fields would inspire important inventions and stimulate development along new and unanticipated lines.[142] Alexanderson asserted that it was unjust and a misuse of the patent system that GE had been put in the position of paying tribute to RCA, especially since the inventions on which RCA had based its monopoly had mostly originated at GE. He recommended a revised agreement that would give the two firms mutually nonexclusive rights in radioelectronics. In conclusion, he voiced his opinion that GE should engage in the development, manufacture, and sale of radio tubes, transmitters, and other types of radio apparatus, as well as undertake to develop new fields of communication.[143]

Alexanderson identified the frequency-modulation (FM) radio system of Edwin H. Armstrong as a promising new approach that RCA had neglected and that GE might well consider. In February 1938, Alexanderson reported that he had visited Armstrong's new FM station and had been favorably impressed. He called FM the most progressive development made in radio broadcasting in several years. He credited Armstrong with the opinion that commercial television would not be established successfully in the absence of an intermediate step, an ultrashortwave broadcasting system. Alexanderson stated that he also viewed a shortwave system as a natural step that should be taken seriously by GE in planning a developmental program in television. He pointed out that the Armstrong FM system provided GE with an opportunity to identify itself with a system that was new and of high quality, an opportunity that RCA had failed to recognize. GE, using a network of Armstrong stations, might add such services as a facsimile newspaper and film and, in the process, might begin to solve some of the difficult problems in television. He explained that the FM facilities would help establish the commercial and financial organization that could become the basis for television in the future.[144]

Alexanderson was well aware of the potential exploitation of television for purposes other than entertainment. In October 1937, he reported that two German scientists who had visited his laboratory had been interested in the development of television screens of great size and brilliance, suitable for viewing by large audiences. He commented that he saw this as a need peculiar to Germany, connected with mass meetings with speeches by Hitler and other government leaders. He conceded that if it proved successful, such a system might appeal to other countries.[145]

In an August 1938 memorandum to Roy Muir, Alexanderson again addressed the issue of the radio license agreement with RCA. He expressed concern that Sarnoff's skillful arguments might lead to acceptance of the view that GE's radio rights were adequate and fair. On the contrary, Alexanderson asserted, GE would be seriously handicapped if it were not permitted to sell radio transmitters. The history of radio demonstrated that new inventions required new business arrangements, he argued, and the original GE-RCA agreement applied to a situation that had existed 20 years ago. He concluded that competition in ideas and initiative was more important than competition in manufacturing.[146]

Like the Armstrong FM broadcasting system, shortwave relay communication seemed to Alexanderson to be identified as an attractive opportunity for development by GE. In a report submitted in November 1938, he stated that work had been started at the Consulting Engineering Laboratory on a directive antenna that might be installed at 10-mile intervals as part of a long-distance communication system that would be almost immune to disturbance. Such a system, he commented, opened the possibility of a nationwide radio system without the use of wire lines. In March 1939, he called attention to the potential use of relay systems in television, stating that he had done an analysis that indicated that the manufacture of relay stations was likely to become both substantial and profitable. GE's goal was to be a leader in new developments, he continued, and he wanted to make certain that GE would not be restricted by RCA in the building of relay stations. He asserted that such a restriction would probably destroy GE's competitive position in television. He also pointed out that a relay system might be very important in case of war.[147]

Alexanderson continued to express his concern over negotiations with RCA. In a June 1939 memorandum, he objected to a paragraph in a draft agreement which would keep GE out of the field of shortwave relay. He stated that he regarded relay development as a new branch of the communications art, in which GE again had an opportunity to assume a pioneering role. He predicted that contributions made by GE would lead development along a different path than it would take if it were left to RCA. Lastly, he asserted that the proposed restriction was shortsighted even from the perspective of RCA, since it would retard development: no major development in any industry could be carried out successfully without some competition. Competition in the power industry had been a great asset, and GE should welcome competition from RCA in power electronics, where GE had actually been handicapped by having to carry too much of the burden alone.[148]

In March 1939, Alexanderson reported that recent shortwave relay experiments had shown that a 25-mile spacing between stations would be satisfactory with a power level of 50 watts. Later the same month, he revealed that he and his colleagues were developing a relay system suitable for television and had already built a frequency-modulated television transmitter for the band between 288 and 294 megacycles per second. He stated that they intended to compare amplitude modulation (AM) with FM and that he did not believe that a larger bandwidth would be needed for FM. He explained that his opinion was based on his earlier experience with high-speed radiotelegraphy and that picture transmission was comparable to telegraphy, with the picture element being like the telegraphic dot. He suggested that the mathematical theory of modulation was less applicable to the analysis of telegraphic signals than it was to sound broadcasting.[149] In his departmental annual report for 1939, Alexanderson asserted that relay experiments using FM had already disproved the theory that a wider band would be required for FM than for AM. He mentioned that a rebroadcast of television programs was about to begin in the Schenectady area and that this would make evaluation of the new method possible and give GE the opportunity to assume a position of leadership.[150]

In addition to investigating the use of frequency modulation for television, Alexanderson explored the possibility of narrow-band FM for radio broadcasting. In a March 1940 memorandum, he reported that recent tests had confirmed his prediction that static could be eliminated without resorting to Armstrong's wide-band system. Alexanderson stated that three receivers had been compared, including one built according to his design. His receiver had demonstrated static suppression equal to that of the Armstrong receiver although the frequency swing was only 3.5 kHz instead of 75 kHz.[151]

A Shortwave Radio Altimeter Project

Late in 1937, Alexanderson began developmental work on a shortwave radio altimeter to be used in military aircraft. In an internal memorandum on the project, he stated that the Consulting Engineering Laboratory had been asked by the U.S. Army Signal Corps to develop an altimeter. He noted that a key problem was the need to measure time intervals of millionths of a second with sufficient accuracy to determine altitude, and he explained that a wavelength of about 3 meters had been selected so that an antenna with small physical dimensions could be employed. In conclusion, he remarked that he and his colleagues did not yet have a suitable broad-band receiver and that one should be built as soon as possible.[152]

·Alexanderson proposed using frequency modulation for the radio altimeter soon after his visit to the new Armstrong FM broadcast station early in 1938. In a March 1938 memorandum to H. E. Dunham of GE's Patent Department, Alexanderson outlined how the FM altimeter would work. He stated that the device would detect reflections from a signal transmitted from a shortwave FM oscillator and that the echo signal would be mixed with the transmitted signal to produce an audible beat that could be calibrated to measure the distance from the reflecting surface. He also commented that a similar arrangement might be used to detect ships in a fog and that it should be possible to determine the distance and direction of ships. In May 1938, Alexanderson reported a successful test of the FM radio echo system using a transmitter wavelength of 3.3 meters and a sweep frequency that was variable from 0 to 250 kHz. A receiver with a bandwidth of 300 kHz had been used, and the output displayed on a cathode-ray tube.[153]

Alexanderson's Views on Creativity and the Interaction between Engineers and Scientists

During the 1930s, Alexanderson continued to develop and express a personal philosophy of engineering creativity and invention. In his annual report for 1930, he stated that he viewed his department as serving a dual function: conducting engineering research, and being a training ground for inventors. He commented that he selected for his staff well-rounded engineers who demonstrated a talent for experimentation. In response to an inquiry from the Highlands Boys Foundation in June 1931, Alexanderson declined to provide an account of his own success. He wrote that the "only kind of success an engineer should have on his mind is the success of the work he [has] undertaken." He did note that an engineer needed to have the judgment not to undertake or persist in a project he could not complete, and that the engineer sometimes required "resourcefulness, skill and courage to carry through in spite of difficulties and opposition." Rather than worry about his own success, he concluded, the engineer ought to preserve his "integrity of character that should be the outstanding trait of the engineer and the professional man generally."[154]

In October 1934, Alexanderson commented on a proposal to place some young scientists in the engineering departments at GE. He stated that the success of such a program would depend on careful selection, and that it would need to be understood clearly that the young physicist who entered the engineering department would function as an engineer and not as a physicist. He added that if the college training of the recruit were in science, it would be important that he begin in the Test Department. Without this background, Alexanderson predicted, GE might end up with individuals who were neither scientists nor engineers and who would do more harm than good. Having expressed these reservations, he concluded with the opinion that the plan was constructive and should be tried.[155]

In a memorandum entitled "Mathematics and Engineering," written in March 1936, Alexanderson stated that an aptitude for mathematics was a good first test to identify potential engineers. He continued, however, that a student might have mathematical talent but lack the creative faculties that were also essential to engineers. He explained that engineering problems were initially solved by the use of clear thinking in terms of physical realities and later were translated into mathematical terms. When that stage was reached, he observed, mathematics was very useful for doing accurate and quantitative analysis. Conceding that the clear-thinking stage of problem solving depended on an attitude of mind that could be cultivated in no better way than in the practice of mathematics, he concluded that there were no simple tests for potential creativity.[156]

During testimony at a Federal Communications Commission hearing in 1937, Alexanderson discussed the interaction among the various laboratories at GE. He stated that the principal objective of the Consulting Engineering Laboratory, which he directed, was to utilize the discoveries of the GE Research Laboratory in solving new problems in engineering. All of the company's laboratories were interrelated, he continued, and the discoveries made in any one were soon available to the others. He noted that his engineers were trained in experimentation and constantly kept in touch with the GE Research Laboratory scientists. He stressed that the importance of this interdepartmental contact depended not on the number but on the quality of the scientists that he and his engineers were in contact with personally.[157] Irving Langmuir expressed a similar view in a paper published in December 1937. He reported that management had determined that 75 percent of the effort of the GE Research Laboratory should be devoted to practical applications. He commented that it was stimulating to the GE scientists to be in contact with those whose primary concern was application, and that it was important for the engineers to interact with those who had a broad scientific perspective.[158]

Alexanderson strongly supported a program of cooperation between GE and university professors. He assessed the importance of such contacts in a memorandum written in February 1939, in which he noted that GE's objective ought not to be the acquisition of rights to the patents of professors who were paid a retainer fee; rather, the company should strive to bring new and constructive ideas into the organization. He went on to say that GE's experience with outside consultants would be evaluated as disappointing if judged solely by the value of patents assigned by these individuals. However, he explained, ideas from outside had been stimulating to GE engineers and had resulted in patents obtained by them, as well as valuable developments in the electrical industry. He stressed that although the consultant's retainer took the form of a patent agreement, it should be understood as being a service agreement for the exchange of ideas. The professors should have independent laboratory facilities, and "above all [each of them] should have a personality with imagination and interests which are likely to be stimulating for our organization." He suggested that a fee of $1,000 per year would interest the best men, and that if GE found 10 good men, it would gain the benefit of the best ideas available, at a modest cost of $10,000 per year. He was certain, he concluded, that in the long run the company would be repaid amply for such an investment.[159]

DURING THE 1930s, as he moved into late middle age, Alexanderson received additional professional honors. He continued to enjoy the recreation of sailing, and he began to develop a hearing problem. In 1934, he was elected a member of the Swedish Academy of Sciences. The following year he nominated William Coolidge of the GE Research Laboratory for a Nobel Prize in physics, citing Coolidge's important contribution on x-ray tubes, but Coolidge failed to win the prestigious award that his colleague Irving Langmuir had received 3 years earlier.[160] In 1938, Alexanderson was awarded an honorary Ph.D. degree by the Royal University of Uppsala, where his father once had taught. Alexanderson had been born in Uppsala 60 years before.[161]

Alexanderson's favorite recreational activity continued to be sailing at Lake George. He had, in fact, introduced the sport at the lake. In 1934 he helped to found the Lake George Yacht Club, which organized weekend races and awarded trophies.[162] In a letter written in 1936, he characterized himself as "boat poor," mentioning that he had acquired a new boat that had cost more than expected and had five other boats to maintain; and in a talk before the Fortnightly Club in Schenectady in April 1937, he extolled the virtues of sailing as a sport in which both the old and the young could participate. He described sailing as a family venture that seemed to be one of the few common interests between the generations. He suggested that the boat tended to become an extension of the sailor's personality.[163] His only trip outside the United States during the 1930s was a recreational cruise to the West Indies in April 1939.[164]

In 1932 a hearing impairment caused Alexanderson to decide to acquire an electronic hearing aid. He contacted GE's Purchasing Department and stated that since he was troubled with deafness, he wished to obtain an aid that had been developed recently by the Western Electric Company. Five years later, he wrote that he had found the Western Electric hearing aid to be as good as any. He commented that he had found that there was a considerable difference in frequency response depending on the shape and size of the hole in the earpiece.[165]

The End of a Decade, and Changes in Management

An analysis of Alexanderson's patents indicates that he applied for a total of 58 patents during the 10-year period 1930–39. During the previous decade he had applied for 81; thus, his annual average declined from 8.1 in the 1920s to 5.8 in the 1930s. Of the applications filed in the 1930s, 43 were in the field of power engineering, including power electronics. This constituted 74.2 percent of the total, as compared with 53 percent for the decade of the 1920s. Thirteen applications in the 1930s covered inventions in the areas of radio, television, and aerial-navigation apparatus; these amounted to 22.4 percent, compared to 44.5 percent in the previous decade. These data appear to confirm that corporate policies that discouraged research and development in telecommunication resulted in a substantial shift in Alexanderson's inventive foci. His most productive years for patent applications during the 1930s were 1931, 1936, and 1939, with 9 patents in each of these years. His least productive year was 1934, when only 1 application was filed, for a "sound reproducing apparatus"—one of two inventions during the 1930s that fell outside the two major categories. By the end of 1939, Alexanderson had applied for a total of 299 U.S. patents that had been or would be granted.

Late in 1939, changes were announced in the top management at GE. Owen D. Young, chairman of the board of directors, and Gerard Swope, the company president, retired from the positions they had held since 1922. Philip D. Reed was selected as the new board chairman and Charles E. Wilson as the new president as of January 1, 1940.[166] Young, who was 65, and Swope, who was 67, were not much older than Alexanderson, who celebrated his 62d birthday in January 1940. He did not look forward eagerly to his own retirement and would, in fact, postpone it as long as possible. Another war had already begun in Europe. It would provide him with new inventive opportunities as an "inventor for victory."[167] This time his efforts would be concentrated on radar and gun-control systems rather than on the development of a transoceanic communication system.

Chapter Nine

"Inventor for Victory"

The Russian invasion of Finland and the fear that Sweden might be next concerned Alexanderson in December 1939. He expressed his apprehension about the threat to Sweden in correspondence with Carl Norman, a professor at Ohio State University who once had been a fellow classmate at the Royal Institute of Technology in Stockholm. Norman suggested that Finland's heroic resistance might save Norway and Sweden.[1] Early in 1940, Alexanderson received a telegram from a Swedish relative urging him to "help Finland." Alexanderson subsequently became chairman of the Finnish Relief Fund in Schenectady County.[2] In his role as a representative of a security-conscious American corporation, responding in August 1940 to an inquiry from a foreign national, he wrote that the man's lack of American citizenship would be a serious obstacle to his finding work at GE due to the "present war conditions."[3]

Radar, Analog Computers, and Gun-Control Systems

The research on gun-control technology that had been under way in Alexanderson's Consulting Engineering Department throughout the 1930s became more urgent in 1940. In April 1940, Alexanderson informed an acquaintance that he would be glad to see him but could not show him much since the radio laboratory was off limits because of work being done for the government.[4] In a July 1940 memorandum to GE's Patent Department, Alexanderson stressed the need to treat certain patent records with strict secrecy because they pertained to a highly confidential project involving a method of varying the direction of a radio beam. He noted, however, that GE should protect patent rights for the future. In a more detailed memorandum filed the same day, he reported that he and his colleagues were testing an antenna that, although physically small, produced a high-intensity beam. He explained that its compactness was important for use on ships, and that zigzag elements were being employed to serve as both an antenna and a transmission line. The ultimate objective was to locate objects by means of radio echoes; the directional beam would be wobbled by using a shaft rotating at 100 rpm. Alexanderson mentioned that a cathode-ray tube would be used as an indicator and that some elements of the system were similar to those developed for television. He cited the work that Albert Taylor of the Naval Research Laboratory had done, including the display of radio echo pulses on a cathode-ray tube.[5]

Alexanderson's annual report on activities at the Consulting Engineering Laboratory for 1940 indicated that a beam antenna system was being developed as a secret project of great importance for national defense. He also reported that GE had received large orders from the navy, the army, and the air corps for amplidynes for use in controlling guns and antennas. In October 1940, he wrote to Walter Baker, a GE executive, concerning work in progress on military projects. Alexanderson stated that his assistant Martin Edwards had the amplidyne control of antennas well under way, while a group under F. G. Patterson was at work on improving antennas.[6]

In a November 1940 memorandum to GE vice-president Roy Muir, Alexanderson discussed a "radio gun detector" that combined a transmitter, receiver, and an antenna control system. The objective was to enable the antenna to follow a moving target automatically. Alexanderson stated that groups of specialists would produce different elements of the complete system and recommended that antenna control be assigned to the radio group at GE. He anticipated that GE would receive the contract to supply all the equipment if they developed the control system that became standard. The same month, he reported that the radio gun detector principle had been demonstrated for the National Defense Committee, which included representatives of other manufacturers. He added that he had discussed the system with Admiral Robinson and other senior naval officers.[7]

In late November 1940, Alexanderson attended a conference at the Naval Research Laboratory; the principal topic of discussion was radio gun control. In a memorandum on the conference, he wrote that pulse radio might be used in several ways to direct guns. First, radio could serve as an accessory to optical systems, since radio echoes could be used to measure range and to detect targets before they could be seen visually. Second, a radio system could also be used to direct guns toward surface targets; that an amplidyne control system adapted to this task had already been tested with a 200-MHz radio beam. Alexanderson mentioned that work was under way to modify this system so that it could track targets automatically. Third, radio detection could be used in the control of antiaircraft guns both in azimuth and elevation. Alexanderson commented that a system of this type would be completed as quickly as possible even if all problems had not been solved completely. He indicated that a frequency between 400 MHz and 600 MHz would be tried with one transmitter and two duplex receivers. He concluded by stating that a separate project to use a much higher frequency, 3,000 MHz, might be undertaken but would require more time.[8]

Alexanderson gave additional information on the design characteristics of a directional antenna suitable for the navy's needs in a January
1941 memorandum. He explained that a method of directing an antenna at a target with an accuracy sufficient to aim a gun was needed.
He noted that the army system employed three antennas but that this
was not feasible for the navy system, and he stated that he had suggested two alternatives to the navy: one would use a mechanical means
to swing the antenna through an arc of about 5 degrees right and left,
while the other would change the electrical length of the transmission
line by means of a rotating condenser. Alexanderson mentioned that
the army was using a frequency of 100 MHz and the navy was using
200 MHz. He explained that GE had proposed an even higher frequency, 400 MHz, to reduce the necessary physical dimensions of the
antenna.[9]

By June 1941, Alexanderson had received a first-hand report on
British experience with the use of radio methods for gun direction. He
discussed the British system with Warren Weaver of the Rockefeller
Foundation, who had just returned from England. Alexanderson
wrote that the British method of tracking by angular velocity was similar to the method he had proposed. However, he noted, the British
were using tracer bullets to determine range, whereas the system being
developed at GE was to employ radio ranging. He mentioned that the
GE engineers planned to compare a method that used motors for
movement in two directions with one that used a motor for one direction and a hand control for the other.[10]

Alexanderson used the terms *radar* and *electric computer* in a memorandum written in May 1941 in which he stated that he had just drafted
a description of a gun director for GE's Patent Department; he urged
that steps be taken to develop an electric computer for use with a 3,000-
MHz radar system that was to be developed at GE. Further details were
revealed in a "Report on Gun Director and Computer," completed the
following month. In it, Alexanderson observed that bombers had introduced a new problem, but that the GE amplidyne offered a means
for rapid control of motion. Radio reflection gave an immediate and
accurate method for determining range, he continued, and the function of the electric computer would be to predict the future position of
the target so that a gun could be directed accordingly. He noted that
the computer prediction would be mathematically based on an inte-
gration and would use a Taylor series, and that tests were planned to
determine the speed and precision with which the computer could adjust itself to the typical motion of target aircraft.[11]

The proposed synthesis of a radar range detector, amplidyne control,
and computer was the subject of another memorandum Alexanderson

wrote in August 1941. He asserted that radar and the amplidyne were two powerful new tools that made it feasible to replace mechanical gun control with an electrical system. He identified three principal functions of the complete system: a radar tracking system to give range and rate data, a computer prediction system, and an amplidyne system to move the antenna and gun. Early in September 1941, he informed GE's Patent Department that tests of the computer and gun-control system were under way. Although this research was secret, he commented, similar methods were applicable in industry and would be described in a separate letter, which would not have to be considered secret.[12]

Alexanderson's nonsecret memorandum dealt with the use of the amplidyne for the control of motors over a wide range of speeds. He reported that a speed range of 1,000 to 1 had been obtained by connecting the amplidyne to the motor through a two-to-one gear; continuous control over the full range was provided by using a rheostat or potentiometer to vary the main control field of the amplidyne. In October 1941, he asked that his position on the issue of separation of secret and nonsecret portions of the speed-control system be assessed for correctness by C. E. Tullar of GE's Patent Department. Alexanderson pointed out that the speed-control system was part of a general research effort on the use of amplidyne control and was apt to be used for many purposes. He noted that the amplidyne-related research had very high priority, was being supported by GE appropriations, and was not charged to the government order. It was for these reasons, he concluded, that the speed-control system had not been designated as secret, even though its immediate use was to control radar antennas.[13]

In October 1941, Alexanderson reported a preliminary test of an analog electric computer designed to solve trigonometric equations encountered in radar range detectors. He explained that the problem was to solve an equation that predicted future changes in the range rate of a target so that the range detector could be controlled accordingly. The radar receiver delivered information in the form of one current representing the range rate of the target and another representing the angular velocity. He noted that if the two currents were designated a and b, a solution, in the form of a current c, could be obtained for the equation $c^2 = a^2 - b^2$ by *feeding currents* a and b into selsyn windings. He concluded by stating that tests of this principle had been successful.[14]

Later the same month, Alexanderson gave further details on the computer, pointing out that it would probably have broad applications for purposes other than gun control. He stated that the computer might be described as a selsyn regulator containing an actuating selsyn, a detector selsyn, and an amplifier. He added that the apparatus could automatically solve an equation with several independent variables. As examples of potential applications, he mentioned that a computer might be used to control the fuel in a diesel engine in a locomotive or in a ship propulsion system. Such an arrangement, he explained, would make optimum economy of operation possible and would maintain certain desired relations between the engine operation and several variables.[15]

He noted that the computer selsyn had several stator windings and two rotor windings. To illustrate the possibilities, he designated the two branches of one stator a and b, the two branches of the second stator d and e, *and the two rotor windings* c and f. He explained that the currents in a and b would act on c to produce a positive torque, while the currents in d and e would act on f to produce a negative torque. When the two torques were equal, the system would satisfy the equation:

$$(a + b)c = (d + e)f. \tag{1}$$

For the condition $c = a - b$ and $f = d - e$, equation 1 becomes

$$a^2 - b^2 = d^2 - e^2. \tag{2}$$

With $e = 0$,

$$a^2 - b^2 = d^2. \tag{3}$$

For the case $b = 0$ and $e = 0$,

$$ac = df. \tag{4}$$

Or with $c = 1$ and $e = 0$,

$$a + b = df. \tag{5}$$

Alexanderson concluded that the selsyn and amplifier could be used to regulate any variable in the equations, with the others serving as independent variables.[16]

In November 1941, Alexanderson reported that the GE engineers had just demonstrated to the army's Signal Corps and Ordnance Department a new system of aided tracking which constituted a radical improvement over any system previously available. He asserted that the GE system had definitely proven superior to competing equipment made by the Sperry Company and by the Bell Laboratories. During the demonstration, however, the GE system had been shown to the other competitors, and they would no doubt modify their future designs. Thus, Alexanderson urged that prompt steps be taken to protect GE's patent rights. He commented that aided tracking had been used extensively in England but that the U.S. Army and Navy had been hesitant to adopt the system because it was known to have certain limitations. He described the English system as employing a position control and a rate control that were linked by a differential gear so that both could be operated by the same hand crank.[17]

Alexanderson went on to say that GE was installing a large number of radar sets on destroyers, and that the navy had decided against the use of aided tracking on these. He concluded by mentioning his contention that rather than discarding aided tracking, GE ought to improve it; and he asserted that the GE engineers had now done so. A few days later, assessing the implications of some tests in which the GE apparatus at Fort Monroe was used to track airplanes, he expressed his conviction that computer control would enable the anticipation of normal variation with considerably greater accuracy than could be obtained by an operator using existing methods.[18]

Alexanderson's proposed strategy for the development of increasingly sophisticated gun control systems was similar to that which had been adopted for the development of television from simpler radio and facsimile systems. In late November 1941, he prepared an outline of a program for the evolutionary development of a new system of gun control. He explained that the program would include several steps, with each having its own usefulness and with succeeding steps taking advantage of experience with simpler forms. His program contained nine stages, marked by incremental increases in automatic control and less reliance on the skill of an operator.[19]

The first stage of Alexanderson's developmental sequence consisted of an amplidyne position control using a hand-crank master controller and a torque amplifier to maintain correspondence between an antenna and the master control. The second stage included both position and rate control, with independent manual control by an operator who could decide whether to use one or both. Alexanderson mentioned that this was the system being provided to the army and navy. In the third stage, aided tracking would be added to position and rate control. Alexanderson observed that this system, with simultaneous operation of position tracking and rate control from a single manual control, had been used in England. The fourth stage would provide aided tracking with manual release; it would permit an operator to select either aided or independent (nonaided) control by changing a hand grip in the manual control. Alexanderson pointed out that the independent mode would be better for making a quick change to pick up a target or for following rapid, unexpected change of direction. The fifth stage would provide aided tracking with an adjustable ratio and would permit the ratio between the position and rate control to be altered as the target speed changed.[20]

The final four stages of the developmental program were to use computers. The sixth stage would employ computer tracking, with an electric circuit capable of controlling speed and azimuth motion linked to a shaft the rotation of which was a function of the range of the target, as determined from radar data. The seventh stage would add another control circuit to produce movement based on the projected range rather than the measured range. In the eighth stage, information from the range detector, azimuth control, and elevation control would be integrated to predict change. The ninth and final stage would achieve detector control by using data obtained automatically, not by an operator. The computer was to maintain the radar detector on a target without manual control. According to Alexanderson, tests had shown that a computer control system could deliver information directly into an amplidyne integrator, and this had opened a new field in ordnance technique. In conclusion, he predicted that a whole system of gun control could be worked out by taking advantage of computer functions already incorporated in the tracking control, and dispensing with mechanical computers.[21]

Alexanderson learned of some high-power British radar tubes when he attended the 30th-anniversary banquet of the Institute of Radio Engineers in January 1942. Soon after the meeting, he reported that Albert Taylor of the Naval Research Laboratory had informed him of the availability of design data on a 100-kw British tube that could be used at a frequency of 600 MHz. According to Alexanderson, Taylor had a number of the tubes, and some might be available for use in the 400-MHz radar equipment being developed at GE. Alexanderson suggested that the tubes might enable testing to begin earlier than had been planned.[22]

In late February 1942, Alexanderson reported that tests of an improved aided tracking system had been carried out by GE engineers in cooperation with the U.S. Army. He stated that he had in mind combining this system with an automatic tracking system of the type being developed at the Massachusetts Institute of Technology Radiation Laboratory. Alexanderson asserted that no discontinuity was necessary in making the transition from manual to automatic control and back. He stressed the need to locate the target quickly and to begin to track smoothly so that the computer would function properly. Later in the year, he made a two-day visit to the Radiation Laboratory at MIT.[23]

Alexanderson's contributions to the war effort attracted some attention from the press during 1942. An article about his exploits, entitled "Inventor for Victory," was published in *Popular Science* in July 1942. The author, Herbert Asbury, asserted that the field of radio communication owed more to Alexanderson than to anyone other than Marconi. Asbury credited Alexanderson's radio inventions with having helped to win World War I and suggested that his latest invention, the amplidyne, was helping to defeat the current enemy. According to Asbury, Alexanderson did not view war as a stimulus to invention; instead, he saw it as providing the incentive and opportunity to develop devices that already were in the inventor's mind.[24]

An article published in the *New York Times* in October 1942 called attention to the anniversary of the ultimatum that had been transmitted to Germany by means of an Alexanderson alternator 24 years before. The reporter commented that Alexanderson was still inventing, although his current activities were covered by wartime secrecy. The writer speculated that Alexanderson's "fruitful days of World War I" might be repeated. Another article published in 1942 described Alexanderson's laboratory as resembling a dozen small factories, with workbenches crowded with an extraordinary variety of tools and apparatus being worked on by his assistants. The writer likened Alexanderson's World War I radio system to the amplidyne that was being used in "new weapons to stop Hitler."[25]

The Consulting Engineering Laboratory during the War

A list of the personnel in Alexanderson's Consulting Engineering Laboratory, dated December 10, 1942, revealed that the staff comprised 16 people, including Alexanderson. They ranged in age from 23 to 64, with nine being 40 or over. Some members of the staff, such as Samuel Nixdorff and Albert Mittag, had been with Alexanderson for many years. Among the others were Martin Edwards, Donald Garr, B. D. Bedford, F. G. Patterson, Earl L. Phillipi, and Gerald A. Hoyt.[26]

A song composed by an unknown writer, filed in Alexanderson's correspondence for 1944, conveys some of the flavor of life in the Consulting Engineering Laboratory during the war years:

We work at the Consulting Lab
We hang around all day
Nothing gives us more a laugh
Than what we do and say

Dr. Alexanderson
From whom the orders flow
The answers to our questions, boys
He always seems to know

Edwards smokes a big cigar
He smokes it all day
He flips the ashes on the floor
And never sweeps away

Chorus:
Psychopathic feller with a
Dissipated look—Consulting Engineer

Don Garr works in the steel mills
With his amplidynes and junk
If one would listen to Westinghouse
You'd think it was the bunk

Defandorff's arriving
So it must be 10 o'clock
Don't offer him a single drink
Tomorrow we need some stock

Jerry Hoyt's from Kansas
But we'll overlook this fact
If his gismos and "de"vices will
Stay on the bastard's track

Nixdorff's in the basement
With his rectifier stuff
He wonders if his helper Jack
Has finally got enough

Choules works on torpedoes
And its gyroscopic brother
The rate that he is travelling
He soon will be another

Mittag sits up at his desk
And figures all the day
Firing currents, synchronizer
And other such array

Patterson and Radar
They reside upon the top
Gremlins and their helpers
Aw Hell, let's stop
                Chorus[27]

Administratively, the work at the Consulting Engineering Laboratory was divided into two sections, with Martin Edwards serving as section head for control systems and F. G. Patterson serving as section head for radio (including radar) systems. Alexanderson was the director of the laboratory.[28]

In the proposed budget for the laboratory for 1943, Alexanderson listed several projects that were under way. He mentioned both automatic and semiautomatic amplidyne control systems for industrial and defense applications, including electric torpedo control. Among power-tube applications he listed a 20-kw frequency changer for the Carnegie Steel Company and the DC power line to Mechanicville. He noted that the laboratory was engaged in several secret radio projects for the army and navy. The proposed expenditures for 1943 included $30,000 for power-tube applications, $90,000 for DC transmission, $50,000 for amplidyne control, and $60,000 for radio development.[29]

The Consulting Engineering Laboratory and the General Engineering Laboratory were consolidated into the General Engineering and Consulting Laboratory in March 1945. The General Engineering Laboratory traced its roots to the Standardizing Laboratory formed in 1895. Alexanderson's unit, which traced its roots to Charles Steinmetz's Consulting Engineering Department in 1910, had undergone several name changes over the years. It had been known as the Radio Engineering Department from 1919 to 1922, the Radio Consulting Department from 1922 to 1928, the Consulting Engineering Department from 1928 to 1933, and the Consulting Engineering Laboratory since 1933. The 1945 consolidation was reported to be in response to the expanding needs of an increasingly decentralized company. The new, combined laboratory was headed by Everett S. Lee. Alexanderson continued to be in charge of the consulting branch, with Martin Edwards as his chief assistant.[30]

In a talk given in November 1945, Alexanderson discussed the history and tradition of in-house consulting engineering at GE. He credited the central principle to Albert Davis of GE's Patent Department, who had recommended that the company follow the lead of the engineers. Alexanderson noted that former GE president Edwin Rice also had believed that it was in the best interest of the company to develop a carefully selected staff of consulting engineers who would be trained for engineering leadership. Alexanderson also spoke of William Emmet, describing him as an outstanding consulting engineer at GE who had shown how the entire works of the company could be used as a laboratory; he asserted that he had applied the Emmet doctrine by having only a small personal staff and drawing on all the talent and other resources of GE in the creation of a radio communication system. He recalled that although GE had allowed him to work part-time for RCA for several years, he had never considered giving up his consulting engineering work for GE. He also commented that he had begun as a power engineer and had never been able to leave the field entirely. In conclusion, he stated his belief that it was vital that GE continue to have the benefit of the leadership of engineers who were free to set their own pace and use their own judgment as consulting engineers.[31]

Amplidyne Applications and the Alexanderson
Radio Alternator Revisited

An October 1942 memorandum described the amplidyne as a "distinctive GE contribution" to the war effort, with more than 120,000 either already built or on order for the army and navy. The applications listed included electric ship propulsion, tank propulsion, coast defense artillery, minesweepers, gun control, and radar antenna control.[32] Fremont Felix of GE's Industrial Engineering Department contributed a short article on the amplidyne to the *GE Review* in 1943. He characterized the amplidyne as "one of the most important electrical developments of recent years." He noted that it could control and regulate speed, voltage, current, or power over a wide range, and had numerous potential industrial applications.[33] Another article on the amplidyne was published in *Scientific American* in January 1944. It credited the invention to Alexanderson and other unnamed GE engineers just before the war. In addition to military applications, the writer mentioned that an amplidyne was used on the world's largest power shovel and on hoists to lift coal from mines.[34]

Alexanderson discussed writing a paper on the amplidyne for the *Proceedings of the IRE* in a June 1943 letter to Alfred Goldsmith, the journal's editor, whom he had known professionally since before World War I. Alexanderson wrote that he was quite interested in doing the paper, since it was like going back to the starting point. He described the amplidyne as being a descendent of the radio tube but a device that demonstrated what could be done without electronics. By contrast, he continued, GE's DC power transmission equipment resembled a "Brobdignagian radio set." In November, Alexanderson informed Goldsmith that the paper was almost completed and that he would have two coauthors.[35] In a January 1944 memorandum, Alexanderson mentioned that a section of the paper had been deleted by a censor. His coauthors were Martin Edwards and Kenneth Bowman, who also had been coauthors of an earlier paper on the amplidyne, published in 1940. The paper was presented in New York City on January 28, 1944, and was published in the *Proceedings of the IRE* in September 1944.[36]

The IRE paper was tutorial and was intended to provide an analytical approach that an applications engineer could use to design an amplidyne system with predictable performance and without excessive instability. Alexanderson and his colleagues explained how graphical characteristics and design principles that commonly were used by designers of vacuum-tube amplifiers could be used with the amplidyne. For example, they demonstrated that the voltage amplification factor of the amplidyne could be calculated by the simple equation $\mu = (f_0/$

$f)^2$, where f_0 was the rotation frequency of the machine and f was the control frequency. Other design parameters, such as the time delay and phase delay as functions of the control frequency, were presented graphically. The authors included sample calculations to show how the characteristics of a given amplidyne machine could be used in designing external antihunt circuits or in matching the amplidyne to a load. They compared the relationship between the amplidyne designer and the application engineer to that between the designer of vacuum tubes and the radio engineer who applied them, noting that the design of circuits for specific purposes was "an art rather than a science." Thus, they observed, in the application of the amplidyne just as in the application of the vacuum-tube amplifier, "the results will largely depend upon the skill of the engineer." They concluded with the hope that the analytical method they had presented would enable the application engineer "to choose amplidyne generators suitable for his purpose and predict the performance of a system before it is completed."[37]

In May 1944, William Brown, a former assistant, wrote to Alexanderson to report that one of the old 200-kw Alexanderson radio alternators had been in operation in Hawaii for the past several months. Brown wrote that he had recently spent 10 days at the transmitter site and had been pleased to observe the reliable operation of the system for which Alexanderson had primarily been responsible.[38] G. J. C. Eshelman, another of Alexanderson's former colleagues, reported in July 1942 that a radio alternator was being used at Tuckerton, New Jersey, for communication with naval ships around the world. According to Eshelman, the alternator was being operated at 15.8 kHz, and the messages "never required repeating."[39]

Reflections on Invention, Engineering, and Creativity

During the 1940s, Alexanderson found numerous opportunities to express his views on creativity as manifest in inventions and engineering design, and on how such creativity might best be encouraged. In March 1940, he wrote to the author of the novel *Utopia, Inc.* with some reactions to the book. Although he had found the novel very entertaining, Alexanderson commented, he questioned whether the idea of gathering the greatest scientists and inventors on Utopia Island would work out well. He observed, "While one may say that chance is the father of invention, it is still true that necessity is the mother of invention and science in isolation becomes sterile." Thus, it would be better not to uproot the inventors and scientists from the surroundings that had provided their inspiration. Alexanderson suggested that a difficulty with the Utopia Island approach would be that many would ask for support of ideas that were either "freakish or unnecessary duplication." He speculated that this problem might be overcome if the funds were controlled by "those few who are acknowledged leaders, each in their own line of endeavour."[40]

In a response to a correspondent in February 1941, Alexanderson wrote that he could not encourage her brother to become an inventor. Few unaffiliated inventors were successful, he noted, while inventions in industry occurred as a normal byproduct of everyday work.[41]

Alexanderson's long and close relationship with the research staff of the GE Research Laboratory was the subject of remarks by Laurence Hawkins, an engineer affiliated with the Research Laboratory, as he introduced Alexanderson as the guest of honor at a meeting of a Schenectady club in May 1941. Hawkins asserted that contacts between the Research Laboratory and Alexanderson had always been helpful and stimulating. He credited Alexanderson with having aroused Irving Langmuir's interest in radio tubes. Hawkins concluded by saying that no other engineer at GE had been quicker than Alexanderson "to appreciate and put to use developments in our laboratory." In his speech, Alexanderson reminisced about his early career at GE and his contacts with such men as Edwin Rice, who had been a sympathetic listener and witness to experiments. Alexanderson suggested that it was GE's tradition of informality and direct approach that had made the company great. He commented that he had probably attracted more publicity than most other GE employees because radio had had more popular appeal than other activities that were equally important to the company.[42]

In a letter written in May 1941 to GE president Charles Wilson, Alexanderson drew a distinction between creative "systems engineering" and "apparatus engineering." He expressed his belief that the best contribution that engineers could make toward filling in the valleys between boom periods in the economic cycle was to engage in systems engineering, while the peak periods would be devoted to the manufacture of apparatus, and to sales. He asserted that the creation of new systems always preceded apparatus business, and that systems for lighting, power, and transportation had been developed before substantial sales of lamps and motors could develop. Alexanderson went on to say that apparatus engineering was more attractive to those with a business orientation because it promised quick cash returns, whereas systems engineering tended to have a time lag of perhaps 10 years. Consequently, he wrote, systems engineering was unprofitable in the short run and had to be financed differently.[43]

He cited the case of amplidyne control, which had been developed in peacetime but only had come to prominence in response to the demands of war. Whenever the amplidyne had been applied to industrial processes, he stated, it had served to accelerate production. Despite this, he noted, GE had experienced difficulty in introducing the system to industry before the war because potential customers had not been interested in speeding up production or spending for new equipment during the economic depression. Alexanderson recommended that systems engineering be given a prominent place in the economics and organization of GE. He concluded that the company ought to be flexible enough that its best engineers could work on apparatus engineering in boom times and on systems engineering during the intervals. In reply, Wilson wrote that he agreed fully with Alexanderson on the necessity of systems engineering and on the "tremendous potentialities of such forward looking work for our company." Wilson promised to discuss with GE vice-presidents Roy Muir and David Prince the elevation of systems engineering to a more prominent place at GE.[44]

During his remarks at an IRE banquet in January 1942, Alexanderson stated that he disagreed with those who thought that engineers ought "to take a holiday" from innovating at times. Engineers, he suggested, were in a position much like that of an airplane pilot and had to keep on carrying out their ideas. He predicted that the war would "bring out so many new ideas that nothing will stay put."[45]

In March 1942, Alexanderson wrote a memorandum to J. F. Young on the subject of how to identify young engineers with exceptional inventive ability. Alexanderson commented that the creative mind did not function in a vacuum and that the beginning of an invention was always stimulated by an observation or discussion that set the creative imagination in motion. It was difficult to measure inventive talent, he continued, since any simple test was apt to pick the clever imitator and reject one who might do more fundamental thinking. He speculated that perhaps the best that could be done was to devise a formula that would select both the clever imitator and the originator. He stated that it was his understanding that Young wanted to study those at GE who were known to have inventive ability. Alexanderson suggested that Young contact GE's Patent Department to identify those who had been awarded the greatest number of patents over the past decade. Alexanderson observed that a good first step in a selection process would be to depend on an individual's self-evaluation. Those who showed a preference for experimentation (which he saw as the source of invention) might then be given a second questionnaire that would have them state in detail what they had done and what they hoped to do.[46]

In June 1942, Alexanderson advised a young correspondent to forget about becoming an inventor but rather to get a job as an engineer. After commenting that even the most experienced inventor had to discard 9 out of 10 ideas, either because they already had been discovered or because there was no demand for them, Alexanderson concluded by saying that an engineer often made inventions in the course of everyday work and also gained the necessary judgment to decide which ideas to follow up.[47]

In November 1942, Alexanderson addressed a meeting of the American Society of Mechanical Engineers on the subject of the engineer-inventor in a corporate environment. He asserted that the inventor at a young age became conscious of an urge to do creative work. He went on to say the young inventor instinctively sought the guidance of older persons who expressed a sympathetic understanding and who could provide guidance from a background of proven inventive ability. The younger person supplied knowledge, Alexanderson suggested, while the older person furnished experience and imagination. He described such a relationship as being intimate and personal and stated that the way conclusions were reached was intuitive. He commented that the best test of an inventor was to observe the individual's reaction when unforeseen difficulties were encountered. When pure analysis indicated that something could not be done, he explained, the successful inventor would use a different approach or restate the problem, and probably would come up with something useful even if it was different than initially envisioned. When an inventor's failures were outweighed by success, Alexanderson concluded, providing adequate financial support for such a person was the best investment that industry could make.[48]

Alexanderson commented on the role of emotion in the creative process in a January 1943 letter, expressing his views that emotion rather than reason served to open the gates of the subconscious mind. Such an emotive response might be stimulated by an intense desire to achieve something that was needed, he continued, or it might result from irritation at someone who claimed that something could not be done or who proposed a poorly reasoned way to do it. "Creativeness cannot be pursued for its own sake," Alexanderson wrote. "It is like happiness, a byproduct of the right way of thinking."[49] The previous day, he had written to C. E. Tullar that the final solution to a problem "has come to my mind in a fraction of a second by a process which is not consciously one of reasoning."[50] Alexanderson's interest in creativity was heightened during this period by his service on the advisory committee of GE's creative engineering program. He was among those who interviewed candidates for the program.[51]

In August 1944, Alexanderson gave a speech entitled "Big Business and Invention," in which he commented that he found it paradoxical that it seemed to require a war to permit Americans' inventive talent to function effectively. He speculated that once the war ended, the "dead hand of regulation" again would stifle invention and progress. He asserted that those who advocated that the government take the initiative in the organization of invention overlooked the fact that industry had a great pool of talent that should be used in industry. He predicted that any effort to transplant these inventors into government would prove futile, and he argued that a more constructive policy would be to relax the pressures of regulation and enable the accumulation of funds to be devoted to new projects.[52]

He cited railroad electrification as an example of a revolutionary technology whose introduction had been hindered by the excessive regulation of rates. As another example, he mentioned that a national distribution system of shortwave communication was needed both for television and other services. He contended that the television industry should not be expected to bear the entire cost of building such a system. He noted that inventions currently were being developed by engineers for war and expressed a feeling of dismay that such a large pool of talent might soon "again be shackled to work on immediate profit." Alexanderson concluded by saying that GE had always provided him with the opportunity to work on new projects "according to my desires."[53]

Alexanderson spoke on the dangers of excessive specialization in electrical engineering at an IRE meeting held in Dayton, Ohio, in October 1944. He suggested that the electrical industry should be viewed as a unit rather than as a number of isolated branches, and that the unity should be reflected in the financial arrangements that controlled the means of expansion as well as in engineering. He commented that engineers with imagination and faith needed to furnish leadership in pointing the direction of technological change, and he expressed concern that without effective leadership from engineers, the future might deteriorate into a virtual Tower of Babel, with each engineering specialty speaking a language not understood by others.[54]

He also stressed the increasing importance of electronics, asserting that more than 10 percent of all the electric energy produced in the United States in 1943 had passed through electronic devices. As an example of the importance of electronic thinking to nonelectronic devices, he cited the amplidyne, which employed the methods of analysis that had originally been developed for electronic amplifiers. Pointing out that electronic tubes had characteristics that forced the electrical engineer to accept a different point of view in design, he predicted that in the future electrical engineers would need to be familiar with the concepts and methods of electronic and radio engineering.[55]

In a November 1944 memorandum, Alexanderson commented on a proposal to award bonuses to GE employees who received patents. He suggested that such a policy would benefit GE by stimulating its engineers to be winning contestants in the "patent race." He speculated that the prospect of substantial rewards would keep the engineers constantly thinking and experimenting in new fields. GE had to take some risks to maintain its position of leadership in the industry, he asserted, and at present, the corporate executives had to push the engineers. Alexanderson wrote that he would like to see the company providing an incentive sufficient to cause the engineering staff to take the initiative in new fields, with the executives having the responsibility for applying the brakes when needed. He concluded by urging that a substantial bonus, on the order of $1,000, should be given when a patent was awarded. Early the following year, he was informed that GE had decided to increase the bonus for assigning a patent to the company—from $1 to $25.[56]

IN JANUARY 1945, on the eve of his 67th birthday, Alexanderson was awarded the Edison Medal, the highest honor given by the American Institute of Electrical Engineers. The award, presented at a joint meeting of the AIEE and the IRE in New York City, was in recognition of his approximately 300 patented inventions and his contributions to the development of "the radio, transportation, marine and power fields." In his remarks during the presentation ceremony, Alexanderson discussed the role of invention in the American social order. He commented that Edison and Marconi had created new industries, and he suggested that what the world needed was more engineer-inventors who could found new industries. He predicted that once the war was over, the new generation of engineers would revitalize the electrical industry with new tools and ideas.[57] Early the following month, Alexanderson was awarded the Cedergren Medal by Sweden's Royal Technical Institute.[58]

Television Revisited

During the 1940s, although he was no longer involved directly with the technical development of television, Alexanderson continued to be an interested observer of trends in the field. He served as a member of the National Television System Committee (NTSC), an industry organization that recommended standards for the television industry, and he attended the committee's meetings during the period from July 1940 to March 1941. As a member of the NTSC, he served on a panel that considered the design specifications of transmitters and receivers as part of an overall system.[59] In a talk given in January 1943, Alexanderson predicted that television would come into its own when it could secure a large enough audience, and that color television would follow. He anticipated that the growth of television would exert indirect effects on other fields. For example, he suggested, the development of microwave-relay techniques for television would lead to the development of airplane navigation systems such that a pilot would have three-dimensional displays to assist his vision.[60] In October 1943, he listed three-dimensional radio vision for airplanes and ships and a shortwave relay trunk-line system as two of his "pet dreams" that seemed ripe for practical development. In another talk, given in January 1944, he stated that an extensive shortwave relay network would enable programs originating in Hollywood to be presented by television in every American home.[61]

Alexanderson described radio as "the new fire which the gods have given to man" in a talk given in June 1944. He speculated that the history of radio seemed to indicate that inventions followed a zigzag course. Initially medium wavelengths had been used to communicate with ships, and later on, long waves had been introduced for trans-oceanic radio. Then medium waves had been adopted for radio broadcasting and short waves had replaced long waves for long-distance communication. Likewise, television first had been tried with medium and short waves, but it did not become practical until the advent of ultrashort waves. Alexanderson concluded by remarking that the beams of the "new fire" might yet open possibilities in power transmission which he hesitated to put into words because they might be misunderstood.[62] In a speech delivered to the Television Broadcasters Association in November 1944, Alexanderson commented on the possibility of international television, asserting that relay techniques should make it easy to link North and South America with television. He suggested that international television would serve to promote business relationships and to bridge language barriers.[63]

Alexanderson received only one patent on a television invention during the 1940s. The patent described a method of producing a simulated color picture by means of a rotating disk situated in front of a black-and-white display. It seems probable that the idea for this invention was stimulated by a sequential color television demonstration by Peter C. Goldmark for the members of the National Television System Committee in September 1940. The Goldmark system also used a rotating-disk filter to produce a color display.[64] Alexanderson sent a memorandum on color television to GE's Patent Department on September 24, 1940, and the application for his patent was filed in May 1941.[65]

Electronic Power Converters and Power Transmission Systems

Even before World War II ended, Alexanderson's principal interest shifted from gun-control systems to electronic power converters, and their applications to long-distance power transmission and to railroad electrification. In June 1944, Alexanderson and Earl L. Phillipi, a co-author, presented a paper on the electronic power converter at an AIEE meeting in Saint Louis. They began by reviewing briefly the 40-year history of technology that had begun with the mercury-arc rectifier, and asserted that finally, "this new industry is now well under way." The pace of development, they suggested, had been retarded by an unwillingness among engineers to accept transient faults in the tubes. Alexanderson and Phillipi contended that it would be much better for designers to accept such faults as almost inevitable and to render them harmless. They mentioned preventive measures that had been used in practical applications of power converters, and they commented that electronic methods of transforming direct current from high to low voltage were "from a technical point of view entirely practical," although there had not yet been a large-scale application.[66]

In February 1944, Alexanderson wrote a memorandum to GE's Patent Department concerning a new method for starting an electric locomotive. He stated that there was renewed interest in railroad electrification and that alternating-current locomotives were being used on the Pennsylvania Railroad. He noted that a defect of the AC locomotive was that it did not produce a high torque at standstill, a limitation that he hoped to remedy. Later he corresponded with Donald R. MacLeod of the Pennsylvania Railroad on the comparative features of DC and AC propulsion. MacLeod wrote that the principal argument in favor of a 3,000-volt DC system was that studies done so far indicated that it would cost less than AC electrification. He added that a very careful study of the entire operation was required before a final selection was made, for railroad managers had learned from experience that any new and untried system always developed unanticipated troubles that had to be overcome before it was put into use on a large scale.[67]

MacLeod continued the discussion of electric locomotives in a letter written in February 1945, in which he commented that an advantage of AC over DC was that several motors could be connected in parallel during acceleration with no sudden increase in tractive effort. He mentioned the difficulty of starting a train when the tracks were slick and stated that the railroad would like a 10-minute rating at standstill without damage to the propulsion motor. Alexanderson replied that GE was attempting to alleviate the handicap of not operating the AC motors at standstill for more than 15 seconds while retaining other advantages of AC. He stated that GE wanted to furnish equipment to test an electronic starting system on an existing Pennsylvania Railroad locomotive, because the experience gained from such a trial would help GE to design the electric locomotive of the future. Later, MacLeod and Alexanderson discussed the design of the regenerative braking system for an AC locomotive that would give an increasing braking effort as the speed increased.[68]

Alexanderson spoke on the new AC locomotive as viewed from the perspective of the long history of railroad electrification at a meeting of the GE Engineering Council in 1945. Recalling that he had started out as a designer of AC railway motors at GE more than 40 years before, he suggested that disputes between the advocates of DC and those of AC and the inability of the AC camp to decide on the best frequency had caused GE to decide to promote DC and to refuse even to bid on AC systems. He reported that Swedish railroads were operating with AC and giving an outstanding performance that was economically viable in spite of long lines and sparse traffic. Alexanderson asserted that GE finally had a well-designed AC locomotive but that AC propulsion would need the backing of management. Thus far, he concluded, the will to create an adequate AC propulsion system for railroads had been lacking within the company.[69]

Alexanderson enthusiastically continued to promote the development of a DC power transmission system. In a speech entitled "Power Electronics as a New Industry," delivered in October 1945, he stated that GE had operated a DC transmission line for several years and that the line had been an excellent proving ground for new tubes and methods. He reported that the New York Power and Light Company had recently become interested in a high-voltage DC system for the Schenectady streetcar system. If this project were undertaken, he continued, it would make possible a full-scale test of a new propulsion system with a 15,000-volt distribution system that employed an electronic power converter on a locomotive. A few months later, Alexanderson told a group of visiting engineers that GE expected to be able to design a DC power line to transfer power for distances of 500 to 1,000 miles through the use of power electronics.[70]

In May 1946, Alexanderson asked the GE Engineering Council to support his proposal for a $500,000 appropriation for the development of DC power transmission. In his remarks before the council, he reviewed the history of the DC line from Mechanicville to Schenectady and the modifications that had been made in it since 1936, and he stated that GE engineers now were ready to take the next step, which was to increase the transmission voltage to 100 kv. He noted that the tubes he and his colleagues were using were designed for 25 kv and that they would need to find out whether they could build tubes that would work at 100 kv in a system with the necessary stability and reliability. However, he asserted, experiments at the GE Research Lab and in Sweden left little doubt that with determination, a GE effort could succeed. He added that the project was needed to help GE to maintain its leadership in power electronics and power transmission. He estimated that half the total cost of the project would be expended on the new high-voltage tubes: $75,000 for research and development, $100,000 for the manufacture of 24 tubes, and $75,000 for engineering and testing. The remaining $250,000 was to cover the cost of transformers and various other apparatus that would be needed for the installation of the DC transmission system.[71]

A few days after the Engineering Council meeting, Alexanderson prepared a memorandum giving further details on the proposed project and the reasons why it should be funded. He mentioned the possibility of a new power plant on the St. Lawrence River and a major electrification of the New York Central Railroad. The two developments were likely to coincide, he commented, since the rail project would create a large demand for power. He added that a DC power line from the St. Lawrence into New York State might prove highly desirable, and estimated the cost of a 50,000-kw installation at only $4 per kilowatt. Conceding that the estimate might prove too optimistic, he pointed out that the facts needed for a sound judgment would not be available until GE had gained the necessary experience with a full-scale demonstration. He wrote that no calculation at present could determine whether long-distance power transmission would become a common feature of the electrical industry of the future, since the question involved imponderables that could not be evaluated until the technical aspects were investigated. He urged approval of the appropriation so that GE might keep pace with the rest of the world in the technology of DC power transmission.[72]

In another memorandum on DC power transmission written in May 1946, Alexanderson conceded that there was little demand for such a system at present but predicted that there would be greater demand in a few years. He noted that several uncertain factors, such as the conservation of natural resources and the state of employment in industry, might affect the future of the system. He stated that the overall objective of his proposed project was to place GE in a position such that it could take orders for a DC power system capable of transmitting 400,000 kw over a distance of several hundred miles.[73]

A few days later, Roy Muir wrote that he understood that Harry Winne (a GE vice-president) and others at GE were in favor of proceeding with a DC power project. He suggested that they get together with Alexanderson and arrange an appropriation. In July 1946, Alexanderson attended a meeting at GE on the development of a high-voltage tube for the DC power project. He urged the selection of a definite goal that could be achieved before the initial appropriation was expended. It would be easier to obtain additional funding, he explained, if there were some completed tests to present.[74] In late August, Winne wrote to David Prince, vice-president in charge of engineering, that GE's application engineers believed it to be problematic whether a DC system would be used but that the project should not entirely be terminated. Consequently, Winne wrote, he could not recommend an appropriation of $500,000. However, he suggested that funding for Alexanderson's project be continued at a reasonable rate.[75]

In addition to working on railroad electrification and DC power transmission, Alexanderson and his associates received a developmental contract from the navy in 1946 for work related to an electronically controlled propulsion system for ships. Alexanderson wrote a memorandum in February 1946 recommending that GE sign a contract with the navy to develop a new type of power-thyratron tube.[76] Later in the year, he suggested that the tube being developed under the navy contract might also be used in a power converter for an electric locomotive operated from a single-phase 60-Hz source. He pointed out that the navy tube was being designed to withstand mechanical shocks due to gunfire and the rolling of ships. The electronic ship propulsion contract was one of three projects listed in a report Alexanderson submitted to David Prince in January 1947. The same month, Alexanderson filed with GE's Patent Department a memorandum on electronic ship propulsion in which he mentioned that one objective of the new system was to reduce substantially the number of tubes required.[77]

In his January 1947 report to David Prince, Alexanderson mentioned that work was continuing on DC power transmission. A higher-voltage tube was under development and preparations were being made to test it, and a small-scale DC system had been set up in the laboratory for experiments. In addition, good progress was being made on a new system of railroad electrification that would combine the superior features of the DC motor with the advantage of power distribution at 60 Hz. Alexanderson predicted that the new locomotive would give better performance than a locomotive operated from a 25-Hz source.[78]

In keeping with his philosophy that alternative approaches should be explored simultaneously, Alexanderson did not neglect the possibility of improving long-distance AC power transmission by electronic methods. In a memorandum to GE's Patent Department written in February 1947, he proposed the use of an electronic regulator located in the power-generating plant to maintain stability in a power line 750 miles long. He stated that an important advantage of the electronic regulator, or stabilizer, was that a new type of generator would not be required for use with very long lines. He gave additional details on the electronic stabilizer in a memorandum written a few days later. Pointing out that AC generators normally had been regulated by controlling the field current but that his proposed method would instead vary the output current, Alexanderson explained that the regulator would simulate a reactive load on the generator which could be altered electronically to maintain the desired voltage and line current. In conclusion, he asserted that both theory and experimental tests indicated that the electronic stabilizer could remove the existing distance limitations on AC power transmission.[79]

In April 1947, Alexanderson reported that an electronic stabilizer had been tested successfully on a simulated 900-mile transmission line. He described the tests as quite interesting, since a condition of resonance between distributed inductance and capacitance had been encountered. He stated that the behavior of such a line was very different from that of a 300-mile line such as was used at present.[80] Tests also were run using the stabilizer with a model of a 600-mile line supplied with three-phase power. Alexanderson filed a patent application covering the electronic stabilizer in May 1947; the patent was issued two years later.[81]

Alexanderson and David Prince prepared a paper on an electronic stabilizer for presentation at an AIEE meeting in the summer of 1947. Several weeks before the AIEE meeting, Prince wrote to Philip Sporn, the eminent power engineer of the American Electric Power Company (formerly AG&E) and enclosed a draft of the paper. Prince admitted that he and Alexanderson were somewhat out of touch with the utility field and would welcome any suggestions. He asked whether Sporn knew where such a system could be tried. It had been tested in the laboratory, Prince explained, but the final proof of its performance would have to await a field test. He acknowledged that the economics of electronically stabilized power systems were still rather vague.[82]

Shortly before the AIEE meeting, Alexanderson wrote to Uno Lamm, a Swedish electrical engineer who he had learned was in the United States. Alexanderson invited Lamm to come to Schenectady and to accompany Alexanderson and his wife to Montreal, where the AIEE meeting was to be held. Lamm, an employee of the Swedish electrical manufacturing company, ASEA, accepted the invitation and discussed his work on the "transductor" with Alexanderson and other GE engineers in early June just before they traveled to Montreal. The contact with Lamm enabled Alexanderson to reestablish his neglected Swedish connection, and he found that Lamm shared his interest in long-distance DC power transmission. *Transductor* was Lamm's term for the magnetic amplifier, a device that Alexanderson had invented before

World War I, which was now being applied not in modulating radio alternators but in electric power applications. His discussions with Lamm would soon cause Alexanderson to renew his enthusiasm for the magnetic amplifier. Where the original application of the magnetic amplifier had been supplanted by the vacuum tube, Lamm's work gave Alexanderson reason to anticipate that the magnetic amplifier might enjoy a renaissance and might in turn supplant electronic tubes in some applications.[83]

In their paper, Alexanderson and Prince stated that they were suggesting a "new approach"—one that was "less radical than d-c transmission"—"to an old problem, the extension of electronic power transmission." They mentioned two methods older than the electronic stabilizer: synchronous condensers and series capacitors. They described their method as analogous to the rotary synchronous condenser, but with the advantages of having no moving parts and having a faster response to transient disturbances. They stated that the electronic stabilizer might be used either to increase the power capacity of a line of moderate length or as an essential additive to a long-line system with inherent instability, and they used graphical analysis to explain how the electronic stabilizer functioned as a "dummy inductive load" to maintain system stability automatically. They also included an example of a laboratory test of their system which had used the stabilizer with an artificial transmission line designed to simulate an actual line more than 600 miles long.[84]

During the discussion of the paper by Alexanderson and Prince, an engineer from the Bonneville Power Administration stated that the problem of stability on lines over 500 miles long was of great interest in the Northwest. He pointed out that comparative cost data on the electronically stabilized system and its possible alternatives were needed, since final decisions inevitably would be based on the comparative cost of transmission per unit of power transmitted.[85]

Uno Lamm, the Swedish engineer who had come to the meeting with Alexanderson, also commented on the paper. He stated that economic studies done in Sweden had shown that DC transmission would be more economical for distances greater than about 350 miles, and that its advantage would increase at a steep rate for still longer lines. He added that plans were being made to transmit 2 gigawatts over distances of up to 900 miles in Sweden. The necessary money should be spent to develop DC systems, he concluded, since so much would be saved in the long run. Lamm mentioned that an experimental DC system operating at 90 kv and transmitting 6,500 kw had been built as a cooperative venture between the Swedish government and ASEA.[86]

During the Montreal meeting, Lamm also presented a paper on the theory of the transductor, or magnetic amplifier. He had completed a doctoral thesis on the transductor in 1943. In his Montreal presentation, he reported that this device had come into wide use in Scandinavia during the past several years as a power amplifier and as a regulator of voltage and current. He pointed out that the transductor stabilizer was rugged and durable, and that power gains of the order of 10 million could be obtained through the use of feedback and with an input of less than 1 microwatt. He commented that former prejudice against the transductor due to its reputation for slow response was being overcome through better understanding of the device.[87]

In an August 1947 memorandum, Alexanderson suggested that an "electronic condenser" might be used as a substitute for the rotary synchronous condenser in AC power systems. He asserted that the electronic condenser would be superior in performance to the synchronous condenser and would be competitive economically. Like the electronic stabilizer that he and Prince had discussed in their Montreal paper, the electronic condenser would have no moving parts and would be almost instantaneous in its response to transients. A further advantage of the electronic condenser, he noted, was that it could be easily modified to serve as a phase balancer at the same time as it was acting as a stabilizer.[88]

In October 1947, Alexanderson recommended that the electronic regulation of power systems be discussed at a meeting of the GE Engineering Council. His proposal was accepted, and he prepared a presentation on the topic for the November meeting of the council. He reported that a Swedish governmental official had informed him that a power line 600 miles long, far longer than any existing line, was being undertaken in Sweden. Alexanderson stated that he and the Swedish engineers had agreed that the best contribution to the project that GE could make was to investigate electronic control using a model of the 600-mile line. Cost data now were available, he concluded, and GE's greatest need was to gain operating experience.[89]

The Magnetic Amplifier Revisited

Shortly after his encounter with Uno Lamm, Alexanderson resumed work on the magnetic amplifier, which he had invented but had not used for many years. Actually, he had revealed a degree of nostalgia concerning the device on at least one previous occasion. In response to a request in 1943 that he articulate some of his "pet dreams" of future developments, he included one "dream in reverse," which referred to the magnetic amplifier. Mentioning that the amplidyne was really an old device he had taken from the shelf and redesigned as a dynamo amplifier, he speculated that perhaps the time was ripe to take another old and reliable device, the magnetic amplifier, from the shelf. He went on to say that it had some advantages over both the electronic tube and the amplidyne because it had no vacuum, hot filament, or moving parts. He was a strong advocate of electronics when there was no better way, Alexanderson wrote, but it seemed that the faithful and reliable magnet had been somewhat neglected ever since Steinmetz had written his famous papers on magnetic hysteresis. Alexanderson concluded his dream in reverse by observing that although magnetics currently was not as glamorous as electronics, GE might someday have a "magnetronics department."[90]

In a September 1947 memorandum to GE's Patent Department, Alexanderson discussed some ideas on multiple-stage magnetic amplifiers with feedback. He commented that there was a resurgence of interest in the magnetic amplifier, noting that he expected it to replace both electronic and amplidyne control for many purposes. After looking over Lamm's recent work, he added, he had concluded that the original magnetic amplifiers used in radio had been designed with the correct principles. Alexanderson proposed a design improvement in the magnetic amplifier used for power frequency control. His improvement took the form of a tuned circuit connected to the control winding and made resonant at the second harmonic of the input frequency. He stated that the innovation would be useful in designing a cascade, or multiple-stage, magnetic amplifier with both high sensitivity and fast response. He noted that a three-stage amplifier with a gain of 1 million would tend to oscillate but that this oscillation might be prevented through the use of negative feedback. His memorandum included a sketch of a three-stage amplifier that used feedback. A few days later, he reported that the proposed improvements had been verified in laboratory tests.[91]

Also in September 1947, Alexanderson read with great interest a paper on recent developments in the field of magnetic amplifiers published in *Electronics*. The paper was written by Walter E. Greene, a naval officer assigned to the Office of Naval Research (ONR).[92] Crediting Alexanderson with having invented the magnetic amplifier and with having received a patent on it in 1916, Greene discussed some German, Japanese, and Swedish improvements in the theory and design of the magnetic amplifier which were causing the ONR and other groups to take a renewed interest in the discarded amplifier. He stated that interrogation of German prisoners had revealed that the German navy had used the magnetic amplifier in gun-control systems and had used it instead of vacuum tubes in electric computers. He mentioned that the Germans had used permalloy as a core and had used dry-disk rectifiers with the magnetic amplifier. In addition, the German army had used magnetic amplifiers in control of the V2 rocket, and some research had been done in Germany on the use of magnetic amplifiers in power transmission.[93]

Greene reported that Japan had developed new magnetic materials that had considerably higher permeability than any made in the United States and that improved the potential of the magnetic amplifier. He cited Uno Lamm's book on the transductor, published in 1943, as being the most complete treatise on the magnetic amplifier available. Greene predicted that the magnetic amplifier would enjoy much greater use in both military and industrial applications and revealed that the ONR was sponsoring a research effort in cooperation with the Bureau of Ships and the Bureau of Ordnance. He mentioned that Lamm had given lectures in the United States that had been sponsored by the navy. Greene concluded that Sweden probably was ahead of any other country in the use of the magnetic amplifier for power control systems.[94]

In a quarterly report submitted in October 1947 to Everett Lee, head of the General Engineering and Consulting Laboratory, Alexanderson stated that the subject of magnetic amplifiers was receiving much attention from the navy and from other organizations. He added that it had been learned that the Germans had used magnetic amplifiers for most purposes for which America had used the amplidyne. He and other GE engineers had begun an intensive study of magnetic amplifiers, he reported, and he expected them to be preferred for many purposes because they had no moving parts and because they were more rugged than electronic tubes. A few days later, Alexanderson informed GE's Patent Department that tests of a magnetic amplifier with regenerative feedback were under way. He explained that GE's objective was to develop an amplifier with a very fast response.[95] In another memorandum, written in late October 1947, he revealed plans to construct a multistage magnetic amplifier that would feature instantaneous positive feedback and delayed negative feedback in order to obtain both a fast response and stable operation.[96]

In a November 1947 memorandum, Alexanderson discussed the use of the magnetic amplifier in gun-control systems and in the control of automatic machine tools. He suggested that magnetic amplifier control should be superior to the amplidyne in both applications. He pointed out that the high-speed response that could be obtained with the magnetic amplifier should enable the design of contour-cutting machines of much greater speed.[97] Later the same month, he recalled that the magnetic amplifier once had been used to regulate the speed of a motor used to drive the Alexanderson radio alternator. After commenting that a limitation of early magnetic amplifiers had been that they had required a large control current, he went on to say that the most significant novelty in current practice was the use of feedback, which enabled the design of magnetic amplifiers with both low control current and quick response.[98]

Alexanderson used the magnetic amplifier as the theme of a talk before a meeting of GE engineers in December 1947. He recalled his "dream in reverse" that the magnetic amplifier would wake up from its 30-year sleep, and stated that his dream now had come true. The magnetic amplifier had been used to telephone across the ocean before electronic amplifiers were available, Alexanderson reminisced, but subsequently engineers had become excited by electronic devices, and the magnetic amplifier had gone to sleep. During the war, however, while American engineers were busy with radar and amplidynes, the magnetic amplifier had awakened in Germany and Sweden. Alexanderson speculated that it had been used in Germany because it had been needed for war purposes and in Sweden because Swedish engineers lacked experience with electronic amplifiers.[99] He reported that GE had now resumed its work on magnetic amplifiers and that he anticipated that magnetic amplifiers would replace electronic and amplidyne control to a great extent. He characterized the magnetic amplifier as the fastest control device known except for the high-vacuum electronic amplifier and predicted that it would stimulate the development of new methods with a broad field of applications. He concluded by urging his audience to support a policy under which new ideas would be given the benefit of the doubt and the company would be willing to take a "gambler's chance." He asserted that decisions were usually not based on an analysis of facts but ultimately rested on the reputation and past record of the company's engineers who sought the backing of management for new projects.[100]

A Reluctant Retirement

On January 1, 1948, as he approached his 70th birthday (January 25, 1948), Alexanderson officially retired from GE after almost 45 years with the company. A retirement dinner was held, and he received a congratulatory message from Owen Young, former chairman of GE's board of directors. Alexanderson responded to Young, who had been retired for several years, that an "antiquarian society" might be a good idea and that perhaps they should hold a "Camp Antiquarian" on Association Island. In reply to a message from GE's president, Charles Wilson, Alexanderson stated that he read into it "permission to continue the work which is my absorbing interest."[101]

In honor of Alexanderson's retirement, Albert Hull, an old friend from the GE Research Lab, composed a poem that read, in part:

How many tubes, that were my children,
Might soon have died at birth,
But for your vision of new circuits
In which they proved their worth

. . . .

How can the pension board be thinking
Your labors should be through?
The place you leave would still be vacant;
There are no more like you!

But I suspect that we shall see you
Continue 'til you drop
Inventing still, you can't be idle,
Inventors never stop![102]

As indicated in Hull's poem and in his own response to Wilson's message, Alexanderson did not really welcome retirement, and he soon arranged a special employment agreement so that he could continue work as a consultant to GE. Initially, he agreed to serve as a consultant for one year for a retainer of $3,000 plus travel expenses in addition to his retirement pension. He still enjoyed vigorous health and would continue as a GE consultant for the next four years.[103]

His most bothersome handicap was his impaired hearing, although he was helped considerably by an electronic hearing aid. He alluded to this in a talk given in 1944, when he included a personal comment on what electronics had meant to him: "It makes it possible for me to hear and that is worth whatever it costs." In another speech, given in 1946, he recounted an experience in which he had picked up a radio signal on his hearing aid and had been able to locate the center of the directional beam. He suggested that a pilot might be able to use this effect as a navigational aid and "hear his way down a beam."[104]

THE FIRST YEAR of Alexanderson's retirement began tragically when Gertrude, his wife of almost 34 years, died on January 25, 1948, his 70th birthday.[105] Alexanderson had eight grandchildren by 1948, including Verner's three-year-old son, who had been named for his grandfather. In addition to the consolation provided by his family, Alexanderson found solace in his favorite outdoor sport, sailing. In March 1948 he wrote to a friend that he expected to make more use of the *Nordic* during the summer.[106]

On June 14, 1949, Alexanderson was married for the third time, having been twice widowed. His third wife was Thyra Isberg Oxehufwud, who was Swedish by birth and still had relatives in Norrkoping.

Alexanderson was a cousin of Anders Oxehufwud, Thyra's former husband, and had served as best man at their wedding in 1925. The Oxehufwuds and Alexandersons had remained friends and exchanged visits over the intervening years. After their marriage, Thyra entered the circle of Alexanderson's friends in Schenectady, which included Larry Hawkins and Irving Langmuir and their wives. Christmas in the Alexanderson home on Adams Road became a particularly festive occasion, celebrated in "the Swedish way," with family and friends joining in traditional songs and enjoying Thyra's Swedish smorgasbord feast.[107]

Chapter Ten

"Inventors Never Stop!"

Alexanderson lived for nearly three more decades after his official retirement from General Electric at the end of 1947. Not surprisingly, he was not content to leave the arena of inventive problem solving, which had occupied him for almost 50 years, and spend all his time sailing at Lake George. As Albert Hull's poem in honor of his retirement suggested, Alexanderson had an addiction to the competitive sport of inventing which persisted into the ninth decade of his life. Despite his retirement, he continued to use GE as a workplace for 4 more years as a consultant until a final and still unwelcome separation occurred at the end of 1951. During this period he often expressed his concern over changes at GE that seemed to him to pose a threat to the continuation of the creative engineering tradition exemplified by Charles Steinmetz.

Although it was not yet public knowledge when Alexanderson retired, a remarkable new semiconductor device, the transistor, had just been invented at the Bell Telephone Laboratories in December 1947. When he learned of the properties of the transistor, Alexanderson soon recognized the potential of solid-state electronic devices in power converters. No longer having access to laboratory facilities at GE, he worked on power applications of semiconductors and magnetic amplifiers in a laboratory in his home, with the aid of his faithful assistant Albert Mittag. Thus the lifelong theme of inventive work at the interface between electrical power engineering and electronics was continued. Alexanderson remained remarkably opportunistic and adaptable in the face of a sequence of changes in electronics technology that affected the means but rarely the ends that were the focus of his professional life.

Views on Changes at GE and the Creative Engineering Tradition

During his four postretirement years as a GE consultant, Alexanderson often addressed the issue of changes at GE which he believed were affecting the creative engineering tradition adversely. In March 1948, in comments on a proposal to provide scholarships for young engineers, he wrote that he had found the environment at GE to be "more stimulating than any university." He did not think that giving someone nothing to do was the best way to stimulate growth, he added, and what GE should avoid was assigning so much to its outstanding engineers that they lacked the time to follow their instincts. Alexanderson recalled that he had been permitted to work on the radio alternator at a time when his regular assignment had been to design railroad motors: this had enabled him to combine valuable practical experience with some imaginative thinking. He mentioned that on another occasion GE's management had allowed him to try out a new idea by building a full-size locomotive to test the idea's practical value.[1]

He went on to say that it was his impression that the current generation of engineers at GE had been well chosen and had creative ability and imagination. However, he suggested, the pressures of routine work and a lack of appropriations for experimental work tended to prevent leading engineers and their junior assistants from adequately developing their potential. He recommended that if GE really wanted to develop and maintain a creative organization, it should "start from the top by investing liberally in the ideas of leading engineers." He urged that these engineers be encouraged to experiment and to make their own mistakes. Not only would this policy encourage junior engineers, he argued, it would also help the present leaders to discover talent. Alexanderson concluded by remarking that there had been a time when GE "was known to its customers through its engineers rather than through its salesmen," and that "a return to this position would be wholesome."[2]

Alexanderson received an opportunity to address the topic of research and engineering at GE publicly when he was invited to deliver a talk at his alma mater, the Royal Institute of Technology in Stockholm, in October 1948. This was a most welcome chance to return to his native Sweden, which he had last visited in 1929.[3] In his speech, Alexanderson recalled that his own career at GE had been spent in an area between engineering and science. He mentioned that Charles Steinmetz had provided him with guidance in the early years at GE, and that Steinmetz himself had become a symbol of the activity known as creative engineering, which linked engineering and science.[4]

Alexanderson asserted that the creative engineer and the scientist were of similar temperament but that the two professions were different enough in character that the organizational structure needed to keep them distinct, with separate responsibilities. Traditionally, he continued, engineering had found its home in industry and science in the university, but GE had taken a radical step in 1900 by establishing a link between science and industry in the form of a laboratory directed by prominent scientists who enjoyed "academic" freedom. He noted that although Steinmetz had played a major role in the founding of the GE Research Laboratory, he had chosen to remain in engineering.[5]

Alexanderson added that although he himself had enjoyed close contact with the GE Research Laboratory in every phase of his work, he had, like Steinmetz, remained in engineering. He asserted that there was good reason for the distinction between creative engineering and science, with the latter having as its objective the discovery of new knowledge, while the principal objective of engineering was the useful application of knowledge. Invention frequently resulted from the exchange of ideas between scientists and engineers, he observed, and the presence of a keen observer who could take advantage of unexpected or accidental occurrences often was important. He commented that the early leaders at GE, such as Edwin Rice and Albert Davis, had understood that the engineer-inventors needed ample freedom to follow their instincts.[6]

Alexanderson explained that many of his own inventions had resulted from the fact that he had been permitted to work both in the field of electric power and in radioelectronics. He pointed out that the evolution of ideas could be traced from AC generators to high-frequency generators and magnetic amplifiers used in radio, and eventually back to methods of electronic control of power transmission. He characterized specialization as the "modern enemy of creativeness," since "new ideas are born when different fields of knowledge are brought together." He argued that the best chance for creative thought occurred when a single mind could combine diversity of experience, and the next best chance was created by bringing about contact between individuals who were somewhat similar in temperament but had considerable diversity of experience.[7]

In April 1949, Alexanderson addressed a conference called to consider GE's policy on inventors. In his remarks, he argued that the apparent lack of understanding of the importance to GE of invention and creative engineering was of recent origin and ought to be corrected. He suggested that the decrease in the appreciation of invention was a result of the trend toward mass production, which was accompanied by regimentation and specialization. He called these "the enemies of creative thinking." Inventors could not be placed in narrow compartments and told what to invent, Alexanderson continued: they required the type of environment that had been traditional in universities and had been established at the GE Research Laboratory. He asserted that a similar lack of restraint had at one time also been company policy with respect to engineering. He stressed that he thought it quite important to GE that such "academic freedom be maintained in at least some corners of the engineering organization." He acknowledged that experimentation was expensive and that it would be a mistake to support every "crackpot who thinks he has made an invention." However, he commented, it was not desirable to scrutinize too closely the engineer's stated objectives when a new line of development was undertaken.[8]

The only thing that really counted was who was asking for an appropriation, Alexanderson continued, since it was a sufficient guarantee if the request came from someone with recognized ability and a record of good judgment and performance. Since "history is made of personalities," he recommended that GE's policy toward engineer-inventors be based on personalities. For the case of the young inventor without a record of achievement, he claimed, this would pose little problem, since the apprentice inventor would inevitably seek the help and guidance of an experienced inventor who would be sympathetic to his ideas. Alexanderson concluded that he had received this kind of assistance from Rice, Steinmetz, Emmet, and Potter, and had in turn helped others in similar fashion.[9]

In a memorandum written in September 1949 to GE vice-president David Prince, Alexanderson stated that he felt that it was essential for GE to maintain a separate group of engineers who had the experience and talent suited for innovation and creative engineering. He argued that these engineers should have facilities for experiments and enough security to be able to devote their time to creative thought without having to find an immediate customer. He noted that the group once known as the Consulting Engineering Laboratory had been organized in that manner. For such a group, he contended, it was very difficult to outline future work in specific detail, and a budget should be flexible enough to allow changes during the year. He pointed out that many possibilities could not be foreseen and that projects tended to be interrelated and overlapping. In conclusion, he urged that a consulting engineering program should be considered as a whole and that the total cost could best be estimated on the basis of the number of consulting engineers.[10]

At a meeting of the GE Engineering Council held in February 1950, Alexanderson gave a presentation on invention and the experimental approach in engineering. He asserted that "industry without invention is dead" and that the experimental approach was still "the only one that leads to new knowledge and new inventions." He used the metaphor of the bicycle, which only remains upright when it is moving, to illustrate the necessity of encouraging invention at GE. He credited scientists with having done an effective job of selling the importance of their profession and argued that "the inventors and creative engineers must do the same." He cited the case of the magnetic amplifier as evidence of the difficulty of predicting the practical applications of an invention. After 20 years of neglect, he noted, the magnetic amplifier was finding new applications and might become the solution to problems in long-distance power transmission.[11]

In a talk given in January 1951, Alexanderson drew a distinction between two types of invention. The object of the first type was to make a product better, cheaper, or more efficient. Such inventions often resulted in immediate profits and were given every encouragement by corporate management. The second type of invention was made with the objective of creating a new field and required imagination or intuition to evaluate initially. Alexanderson noted that the second type of invention often required 15 years or more to become important commercially. He suggested that the best protection against the frustration that affected many inventors was to have numerous interests. He cited some of his own experiences with inventions in the fields of railroad electrification and radio whose importance had not been recognized for many years.[12]

Alexanderson continued his critique of corporate policy on the support of the creative engineer-inventor in a letter to the business and financial editor of the *New York Herald* in September 1951. He again distinguished between two classes of invention and asserted that it had become difficult for the inventor to obtain backing for basic inventions that did not promise an immediate payoff. He speculated that this might be the result of increased emphasis on the efficiency of industrial production, with the decisions being dominated by sales policy. In contrast to the situation when he had first come to work at GE, he noted, creative engineering no longer was represented in top corporate management.[13]

He explained that although a person with creative ability had an inner urge to create which did not require artificial stimulation, financial backing and facilities were needed. A system designed to encourage invention should not try to judge the value of a specific invention, he added. Rather, it should attempt to locate and encourage the right kind of inventor. Management should only decide what it could afford to spend on pioneering development and invention and should not confine its support to a specific project. Acknowledging that any appropriation for research and development was somewhat of a gamble, Alexanderson asserted that it was a good gamble if the effort was led by a person of proven ability. He insisted that it took an inventor to identify and understand an inventor, since the creative mind functioned by means of shortcuts and hunches that were partly subconscious. As a result, he continued, inventors often were not able to give a clear explanation of what they had done and why. He recommended that a group of scientists and engineers with inventive credentials be established in each large corporation to take responsibility for invention and its encouragement.[14]

In a November 1951 memorandum, Alexanderson wrote what amounted to a valedictory at the end of his long affiliation with GE. He began by asserting that during the past 50 years, GE had changed from an organization dominated by pioneers in electrical engineering to an organization oriented toward manufacture and sales. He conceded that this evolution had been caused by changing times, but he called attention to the danger that the pioneering spirit would be extinguished. In the present context, he stated, an engineer was not considered a success until he had left engineering to become a manager, and as a result those who would have become engineering leaders 50 years ago were being diverted into other channels. He contended that present-day managers had little time to be concerned with pioneering work that might result in business in 10 or 20 years.[15]

Alexanderson wrote that he saw an opportunity for the "General Engineering Laboratory" to become an organization of engineer-inventors which could keep alive the pioneering tradition at GE. Such an organization, he noted, could not be created by executive action and money alone but had to be grown from the grass roots. He stressed that appropriations for such a unit should not be restricted to an objective recognized at the start but should be broad and flexible enough to accommodate changes in plan as work progressed. He contended that executive action should be limited to deciding how strong a pioneering organization the company was willing to support, and that the specific uses of resources should be left to the laboratory director and a staff of assistants or counselors. Pointing out that Bell Telephone Laboratories and the GE Research Laboratory had become great through such an arrangement, he speculated that if the General Engineering Laboratory were to be financed on the same principle as the GE Research Laboratory, the two in concert could well occupy the same position in the electrical industry that the Bell Telephone Laboratories did in the communication industry.[16]

He acknowledged that it often was difficult to draw a line between science and engineering and that some people might believe that the GE Research Laboratory filled the need for pioneer work at GE. However, he contended that the scientists had enough to occupy them, and he recalled that Langmuir and Coolidge had often asked him to take over responsibility for a new subject at a certain point. Alexanderson felt that the GE scientists did not want to take on engineering responsibilities, and he expressed the hope that top management would recognize the need for a pioneer organization in engineering. If his recommendation were implemented, he suggested, it should be done gradually, and the limit would not be what budget the company could afford but the limited availability of first-class talent. However, he predicted, when the opportunity became known, talented young engineers from the operating departments and from outside would present themselves. In the selection of staff for the engineering laboratory, he commented, the rule should be to "Find the right man and he will create his own job." Alexanderson asserted that there existed a need for an organization of pioneers in electrical engineering which would be equal in status and privileges to the GE Research Laboratory and the Bell Telephone Laboratories. He concluded by expressing his hope that the General Engineering Laboratory "[could] be developed to fill this place."[17]

A Regulator for a Swedish Transmission Line

Alexanderson spent approximately two months in Sweden in the fall of 1948, during which time he discussed with Swedish engineers their plan to install a long power transmission line from the northern part of the country to the more populated southern section. Soon after his return to Schenectady, he informed a GE executive that a 600-mile line, operated at 380 kv and rated at 285,000 kw, would be completed in Sweden by 1951. Alexanderson reported that he had given Swedish engineers the results of the GE tests on stabilized power transmission, and that they had accepted his recommendations as the most promising solution to their problem. Stabilizing apparatus would be installed at the midpoint of the Swedish line and would produce experimental data that could be used on future power lines; and Alexanderson, noting that credit might be available through the Marshall Plan, suggested that GE furnish the needed stabilizing equipment. He concluded by stating that he would like to begin work immediately on the project.[18]

In a January 1949 memorandum, Alexanderson wrote that he was convinced that he had a valid solution to the problem of long-distance power transmission but that the ultimate proof would come from the full-scale trials in Sweden. He stated that GE's principal incentive to secure the contract was the opportunity to make an important engineering contribution. He commented that the evolution of thought was slow, and he recalled that the evolution of railroad electrification had been killed by the controversy between the advocates of AC and those of DC. He urged patience with regard to the Swedish proposal.[19]

In March 1949, Alexanderson wrote to a Swedish official to report that since his return from Sweden he had been running tests on a model that simulated the Swedish line, and that the proposed design seemed quite practical.[20] And in February 1950, he wrote a memorandum on a "magnetic regulator" that would be capable of causing a 600-mile power line to operate like a line of half the length if the regulator was inserted at an intermediate location.[21]

In October 1950, he reported that the new magnetic regulator was capable of maintaining stable operation of a model line equivalent to a 1,200-mile line with a phase displacement of 120 degrees. The system's tendency to oscillate had been overcome, he stated, and a level of stability that previously had seemed feasible only with DC transmission was now possible for a long AC power line. He also reported the results of the magnetic regulator experiments to a contact in Sweden. He recommended that the Swedish engineers equip the 380-kv line with stabilized excitation at the generator, a magnetic amplifier stabilizer at the midpoint of the line, and a synchronous condenser with stabilizer at the receiving end. He asserted that with this design they should be able to operate at a level of 500,000 kw. In February 1951, Alexanderson and R. W. Kuenning filed for a patent on a power system stabilizer that employed the magnetic amplifier.[22]

As the year 1951 ended, Alexanderson reluctantly ended his consulting work at GE, which had been extended four years beyond his official retirement. In December 1951 he wrote to an old friend, John Hammond, Jr., that he had been permitted to continue work on special problems at GE for four years since his nominal retirement, but that he was now winding up his work because of GE's rigorous rule on age. He added that he expected to find other interests. A few weeks later, the president of the Mohawk Development Services (MDS), Lawrence A. Hawkins, who had retired from GE, wrote to Edwin H. Armstrong, an independent inventor, that Alexanderson had joined the staff of MDS as a consultant and that his services now were available to clients in areas related to his past experience.[23]

A Paper on Power Applications of the Transistor

Beginning in 1951, Alexanderson developed a strong interest in the transistor and its possible uses for the control of power machinery. He clipped an article from the *New York Times* on the junction transistor and read a tutorial article in the July 1951 issue of the *Proceedings of the IRE* which discussed the characteristics of the transistor and various circuit configurations suitable for its use.[24] He also obtained some technical information on the transistor from Ralph Bown of the Bell Telephone Laboratories. In April 1952, Alexanderson sent Bown a draft of a paper on power applications of the transistor for possible publication in the *Bell System Technical Journal*. Bown replied that he was uncertain whether a paper on power applications by a non-Bell-affiliated author would be accepted for this periodical, although otherwise he would have been glad to see it published.[25]

Alexanderson then decided to send his manuscript on the transistor to Alfred Goldsmith, another long-time friend, who was still the editor of the *Proceedings of the IRE*. In his cover letter, Alexanderson told Goldsmith that he was now retired from GE and had been studying the literature on transistors. He had written the enclosed paper, he continued, and was sending it to Goldsmith, who he presumed was still running the IRE. Alexanderson stated that the "transistor looks to me like the beginning of a new industry, like electronics was when you and I were getting into radio." Goldsmith responded that he would present the paper with a favorable recommendation to the board of editors. He commented that he had great faith in Alexanderson's intuition and believed that he again had taken a correct stand.[26]

Alexanderson's paper, "Control Applications of the Transistor," which was to be his last published professional paper, appeared in the *Proceedings of the IRE* in November 1952. He acknowledged that publications from the Bell Laboratories were his principal sources for the paper. However, he stated, he had extrapolated beyond published data to examine the possibility of using transistors as a substitute for controlled power rectifiers. On the basis of his analysis, he estimated an overall efficiency of 95 percent—which he termed "highly promising"—for a transistor used as a control device. He also worked through some examples in which the transistor was used to control the speed and direction of rotation of a DC motor operated from an AC source and to operate a variable-speed AC motor. He pointed out that radio signals might be used to achieve remote control of transistorized motors. Noting that the transistor had desirable properties that were not possessed by other control devices, he wrote, "The possibilities of polyphase and composite transistor circuits are too numerous for exploration at this time." He speculated that "a new branch of industry and research" would emerge if transistors capable of fairly high power could be developed. He expressed the hope that his paper might provide "an incentive for such a development."[27]

Magnetic Amplifier Applications

In September 1952, Alexanderson signed an agreement to act as a consultant to RCA, the company he once had served as chief engineer. In June 1953 he contacted Charles B. Jolliffe of RCA concerning a patent application on a push-pull magnetic amplifier used with an auxiliary transistor amplifier. Alexanderson explained that with the addition of the transistor, the magnetic amplifier could compete with the amplidyne and would offer the advantage of having no moving parts.

During 1953, he filed successful patent applications for the push-pull magnetic amplifier and for a magnetic amplifier used for motor control.[28] He also discussed with RCA's Patent Department an invention that would use a pictorial display to aid an aircraft landing in fog.[29]

In April 1953, Alexanderson informed a Swedish engineer that he no longer worked for GE and had no test facilities. He was now a consultant to RCA, he continued, and had returned to his interest in radio and electronics. However, he added, he could not keep from thinking about problems in power transmission.[30]

During 1953, Alexanderson was invited to write a brief account of the early use of the magnetic amplifier in radio for inclusion in a proposed book on the theory and applications of magnetic amplifiers.[31] The principal author of the book was Herbert F. Storm of the GE General Engineering Laboratory, although chapters were contributed by several other GE authors. In the preface to the book, published in 1955, Storm credited Alexanderson with the first significant application of the magnetic amplifier. Storm attributed the subsequent "eclipse of interest" to vacuum tubes and the amplidyne. He mentioned that there had been two important developments during the 1930s: improved magnetic materials and metallic rectifiers. He reported that these innovations had elevated the performance of magnetic amplifiers to "astonishing levels" and permitted applications that had not been feasible before. He pointed out that the new magnetic amplifiers could be made smaller than vacuum-tube amplifiers, required no warm-up time, could withstand severe vibration and shock, and had no moving parts. He stated that magnetic amplifiers were now widely used both in industry and in military applications, although they posed challenging analytical difficulties because of their nonlinear behavior.[32]

In his chapter, entitled "Magnetic Amplifiers for Radiotelephony," Alexanderson recalled that he had originally used the magnetic amplifier between the microphone and a 2-kw radio alternator before practical electronic amplifiers were available. He included and discussed a circuit diagram of the arrangement that had been used in 1915 to transmit radiotelephone signals between Schenectady and Alfred Goldsmith's laboratory in New York City. He explained how tuning capacitors had been combined with the magnetic amplifier to increase its speed of response. Alexanderson then went into the design of his 200-kw alternator transmitter, which had utilized both electronic and magnetic amplifiers between the microphone and the alternator. He also gave a brief discussion of how the frequency transmitted had been controlled with great precision with a circuit that had used the magnetic amplifier and a mercury-arc rectifier.[33]

During 1954 Alexanderson filed several successful patent applications, and all of the inventions were applications of magnetic amplifiers. In March 1954, he applied for a patent on a motor-control system that employed a two-stage magnetic amplifier as well as two field windings that could change the direction of rotation of the motor.[34] In July 1954, he filed applications for four inventions, all of which utilized the magnetic amplifier to control AC or DC motors. His final two applications for 1954 were filed in December, with the first covering magnetic amplifier control of a polyphase motor.[35]

On Christmas eve 1954, Alexanderson applied for a patent on a simple "magnetic computer," which used magnetic amplifiers to combine signals that were related mathematically. As an example of how such a computer might be used, he suggested that it could be part of a collision avoidance system for aircraft. He explained that a voltage x could be differentiated by the computer to provide an indication of the rate of approach, given in terms of dx/dt.[36] The magnetic computer patent application was Alexanderson's last successful patent application for almost five years.

A Talk at the Babson Institute and a Television Patent

In October 1954, Alexanderson accepted an invitation from Roger W. Babson to speak at the Babson Institute on the theme "Looking Ahead 50 Years." In his letter of invitation Babson mentioned that John Hammond, Jr., had recommended Alexanderson for the talk.[37] In his speech, given on November 6, 1954, Alexanderson stated that social factors as well as technical possibilities needed to be considered in speculations about the future, for science and invention had altered the social system and had taught the lesson that "the inventor has a social responsibility." He asserted that industries established during the past 50 years fell into two main categories. The first included the airplane industry and the electric power industry, both based on scientific principles discovered in the nineteenth century, while the second included radio and electronics, which were largely based on scientific principles discovered in the twentieth century. Alexanderson predicted that the dominant inventions of the future would be in the fields of electronics and chemistry. He singled out the transistor as an important recent invention whose "full significance is still obscure," although he anticipated that it would "undoubtedly make possible radical improvement and simplification of the computer or 'electronic brain.'"[38]

He went on to say that the inventor was "at heart an individualist but inventing is becoming more and more a cooperative effort." He observed that the inventor who worked for a corporation was not apt to receive great financial rewards but might "find congenial associates and inspiration as well as peace of mind." He cited Fessenden and Armstrong as examples of inventors whose individualism had perhaps been "too dominant." Alexanderson forecast that during the next half century the process of invention would be even more a cooperative effort, although all inventive initiative would not come from big business. He termed "depressing" the thought that wars and weapons had always served as great incentives to science and invention, but noted that he could see no good solution to the problem. He ended the talk by speculating on whether the existing social system and government could adjust sufficiently to the great changes that were taking place, and whether the scientist and the inventor could adapt to the "psychological and ethical demands of human nature."[39]

A few weeks after his talk at the Babson Institute, Alexanderson wrote to Roger Babson to comment on the latter's idea that television might be used for shopping. Alexanderson expressed agreement and predicted that it probably would not be long before many telephones would have a television attached, to be used for shopping and other purposes.[40]

In February 1955, the *New York Times* reported that Alexanderson had just been granted a patent on a color television receiver. The article stated that the invention was intended for use with the RCA system and would receive black-and-white programs as well as color. Some biographical information on the inventor was included. Alexanderson was credited with the first home reception of television in 1927, and with having received a patent on an average of every seven weeks throughout his career at GE. The article also mentioned that he was the father of yachting at Lake George and an enthusiastic canasta player.[41]

Alexanderson revealed his keen sense of humor in a tongue-in-cheek characterization of the plots of popular television programs in a November 1960 letter to John Crosby of the *New York Herald Tribune*. The plots of the four most popular shows, Alexanderson suggested, could be summarized as follows:

Fist fight and female and guns with horse and wagon as a
    background.
Fist fight and female and guns with a river boat as a
    background.
Fist fight and female and guns with Indians and horses as a
    background.
Fist fight and female and guns with horses and cows as a
    background.

He speculated that the popularity of these programs probably was due to a desire to escape from modern art, modern music, and the cold war, and return to fundamental emotions. He concluded that future television plots might shift, for example, to feature Vikings and find a substitute for fist fights, but he predicted that no substitute would be found for females.[42]

Another Trip to Sweden: A DC Transmission Project

Alexanderson and his wife returned to Sweden for a visit in the summer of 1955. He took the opportunity to observe a pioneering Swedish project for the long-distance DC transmission of power.[43] Uno Lamm, the Swedish engineer who had visited Alexanderson in Schenectady in 1947, had played a major role in the project, which used a 60-mile-long submarine cable to carry DC power from the Swedish mainland to the island of Gotland. A mercury-arc electronic power converter was used at the mainland end of the cable to convert from alternating current to 100-kv direct current. Another electronic converter, located at the island terminal of the cable, changed the DC back to AC for local distribution on the island, which had a population of about 60,000 and lacked adequate local sources of electric power. In a paper on the design of the project, Lamm described it as the first large-scale DC link between two remote AC systems.[44] Developmental work on the electronic converters for the Gotland cable had been carried out during the 1940s as a joint venture of the Swedish government and ASEA, the company that employed Lamm. The cable had been laid in June 1953 by a British cable ship, and transmission of power to Gotland had begun in March 1954.[45]

Soon after his return from Sweden, Alexanderson submitted a report to Charles Jolliffe of RCA on what he had learned. Alexanderson described the Gotland power system as the "largest and most modern application" yet made of power electronics. He mentioned that the frequency regulation was maintained by a radio link, and termed the whole project a "striking example of how communications and power engineering are becoming more and more interwoven by the use of remote automatic control." He also discussed Sweden's 1,000-mile-long AC power line, which was operated at 380 kv and could carry up to 600,000 kw. He noted that a technique had been developed to clear faults on the line by remote radio control. He asserted that Sweden now led the world in the technology of electric power transmission and that Swedish engineers believed that "electronics is the key to future development and is shaping the thought in power engineering." Alexanderson reported that he had told the Swedish engineers that he believed that great improvements in the design of AC traction motors could be made through the use of silicon diodes. He wrote that recent developments in the area of solid-state rectifiers were not yet known to the European engineers.[46]

Some Hearing-Aid Experiments

As his hearing loss became increasingly severe, one application of the transistor that proved of great personal interest to Alexanderson was the transistorized hearing aid. In November 1950, he wrote to an engineer at the Zenith Radio Corporation about his experience with a Zenith hearing aid. Alexanderson stated that although he suffered from an extreme loss and could not understand speech from a radio loudspeaker no matter how high the volume was set, the aid enabled him to understand the speech of a person a few feet away, albeit with difficulty. He suggested that he might be helped if a suppression circuit that could be tuned to select the frequency suppressed were added to the aid. During 1955 he corresponded with Zenith about transistorized aids that were under development, and he was among the first to use one. In October 1956 he complained to a Zenith vice-president that he had to set the aid at maximum power.[47]

In 1959, Alexanderson wrote to the Zenith engineering department that he was wearing two aids because one ear had a defective frequency spectrum. He mentioned that he would like to try a powerful bone conduction instrument. A few days later, he was provided with a Zenith bone conduction aid.[48] In June 1960, he provided a local hearing-aid representative with the specifications for a modified cord that incorporated a 500-ohm resistor in one lead. He had measured his hearing characteristics, he explained, and the added resistance would serve as an equalizer and as a high-pass filter for the left ear. After receiving the modified hearing-aid cord, he wrote that it had performed exactly like his experimental cord and provided sufficient sound without squealing, together with longer battery life. He suggested that others with problems like his might benefit from such a modification and explained how a cord could be modified to fit the needs of each individual by a simple test using a calibrated variable resistor in one lead.[49]

Alexanderson undertook an experimental self-analysis of his hearing problem in 1963. He wrote that he had done some tests that were more revealing than a conventional audiogram. One test had been done by using a piano and no hearing aid; he had heard no sound except three octaves above middle C. When all the keys were struck hard, he had heard only a pitch of 2,400 Hz. In a second test he reported having heard the sound of running water from a faucet without a peak at 2,400 Hz. In a third test he had listened to the sound of a grandfather clock, which he reported had sounded natural without a 2,400-Hz tone. On the basis of his tests, Alexanderson deduced that his left ear behaved much like a radio tuned to 2,400 Hz with a detector that responded to amplitude modulation of the carrier frequency. He concluded by remarking that he would like to try using a filter that cut off sharply at 2,000 Hz or a wave-trap circuit tuned to 2,400 Hz.[50]

Research on the Use of Semiconductors in Motor Control

Alexanderson expressed a pragmatic philosophy on the role of theory in science in a letter to his old friend Samuel Robinson, a retired admiral who was living in Houston, Texas. Recalling that "in the good old days" a theory of ether wheels had been used to explain electromagnetic waves, Alexanderson suggested that greater scope was left for the imagination if it was assumed that ideas did not represent reality but were only "convenient interpretations." He remarked that his laboratory work now went at a more "pedestrian" pace but that he was "keeping at it" and enjoyed having some real interests. In January 1958, he wrote to Robinson about his research on the use of semiconductors in power applications. His objective, he stated, was to use semiconductor diodes to do many things that had formerly been done with vacuum tubes and amplidynes. He mentioned that he had been experimenting with an electronic torque amplifier that worked on the principle of a DC motor but used semiconductor diodes instead of a commutator.[51] In reply, Robinson recalled an occasion when he had been visited by Alexanderson and they had discussed "transistors by the hour."[52]

In the fall of 1960 Alexanderson suffered another personal loss with the death of his most faithful assistant, Albert Mittag, who had worked for Alexanderson during most of his career of over 40 years at GE and in Alexanderson's private laboratory after their retirement. In early October, Alexanderson wrote to Samuel Robinson that he was still working regularly in his laboratory but that Mittag had died of liver cancer only two months after coming to help in the lab as usual on July 1.[53] In a letter to Mittag's widow, Alexanderson wrote that he and Mittag had been close associates for most of their professional lives. He praised Mittag's skill and "penetrating accuracy" as having been very valuable, and he recalled that he and Mittag had accompanied Samuel Robinson on the first voyage of the *New Mexico* after they had collaborated on the design of its electric propulsion system.[54]

Alexanderson's final cluster of inventions, made during the 1960s, were in effect adaptations of some of his earlier inventions involving electronic control of electric motors. In essence, he undertook to employ recently developed semiconductor devices, such as the silicon-controlled rectifier, for functions that had previously been performed by thyratrons or vacuum power tubes. He had acquired a half-dozen semiconductor triodes in July 1959, and by August of the following year he was ready to begin the process of filing patent applications on some modified control circuits. In a letter to his patent attorney, he discussed filing an application covering a "system of position control." Explaining that the invention was intended to achieve position control by means of electronics and a motor that would not have a commutator or other contacts that could cause sparks, Alexanderson pointed out that there was a need for a nonsparking position control in such fields as aviation, atomic reactors, and the chemical industry.[55] In November 1960, he reported that he was constructing a prototype of the position-control system and that it might be used in tracking earth satellites as well as in industrial applications.[56]

In an October 1960 letter to John H. Clough, a GE engineer, Alexanderson commented on recent achievements in the design of electric locomotives using mercury arc rectifiers. He predicted that the next step would be to replace the mercury-arc devices with semiconductor devices, and he speculated that his own simplified system of control might then be looked on with favor.[57] In December, Alexanderson recorded that his position-control system had been tested successfully. He mentioned that he had encountered a problem—some of the semiconductor triodes had tended to fire out of proper sequence—but he had finally isolated the cause of the difficulty as a transient pulse in the control transformer and had eliminated the effect by means of a capacitor. He wrote to Clough to complain that he was running out of semiconductor triodes and to request assistance in getting replacements.[58]

By January 1961, Alexanderson had decided to apply for two additional patents. He wrote to Clough that he believed that the three inventions constituted a "major achievement." Alexanderson described the first invention as a system of reversible torque and speed control; the second as a system of firing semiconductor triodes which was generally applicable in power electronics; and the third as part of a system for control of speed, torque, and position. He commented that he had overcome several hurdles and had only recently managed to convince himself that he was not trying to do something impossible. Now, however, he could demonstrate a smooth, stable control system of a new type and could show the equivalent of tracking a radar target or earth satellite. He characterized the three inventions as being a "definite step ahead in the art of electronic motor control" and expressed his belief that GE certainly should be interested in them.[59]

In April 1961, Alexanderson filed two patent applications, one of which was later abandoned. A patent that was issued in August 1962 was described as a further development of an invention described in a patent application filed in 1959. Where the earlier invention had employed a magnetic amplifier as part of the power circuit of a motor, the

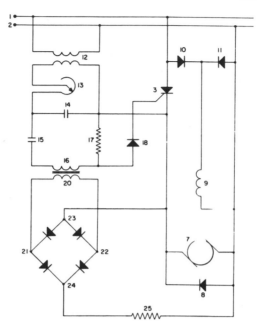

Figure 10.1. Alexanderson's last patented invention: an adjustable-speed motor-control system using semiconductor devices. A silicon-controlled rectifier triode is designated *3* in the patent drawing, and the motor armature is shown as *7*. From Ernst F. W. Alexanderson, U.S. Patent no. 3,736,481, issued May 29, 1973.

later one substituted semiconductor triodes. However, a magnetic amplifier was still used in the control circuit.[60]

In June 1961, Alexanderson filed a patent application entitled "Electric Motor Control System." This invention, he explained, was related closely to an invention for reproducing position covered by his patent filed in 1932 and used during World War II, although the earlier system had used thyratrons. He added that semiconductor triodes of the "thyratron type" had certain characteristics that made them more desirable than actual thyratron tubes. He also mentioned that as an alternative to thyratrons he was using devices known as silicon-controlled rectifiers, which were made by GE, and that he was using magnetic amplifiers to time the firing of the rectifiers.[61]

Alexanderson's final patent application, entitled "Adjustable Speed Motor Control System," was filed in March 1968, during his ninetieth year, and the patent was issued in May 1973, only two years before his death. This invention also used silicon-controlled rectifiers and a magnetic control circuit. As a synthesis of one of his first inventions—the variable-speed motor—and the latest semiconductor electronic devices, it constituted a virtual paradigm of Alexanderson's long inventive career.[62] It is a testament to his creative imagination and his determination that he was able to master the idiosyncrasies of silicon-controlled rectifiers and achieve this synthesis while in his eighties.

In February 1961, Alexanderson wrote to John Hammond, Jr., about his work on motors controlled by semiconductors and implied that GE did not seem interested in it. Alexanderson concluded that the real problem for the outside inventor was to find a way to penetrate the "hard shell with which corporations surround themselves."[63] Later in the year, Hammond asked Alexanderson to recommend someone qualified to work on a system of very-low-frequency communication with Polaris submarines. Alexanderson suggested two of his former associates, Earl Phillipi and Harold Beverage, as possible candidates. He mentioned that Phillipi had spent four years at a naval research facility at Key West, where he had experimented with using a frequency of 42 kHz to communicate with submarines. Hammond promised to get in touch with Phillipi.[64]

Some Thoughts on Inventors and Future Developments

Early in 1962, Carter Davidson, president of Union College, asked Alexanderson to comment on a proposal to establish a chair of mechanical engineering which would focus on invention. Alexanderson replied that he found the idea interesting and that he thought a radically different approach was needed in education for creativity and invention. However, he observed, practical invention generally was the result of a team effort, and the "ideal team" should include inventors, scholars, and engineers. He explained that the scholarly type was strong in analysis while the inventor tended to proceed synthetically and intuitively. He used Edison as an example of a synthesizer and Steinmetz as an example of one who had been outstanding in analysis. He concluded by stressing that it was important to have facilities for the construction of experimental models.[65]

In 1968, Alexanderson wrote a short essay entitled "Inventors I Have Known," which was published in 1972 in a book by Philip Alger. Alexanderson began his essay by characterizing invention as "one of the creative arts," but he noted that unlike the painter or composer, the inventor needed to learn to cooperate with the business executive, for "industrial organizations need the services of both." He reminisced about some engineer-inventors such as Edwin Rice, William Emmet, Reginald Fessenden, and Guglielmo Marconi, and recalled some highlights of his own career. He also commented that some technological

Figure 10.2. Alexanderson attending the Charles P. Steinmetz centennial luncheon in April 1965. Alexanderson is seated in front of a portrait of Steinmetz. Courtesy of the General Electric Company, Schenectady, N.Y.

systems on which he had worked, such as DC power transmission and railroad electrification, had been quite slow to develop, especially in the United States.[66]

However, Alexanderson pointed out, these systems had begun to be used elsewhere. He mentioned the DC cable from Sweden to Gotland which he had seen go into service in 1955, and a subsequent DC cable across the English Channel. Although the diesel-electric locomotive had dominated in the United States, Alexanderson predicted that "when the oil gets scarce the Diesel engines may be replaced by electronic converters." He speculated that electronic converters might also come into use to make possible the use of such exotic sources of energy as the ocean tides and solar power. He concluded the essay by commenting on the patent system. He felt that it probably had failed to fulfil its original purpose of protecting the lone inventor. However, he argued, it would be a mistake to abandon the system, because it protected "the institutions which favor invention." It was his view that "the creative mind needs a safe and resourceful place for action and co-operation."[67]

The Final Years of an Engineer-Inventor

During the last few years of his life, Alexanderson suffered from pernicious anemia as well as deafness and failing eyesight. The latter afflictions must have been especially frustrating to one who had spent much of his life in the development of communications systems. When I interviewed him in October 1972, he could only respond to questions presented in the form of enlarged block-letter printing, which he read with the aid of a magnifying glass. He used a combination of lenses to read a clock face across the room. His mind remained lucid, and he retained an extraordinary recall of details of his life and of episodes that he could "remember so well." On one occasion he recited several stanzas of Latin verse he had apparently learned as a child as the son of a classical-language scholar. At the time, he was eagerly awaiting the issuance of his final patent.

Alexanderson died in his 97th year on May 14, 1975, and was survived by his wife, Thyra, four children, nine grandchildren, and five great-grandchildren.[68] Thyra Alexanderson recalled that their life of over a quarter-century together had been "happy years." She also wrote: "Alex was a remarkable man. When he left us, we all realized it was the end of an era."[69]

Epilogue

One may well ask, as did an early reader of this biography, why Alexanderson is less well known than some other inventors such as Marconi, de Forest, Zworykin, and Farnsworth. The reader went on to speculate that a partial answer might be that some of Alexanderson's major technological systems, such as mechanical-scan television and the radio alternator, were supplanted by other technologies that were more similar to current systems. This obsolescence effect may be seen to operate in technology as it does in science, where the creators of supplanted paradigms tend to receive less attention, but there probably are other factors at work as well.

Although most people probably could identify Morse, Edison, Bell, and Marconi as heroic inventors, I am less certain than the above-mentioned reader about the public's cultural literacy about such inventors as Farnsworth and Zworykin. At least among professional engineers, Alexanderson still has name recognition, as indicated in a survey of readers of the *IEEE Spectrum* published in 1984.[1] In the survey, Alexanderson was tied with Zworykin for sixth place among the top 10 technical contributors in the period 1900–1939. Farnsworth did not make the list, but it did include de Forest, Steinmetz, Armstrong, Marconi, and Edison. It would be interesting to compare these results with those of a survey of a more representative cross-section of the public.

At an Edison Centennial Symposium in 1979, David A. Hounshell presented a paper entitled "The Inventor as Hero in American History," in which he discussed why twentieth-century inventors are less apt to gain public acclaim and "heroic" status than those who came before. Hounshell suggested that the changing perception reflected qualitative changes in Americans' values and attitudes toward technology. These changes seem connected to changes in the inventors' environment as many of them became corporate employees. The rewards they sought no longer were the national fame and honor that had motivated earlier inventors. Hounshell concluded, "The end of the heroic age of invention suggests that Americans began to realize that technology had become a corporate affair which precluded the identification and, therefore, celebration of the dominant individual technologist."[2]

Even Edison's public image has changed substantially during the twentieth century, as Wyn Wachhorst explained in his book on the mythology of Edison. Wachhorst found that the amount of printed material on Edison reached a pronounced peak in the 1920s and then declined steadily, and he commented that the "descending slope . . . closely approximates the fate of all traditional heroes in a modern, fragmented, demythologized culture." This seems to be a reflection of a change from the "age of Edison," when most Americans admired "idols of production" such as industrial leaders and inventors. More recently, there has been a shift to "idols of consumption": celebrities who are more likely to be stars of sports, music, or television than inventor-engineers.[3]

In a paper on long-term patterns of research and development at General Electric, George Wise gave a perceptive analysis of how Steinmetz and Alexanderson sought to perpetuate the "Edisonian style" in a corporate environment that tended to favor a style associated with the name of Elihu Thomson. According to Wise, these two styles became "the thesis and antithesis from which GE R&D would be synthesized." The Thomson style tended to place more stress on teamwork and incremental improvement of product and process, whereas the Edison style was more independent and risk oriented. The Edison approach tended to be more conducive to major new initiatives. Wise credits Alexanderson's developmental group with being "one of the most productive since Edison's" among the several research and development organizations formed at GE. Wise credits Steinmetz and Whitney for their role in providing protection and support for certain "subordinates who displayed Edisonian originality and independence, such as Alexanderson and Langmuir."[4] In turn, Alexanderson assumed this role for those who worked on projects that he defined after the passing of Steinmetz.

Even in the post–World War II climate, Wise notes, "the predominant Thomson style of GE technology was balanced by an element of independence based on technical prestige." However, by the late 1970s, an in-house study of research and development policy at GE "clearly indicated that the light of the Edison-Steinmetz-Alexanderson-Emmet-Langmuir heritage had dimmed in an era of decentralization and strategic planning." Wise concluded with the contention that many of the more outstanding achievements at GE occurred "only because occasional injections of the Edison spirit have set up creative tensions between the methods of today and the opportunities of tomorrow."[5]

Developments in the technology of power electronics since Alexanderson's death have confirmed his confident expectations of a bright future for solid-state power converters in such fields as DC power transmission, electric locomotives, and variable-speed industrial motors. In 1976, the year after his death, the first permanent tie between the eastern and western power grids was made at "the electrical continental divide" in Stegall, Nebraska. The nonsynchronous DC link used a solid-state power converter designed at GE and used silicon-controlled rectifiers, otherwise known as thyristors.[6]

The first thyristor-controlled electric locomotive began regular operation in Sweden in 1964. This type of design has demonstrated superior characteristics of adhesion and smooth acceleration in comparative tests and has been quite successful in Europe.[7] As Alexanderson had anticipated, increases in the cost of diesel fuel made the replacement of diesel-electric locomotives by thyristor locomotives seem more attractive by the early 1970s.[8]

The development of solid-state-controlled industrial motors of the type covered by Alexanderson's last cluster of patents also has been stimulated by efforts to conserve energy. The use of solid-state power converters and microprocessors in the control of electric motors makes possible a substantial reduction in the motor's energy consumption. (Electric motors account for almost two-thirds of American electric energy consumption.)[9]

In his recent book *American Genesis*, Thomas P. Hughes characterized the century from 1870 to 1970 as having been "an era of technological enthusiasm in the United States." He anticipates that future historians may decide that this century "was the most characteristic and impressively achieving century in the nation's history, an era comparable to the Renaissance." Hughes wrote about some of the inventors and engineers whose "values of order, system, and control . . . have become the values of modern technological culture."[10] Alexanderson's long life—almost a century—closely conforms to the century examined by Hughes, and Alexanderson's professional career was long enough to cross most of the chronological boundaries in Hughes's analysis. I hope that this biography will serve to add Alexanderson to the list of major contributors to electrotechnology during America's century of "technological enthusiasm."

Appendix

A. Alexanderson's Patent Applications by Type, 1903–1968

Year	Electric Power Machinery and Systems	Radio Devices, Circuits, and Systems	Power Electronics Circuits and Systems	Facsimile, Television, and Radar	Miscellaneous	Total
1903	2					2
1904	7					7
1905	12	1				13
1906	7					7
1907	16	1				17
1908	10	1				11
1909	4	2				6
1910	4	1				5
1911	5	3			1	9
1912	2	3				5
1913	6	3			2	11
1914	9	3				12
1915	6	2				8
1916	7	9				16
1917	2	10				12
1918	1	12				13
1919	1	5				6
1920	6	1				7
1921		6	1			7
1922	1	4	1			6
1923	2	3	3	1		9
1924	3					3
1925	2	4	7			13
1926	2	1	7	3	1	14
1927	2	4	3			9
1928	2	3	1	4		10
1929	2	1				3
1930				2		2
1931		3	4	2		9
1932	1		3			4
1933			4	2		6
1934					1	1
1935	2					2
1936			8		1	9
1937		1	7			8
1938		1	7			8
1939	2	2	5			9
1940	2		2			4
1941			3	1		4
1942	6					6
1943	1			2		3
1944	2					2
1945				1		1

Year	Electric Power Machinery and Systems	Radio Devices, Circuits, and Systems	Power Electronics Circuits and Systems	Facsimile, Television, and Radar	Miscellaneous	Total
1947			2			2
1949	1		1			2
1951			5			5
1952			1	1		2
1953	3					3
1954	6			1		7
1959	1					1
1961			2			2
1968			1			1
Total	150	90	78	20	6	344

B. Patents Issued to Alexanderson

Date Filed	Title	Date Issued	Patent No.
July 16, 1903	Protective Means for Alternating-Current System	May 1, 1906	819,627
July 23, 1903	Motor Winding	Jan. 15, 1907	841,609
Feb. 15, 1904	Alternating-Current Motor	Dec. 17, 1907	873,702
Aug. 12, 1904	Synchronous Motor	Mar. 21, 1905	785,532
Aug. 13, 1904	Two-Speed Winding for Three-Phase Motors	Mar. 21, 1905	785,533
Aug. 19, 1904	Winding for Three-Phase Motor	Mar. 28, 1905	785,995
Aug. 19, 1904	Alternating-Current Generator	May 9, 1905	789,476
Sept. 14, 1904	Alternating-Current Generator	May 19, 1905	789, 404
Sept. 14, 1904	Ground Detector and Cut-Out	July 10, 1906	825,286
Feb. 6, 1905	Alternating-Current High-Frequency Generator	Dec. 1, 1908	905, 621
Mar. 10, 1905	Voltage-Regulator	Nov. 21, 1905	805,253
Mar. 10, 1905	Voltage-Regulator	Aug. 28, 1906	829,826
May 5, 1905	Alternating-Current Generator	Dec. 25, 1906	839,358
May 5, 1905	Synchronous Motor	Sept. 1, 1908	897,507
May 15, 1905	Overload Protective Device	Apr. 12, 1910	954,845
June 29, 1905	Dynamo-Electric Machine	Apr. 9, 1907	849,713
Aug. 19, 1905	Alternating-Current Motor	July 9, 1907	859,358
Aug. 20, 1905	Means for Controlling Self-Exciting Generator	Aug. 21, 1906	829,133
Aug. 24, 1905	Frequency-Changer	July 9, 1907	859,359

Date Filed	Title	Date Issued	Patent No.
Oct. 16, 1905	Alternating-Current Motor	Sept. 1, 1908	897,508
Nov. 9, 1905	Alternating-Current Motor	Jan. 15, 1907	841,610
Nov. 9, 1905	Induction Motor	Dec. 3, 1907	872,550
Feb. 19, 1906	Self-Exciting Alternator	Mar. 10, 1908	881,647
Mar. 7, 1906	Induction-Motor Control	Mar. 24, 1908	882,606
May 19, 1906	Alternating-Current Motor	May 4, 1909	920,896
June 4, 1906	Braking Alternating-Current Motors	Aug. 16, 1910	967,295
July 12, 1906	Induction-Motor Control	Oct. 20, 1908	901,513
July 19, 1906	Compensated Motor	Jan. 21, 1908	876,924
Dec. 10, 1906	Alternating-Current Motor	July 23, 1907	861,012
Jan. 12, 1907	Dynamo-Electric Machine	July 23, 1907	861,072
Jan. 12, 1907	Electric Motor	Apr. 27, 1909	919,414
Jan. 26, 1907	Motor Control	Apr. 21, 1908	885,128
Mar. 1, 1907	Electric Motor	June 1, 1909	923,311
May 1, 1907	Alternating-Current Motor	June 1, 1909	923,753
May 31, 1907	Starting-Motor Converter	Jan. 21, 1908	876,923
July 15, 1907	Alternating-Current Motor	June 1, 1909	923,754
Aug. 3, 1907	Induction Generator	May 4, 1909	920,809
Aug. 3, 1907	Method of Manufacturing Rotating Bodies for High-Speed Machinery	July 25, 1911	998,734
Aug. 10, 1907	Motor Control	May 10, 1910	957,454
Sept. 11, 1907	Motor-Control System	Nov. 16, 1909	940,112
Oct. 14, 1907	Dynamo-Electric Machine	Feb. 15, 1910	949,345
Oct. 31, 1907	Single-Phase Motor Control	Feb. 15, 1910	949,346
Nov. 21, 1907	Self-Exciting Generator (with E. H. Widegren)	Aug. 11, 1908	895,933
Dec. 17, 1907	Motor Control	Oct. 5, 1909	936,071
Dec. 26, 1907	Single-Phase Motor	Oct. 5, 1909	935,881
Dec. 26, 1907	Single-Phase Motor	May 4, 1909	920,710
Jan. 4, 1908	Single-Phase Motor Control	July 6, 1909	927,397
Jan. 4, 1908	Single-Phase Commutator Motor	Mar. 29, 1910	953,366
Jan. 7, 1908	Single-Phase Commutator Motor	Mar. 8, 1910	951,357
Jan. 7, 1908	System of Electric Distribution	May 18, 1909	921,786
Jan. 25, 1908	Telephone Relay	Oct. 27, 1908	902,195
Feb. 19, 1908	Electric Braking	Aug. 16, 1910	967,296
Apr. 17, 1908	Alternating-Current Motor	July 6, 1909	927,398
June 1, 1908	Alternating-Current Motor Control	July 6, 1909	927,399
June 1, 1908	Alternating-Current Motor Control	Jan. 18, 1910	946,751
June 1, 1908	Commutator Motor	Feb. 22, 1910	949,992
Aug. 6, 1908	Single-Phase Motor Control	June 1, 1909	923,312
Apr. 26, 1909	High-Frequency Alternator	Nov. 14, 1911	1,008,577
Sept. 2, 1909	Single-Phase Motor	Mar. 26, 1912	1,021,289
Sept. 22, 1909	Method of and Apparatus for Indicating the Speed of Electric Motors	June 16, 1914	1,100,280
Oct. 9, 1909	Electric Braking	Nov. 1, 1910	974,225

Date Filed	Title	Date Issued	Patent No.
Nov. 11, 1909	Electric Locomotive	Feb. 15, 1910	949,347
Nov. 11, 1909	Telephone Relay and System	June 27, 1911	996,445
Jan. 5, 1910	Motor Control	June 27, 1911	996,390
Jan. 6, 1910	Reinforcer for Telephone Circuits	June 27, 1911	996,391
Jan. 6, 1910	Alternating-Current Motor Control	June 2, 1914	1,098,656
Feb. 4, 1910	Alternating-Current Motor	July 29, 1913	1,068,404
July 7, 1910	Dynamo-Electric Machine	Mar. 31, 1914	1,091,613
June 9, 1911	Method of and Apparatus for Relaying High Frequency Currents	Oct. 22, 1912	1,042,069
June 19, 1911	System of Electrical Distribution	Aug. 17, 1915	1,150,652
July 7, 1911	Dynamo-Electric Machine	Mar. 31, 1914	1,091,614
Aug. 21, 1911	Air-Inlet Valve for Carburetors	Jan. 6, 1914	1,083,789
Oct. 19, 1911	Alternating-Current Motor	Mar. 5, 1912	1,019,042
Oct. 19, 1911	Alternating-Current Motor	May 6, 1913	1,060,731
Oct. 19, 1911	Electrical Braking with Alternating Current Motors	Mar. 3, 1914	1,089,384
Oct. 19, 1911	High-Frequency Alternator	Sept. 8, 1914	1,110,028
Oct. 19, 1911	Bearing	Sept. 8, 1914	1,110,030
June 20, 1912	Phase Balancer	Apr. 21, 1914	1,093,594
July 17, 1912	High-Frequency Alternator	Jan. 26, 1915	1,126,334
Aug. 28, 1912	Means for Compensating Polyphase Alternating-Current Commutator Motors	Dec. 2, 1913	1,080,403
Dec. 7, 1912	High-Frequency Alternator	Sept. 8, 1914	1,110,029
Dec. 7, 1912	Controlling Alternating Currents	Nov. 28, 1916	1,206,643
Apr. 2, 1913	Neutralizing Inductive Disturbances	Dec. 15, 1914	1,120,992
Apr. 14, 1913	Telephone System	July 7, 1914	1,102,628
Apr. 26, 1913	System of Ship Propulsion	Feb. 6, 1917	1,215,094
June 19, 1913	Dynamo-Electric Machine	Apr. 7, 1914	1,092,420
July 10, 1913	Induction Motor	Dec. 1, 1914	1,119,741
July 29, 1913	Method of Frequency Transformation	March 7, 1916	1,174,793
July 29, 1913	Internal Combustion Hydraulic Pump	March 12, 1918	1,259,338
Oct. 29, 1913	Selective Tuning System	Feb. 22, 1916	1,173,079
Nov. 11, 1913	Boiler and Heating Arrangement Therefor	Mar. 7, 1916	1,174,375
Dec. 1, 1913	Induction Motor	May 30, 1916	1,185,461
Dec. 18, 1913	Motor Excitation	Dec. 31, 1918	1,289,592
Jan. 29, 1914	Split-Phase System	Feb. 1, 1916	1,170,211
Jan. 29, 1914	Series-Multiple Control	Apr. 15, 1919	1,300,542
Jan. 29, 1914	Starting Phase-Converters	Apr. 15, 1919	1,300,545
Apr. 18, 1914	Automatic Control of Phase-Converters	July 17, 1917	1,233,952
Apr. 30, 1914	Wireless Signaling System	July 15, 1924	1,501,830
June 15, 1914	Method of and Means for Controlling Electric Energy	May 11, 1920	1,340,101

Date Filed	Title	Date Issued	Patent No.
Aug. 22, 1914	Apparatus for Producing an Electromotive Force of Special Wave Form	Dec. 18, 1917	1,250,752
Sept. 11, 1914	Alternating-Current Commutator Motor	Aug. 15, 1916	1,194,923
Oct. 8, 1914	High-Speed Rotating Body	Dec. 12, 1916	1,208,441
Oct. 24, 1914	Alternating-Current Motor	Aug. 8, 1916	1,194,265
Nov. 20, 1914	Method of and Means for Controlling Alternating Currents	Jan. 6, 1925	1,522,221
Nov. 20, 1914	Method of and Means for Controlling Alternating Currents	July 5, 1927	1,634,970
June 22, 1915	High-Frequency Alternator	June 12, 1917	1,229,856
June 22, 1915	System of Electrical Distribution	Oct. 9, 1917	1,242,632
Aug. 3, 1915	Phase Balancer	May 16, 1916	14,133 (reissue)
Aug. 5, 1915	Method of and Apparatus for Producing and Distributing Electric Current Waves of Radio Frequency	Aug. 9, 1921	1,386,830
Sept. 3, 1915	System of Ship Propulsion	Dec. 31, 1918	1,289,593
Nov. 26, 1915	Means for Controlling Alternating Currents	Jan. 20, 1920	1,328,797
Nov. 26, 1915	Condenser	May 14, 1918	1,266,377
Dec. 17, 1915	Multi-Speed Alternating-Current Motor	Apr. 23, 1918	1,263,992
Jan. 10, 1916	Wireless Signaling System	July 15, 1924	1,501,831
Jan. 21, 1916	Method of and Means for Controlling High-Frequency Alternating Currents	Jan. 20, 1920	1,328,610
Mar. 11, 1916	Electrical System of Power Transmission	May 20, 1919	1,304,239
Mar. 22, 1916	Electric System of Ship Propulsion	May 20, 1919	1,304,240
Apr. 1, 1916	Motor-Control System	Oct. 8, 1918	1,280,624
Apr. 19, 1916	Wireless Signaling System	Aug. 28, 1923	1,465,962
Apr. 19, 1916	Wireless Signaling System	Aug. 28, 1923	1,465,961
Apr. 19, 1916	Regulator for Phase-Balancers	July 17, 1917	1,233,953
Apr. 19, 1916	Wireless Signaling System	Sept. 9, 1924	1,508,151
Apr. 22, 1916	Electrical System of Power Transmission	June 4, 1918	1,268,662
July 28, 1916	Rotor for Dynamo-Electric Machine	Apr. 2, 1918	1,261,673
Aug. 7, 1916	Power-Transmitting Mechanism	Apr. 24, 1917	1,223,924
Aug. 24, 1916	Wireless Signaling System	Aug. 12, 1919	1,313,042
Sept. 23, 1916	High-Frequency Alternator	Aug. 22, 1922	1,426,943
Oct. 6, 1916	Means for Frequency Transformation	June 28, 1921	1,382,877
Oct. 13, 1916	System of Ship Propulsion	Feb. 6, 1917	1,215,095
Jan. 20, 1917	Wireless Signaling System	Aug. 24, 1920	1,350,911

Date Filed	Title	Date Issued	Patent No.
Feb. 10, 1917	Series-Multiple Control	Jan. 11, 1921	1,365,441
Feb. 23, 1917	System of Phase Modification	Apr. 15, 1919	1,300,543
Feb. 23, 1917	System of Phase Modification	Apr. 15, 1919	1,300,544
Mar. 17, 1917	Speed Control of Induction Motors (with D. C. Prince)	Apr. 22, 1919	1,301,632
Aug. 25, 1917	Radio Receiving System	Aug. 24, 1920	1,350,912
Sept. 4, 1917	Radio Signaling System	Aug. 22, 1922	1,426,944
Sept. 13, 1917	Antenna	Nov. 23, 1920	1,360,167
Oct. 6, 1917	System of Electrical Distribution	Apr. 26, 1921	1,375,991
Oct. 16, 1917	Radio Signaling System	Nov. 4, 1919	1,320,959
Nov. 26, 1917	Wireless Signaling System	Aug. 12, 1919	1,313,042
Dec. 6, 1917	System of Frequency Transformation	Sept. 23, 1919	1,316,995
Feb. 15, 1918	Antenna	Nov. 23, 1920	1,360,168
Apr. 2, 1918	Automatic Control of Phase-Converter	Aug. 27, 1918	14,510 (reissue)
Apr. 11, 1918	Means for Controlling Alternating Currents	Mar. 16, 1920	1,334,126
Apr. 20, 1918	Means for Controlling Alternating Curents	Jan. 20, 1920	1,328,473
May 4, 1918	Radio Signaling System	Dec. 8, 1925	1,564,807
Sept. 16, 1918	Unidirectional Radio Receiving System	Aug. 14, 1923	1,465,108
Oct. 17, 1918	Regulating System for Alternating-Current Circuits	Apr. 20, 1920	1,337,875
Nov. 4, 1918	High-Frequency Alternator	Feb. 22, 1921	1,369,601
Nov. 26, 1918	Bearing	Sept. 2, 1919	1,315,069
Nov. 26, 1918	Cooling Dynamo-Electric Machines	June 28, 1921	1,382,878
Dec. 6, 1918	High-Frequency Alternator (with S. P. Nixdorff)	Jan. 25, 1921	1,366,627
Dec. 24, 1918	Radio Signaling System	Nov. 23, 1920	1,360,169
Dec. 31, 1918	System of Radio Communication	Dec. 20, 1921	1,400,847
Jan. 14, 1919	Removing Sleet From Antennae	Jan. 31, 1922	1,404,726
Feb. 13, 1919	Amplifying System	June 13, 1922	1,419,797
Apr. 18, 1919	Radio-Receiving System	Apr. 26, 1921	1,375,992
June 5, 1919	Radio-Receiving System	Apr. 5, 1921	1,373,931
Aug. 23, 1919	Rotary Transforming Apparatus	June 20, 1922	1,420,398
Sept. 15, 1919	Signaling System	Aug. 18, 1925	1,549,737
Feb. 16, 1920	Electric Ship Propulsion	Jan. 29, 1924	1,481,853
Feb. 16, 1920	Electric Ship Propulsion	Jan. 29, 1924	1,481,882
Feb. 16, 1920	Electric Ship Propulsion	Feb. 5, 1929	1,701,350
Mar. 23, 1920	Control Ssytem	July 10, 1923	1,461,571
Mar. 23, 1920	Double Squirrel-Cage Synchronous Motor	May 27, 1924	1,495,969

Date Filed	Title	Date Issued	Patent No.
Sept. 28, 1920	Electron-Discharge Device	Apr. 21, 1925	1,535,082
Dec. 11, 1920	Electric Ship Propulsion	Mar. 30, 1926	1,579,051
June 7, 1921	Radio Receiving System	July 21, 1925	1,546,878
June 24, 1921	Radio Transmitting System	Dec. 2, 1924	1,517,816
June 24, 1921	Radio Receiving System	Dec. 11, 1923	1,477,413
Aug. 29, 1921	High-Frequency Signaling System	July 8, 1924	1,500,785
Sept. 17, 1921	High-Frequency Signaling System	Nov. 8, 1927	1,648,711
Oct. 28, 1921	System of Distribution	May 12, 1925	1,537,055
Oct. 28, 1921	Radio Receiving System	Apr. 22, 1924	1,491,372
Jan. 9, 1922	System of Motor Control	Apr. 15, 1924	1,490,720
Jan. 17, 1922	Radio Signaling System	Mar. 17, 1931	1,797,039
Apr. 10, 1922	Antenna	July 17, 1928	1,677,698
Apr. 10, 1922	High-Frequency Signaling System	Dec. 7, 1926	1,610,073
Oct. 16, 1922	System of Electric Ship Propulsion	July 10, 1928	1,676,312
Jan. 2, 1923	System of Ship Propulsion	Aug. 6, 1929	1,723,906
May 11, 1923	Signaling System	Aug. 6, 1929	1,722,998
May 11, 1923	Radio Receiving System	Aug. 6, 1929	1,723,907
May 11, 1923	System of Distribution	Mar. 5, 1935	1,993,581
May 22, 1923	System of Distribution	Dec. 8, 1931	1,835,131
May 25, 1923	Radio Receiver	Aug. 28, 1934	1,971,762
July 13, 1923	System of Distribution	Apr. 7, 1931	1,800,002
Oct. 20, 1923	Method and Apparatus for Picture Transmission by Wire or Radio (with R. H. Ranger)	Jan. 6, 1931	1,787,851
Dec. 22, 1923	Surge Preventer	Sept. 22, 1925	1,554,698
June 12, 1924	Method of and Means for Operating Motors	Nov. 24, 1925	1,563,004
Nov. 1, 1924	System of Distribution	Mar. 8, 1927	1,620,506
Nov. 28, 1924	Means for Transmitting Angular Motion	Sept. 24, 1926	1,600,204
Jan. 8, 1925	Method of and Apparatus for Multiplex Signaling	July 29, 1930	1,771,700
Jan. 17, 1925	Regulating System	Jan. 3, 1928	1,655,035
Feb. 11, 1925	System of Distribution	Aug. 14, 1928	1,680,758
Mar. 11, 1925	Electric Motor	Apr. 24, 1928	1,667,647
Apr. 25, 1925	Control of Electric Power (with A. H. Mittag)	Jan. 3, 1928	1,655,036
May 9, 1925	Radio Signaling System	Feb. 3, 1931	1,790,646
May 25, 1925	Locomotive Control	Jan. 3, 1928	1,655,037
Aug. 5, 1925	Power Factor Control	Jan. 3, 1928	1,655,038
Sept. 2, 1925	Speed-Control System (with A. H. Mittag)	Jan. 3, 1928	1,655,039
Oct. 1, 1925	Power-Amplifying Means	Mar. 19, 1929	1,706,094
Oct. 17, 1925	Oscillation Generator	Jan. 8, 1929	1,698,290

Date Filed	Title	Date Issued	Patent No.
Nov. 30, 1925	Radio Signaling	July 14, 1931	1,814,813
Dec. 15, 1925	Control of Electric Power (with A. H. Mittag)	Jan. 3, 1928	1,655,040
Jan. 15, 1926	Regulating Apparatus	Jan. 3, 1928	1,655,041
Jan. 15, 1926	Control System	Dec. 30, 1930	1,787,299
Mar. 17, 1926	Voltage Regulator	Oct. 13, 1927	1,652,923
Mar. 17, 1926	Rectifying Apparatus	June 25, 1929	1,718,515
Mar. 29, 1926	Alternating-Current Motor	Oct. 12, 1926	1,603,102
May 17, 1926	Control System (with S. P. Nixdorff)	May 8, 1928	1,669,153
May 20, 1926	Circuit-Control Apparatus (with A. H. Mittag)	Nov. 13, 1928	1,691,423
June 14, 1926	Ignition System	Aug. 6, 1929	1,723,908
July 14, 1926	Speed-Control System	Jan. 3, 1928	1,655,042
Aug. 9, 1926	Electrical Apparatus	Dec. 4, 1928	1,694,244
Aug. 9, 1926	Transmission of Pictures	Apr. 1, 1930	1,752,876
Aug. 9, 1926	Radiotelegraph	June 24, 1930	1,768,433
Aug. 9, 1926	Transmission of Pictures	Nov. 3, 1931	1,830,586
Oct. 19, 1926	Electrical Transmission of Pictures	Dec. 4, 1928	1,694,301
Mar. 9, 1927	Control of Electric Power	July 9, 1929	1,719,866
Mar. 30, 1927	System for Producing High Frequency Oscillations	July 5, 1932	1,866,337
May 13, 1927	System of Electrical Distribution	July 17, 1928	1,677,699
May 13, 1927	System of Electrical Distribution	July 17, 1928	1,677,700
May 13, 1927	Electrical System	Feb. 7, 1933	1,896,534
Nov. 15, 1927	Means for Eliminating Fading	Apr. 12, 1932	1,853,021
Nov. 15, 1927	Radio Signaling System	Sept. 16, 1930	1,775,801
Nov. 16, 1927	Electric-Discharge Device	Dec. 30, 1930	1,787,300
Dec. 9, 1927	Signaling by Phase Displacement	Oct. 18, 1932	1,882,698
Jan. 23, 1928	Transmission of Pictures	Dec. 4, 1928	1,694,302
Feb. 15, 1928	Radiant Energy Guiding System for Airplanes (with J. H. Hammond, Jr.)	July 4, 1933	1,917,114
Mar. 26, 1928	Speed-Control System	Nov. 19, 1929	1,736,689
Mar. 26, 1928	Transmission of Pictures	Feb. 10, 1931	1,792,264
May 9, 1928	Stabilization of Tuned Radio Frequency Amplifiers	Sept. 9, 1930	1,775,544
June 16, 1928	Signaling	Jan. 5, 1932	1,839,455
July 24, 1928	Transmission of Pictures (with R. D. Kell)	Nov. 25, 1930	1,783,031
Oct. 17, 1928	Amplifying Electrical Impulses (with R. D. Kell)	May 2, 1933	1,906,441
Oct. 27, 1928	Speed-Indicating System	Feb. 11, 1930	1,747,041
Nov. 17, 1928	Method and Means for Determining Altitude from Aircraft	Aug. 7, 1934	1,969,537
Apr. 8, 1929	Method and Means for Indicating Altitude from Aircraft	June 6, 1933	1,913,148
Nov. 30, 1929	Alternating-Current Commutator Machine	July 12, 1932	1,867,396

Date Filed	Title	Date Issued	Patent No.
Dec. 19, 1929	Electrical Transmission System	Aug. 8, 1933	1,921,718
Apr. 23, 1930	Picture Transmission Apparatus	July 5, 1932	1,866,338
June 5, 1930	Picture Transmission	May 10, 1932	1,857,130
July 15, 1931	Automatic Steering System	May 18, 1934	1,958,258
Aug. 20, 1931	High Frequency Transmission System for Railways	May 14, 1935	2,001,514
Sept. 19, 1931	Television Receiver	Nov. 29, 1932	1,889,587
Oct. 16, 1931	Television Apparatus	Nov. 14, 1933	1,935,427
Nov. 16, 1931	Indicating System for Aircraft	May 9, 1933	1,907,471
Dec. 1, 1931	Airplane Landing Field Using Directional Radio Beams (with J. H. Hammond, Jr.)	Apr. 13, 1937	2,077,196
Dec. 16, 1931	System of Electric Power Transmission	July 4, 1933	1,917,081
Dec. 16, 1931	System of Electric Power Transmission (with P. L. Alger and S. P. Nixdorff)	July 4, 1933	1,917,082
Dec. 16, 1931	Regenerative Electric Regulator (with S. P. Nixdorff)	July 3, 1933	1,917,146
Mar. 12, 1932	Electric Valve Converting System	Nov. 28, 1933	1,937,377
May 19, 1932	System for Reproducing Position	Apr. 24, 1951	2,500,514
Nov. 2, 1932	Torque Amplifying System	Jan. 7, 1936	2,027,140
Dec. 1, 1932	Electric Valve Excitation Circuits (with A. H. Mittag and E. L. Phillipi)	Apr. 10, 1934	1,954,661
May 2, 1933	Colored Television Apparatus	Jan. 22, 1935	1,988,931
May 2, 1933	Sound–Motion Picture Producer	Nov. 28, 1933	1,937,378
May 9, 1933	Electric Valve Converting System	Dec. 12, 1933	1,939,428
May 9, 1933	Electric Translating System	Aug. 7, 1934	1,969,538
May 9, 1933	Electric Valve Converting System	Dec. 12, 1933	1,939,429
Nov. 28, 1933	Electric Valve Converter	May 18, 1937	20,364 (reissue)
Jan. 27, 1934	Sound Reproducing Apparatus	Oct. 23, 1934	1,978,183
Jan. 31, 1935	Control System	June 4, 1946	2,401,450
Sept. 21, 1935	Follow-Up Control System	Jan. 28, 1947	2,414,919
Feb. 29, 1936	Electric Valve Translating Circuit	Sept. 7, 1937	2,092,545
Feb. 29, 1936	Electric Translating Circuit	Sept. 7, 1937	2,092,546
Feb. 29, 1936	Electric Valve Translating Circuit	Nov. 2, 1937	2,098,023
Feb. 29, 1936	Electric Power System (with A. H. Mittag)	June 15, 1937	2,084,177
Feb. 29, 1936	Electric Valve Converting System	June 28, 1938	2,122,271
Apr. 15, 1936	Electric Valve Circuit	Jan. 4, 1938	2,104,633
May 23, 1936	Control System	Dec. 3, 1946	2,412,027
Oct. 31, 1936	Electric Valve Circuit	Feb. 20, 1940	2,190,759
Oct. 31, 1936	Electric Valve Circuit	Feb. 20, 1940	21,919 (reissue)
Mar. 19, 1937	Cable for Transmitting Electric Power	Dec. 27, 1938	2,141,894

Date Filed	Title	Date Issued	Patent No.
Mar. 27, 1937	Discharge Lamp System	Nov. 1, 1938	2,135,268
May 15, 1937	Electric Valve Converting System	Mar. 19, 1940	2,193,912
May 29, 1937	Course Guiding System	Dec. 26, 1939	2,184,267
Oct. 19, 1937	Electric Power Transmission System	Sept. 10, 1940	2,213,945
Oct. 19, 1937	Electric Motor Control System	Apr. 1, 1941	2,236,984
Oct. 19, 1937	Electric Motor Control System	June 2, 1942	2,285,182
Dec. 14, 1937	Asynchronous Electric Power Transmission System	July 9, 1940	2,207,570
Feb. 17, 1938	Electric Valve Frequency Converter System	Mar. 19, 1940	2,193,913
Feb. 17, 1938	Electric Valve Frequency Converter System	Mar. 19, 1940	2,193,914
Apr. 7, 1938	Protective System for Electric Valve Translating Apparatus	Jan. 9, 1940	2,186,815
Apr. 28, 1938	Frequency Controlling System	Apr. 22, 1941	2,239,436
June 1, 1938	Protective System	Nov. 26, 1940	2,222,696
Aug. 24, 1938	Navigation and Landing of Aircraft in Fog	June 10, 1941	2,245,246
Nov. 5, 1938	Electric Transforming Apparatus	Apr. 29, 1941	2,240,201
Nov. 5, 1938	Electric Power Transmission System	July 16, 1940	2,208,183
Jan. 21, 1939	Follow-Up Control System (with M. A. Edwards and K. K. Bowman)	Jan. 21, 1947	2,414,685
Mar. 18, 1939	Excitation Control System for Synchronous Machines	Sept. 17, 1940	2,215,312
Apr. 26, 1939	Electric Power System	Apr. 8, 1941	2,237,384
Apr. 26, 1939	Method of and Apparatus for Starting and Operating Thyratron Motors	Nov. 11, 1941	2,262,482
May 26, 1939	Electric Power Converting Apparatus	Dec. 17, 1940	2,225,328
June 24, 1939	Dynamoelectric Machine (with M. A. Edwards)	Jan. 7, 1941	2,227,992
Sept. 7, 1939	Radio Distance Meter	July 8, 1941	2,248,599
Sept. 29, 1939	Electric Valve Converting System and Control Circuit Therefor	Sept. 17, 1940	2,215,313
Sept. 29, 1939	Radio Distance Meter (with F. G. Patterson and C. A. Nickle)	Oct. 21, 1941	2,259,982
May 3, 1940	Electric Valve Circuits (with A. H. Mittag)	July 8, 1941	2,248,600
July 20, 1940	Speed Regulating System	Sept. 23, 1941	2,256,463
Nov. 20, 1940	Electric Drive	Feb. 23, 1943	2,312,061
Nov. 20, 1940	Electric Drive	Feb. 23, 1943	2,312,062
Jan. 16, 1941	Electric Control Circuits	Dec. 2, 1947	2,431,903
May 27, 1941	Television System and Method of Operation	Nov. 9, 1943	2,333,969
Aug. 29, 1941	Electric Valve Translating System	Aug. 29, 1944	2,357,067
Sept. 12, 1941	Control Circuit for Electric Valve Apparatus	Mar. 7, 1944	2,343,628

Date Filed	Title	Date Issued	Patent No.
Apr. 1, 1942	Electric Drive	Apr. 6, 1943	2,315,489
Apr. 1, 1942	Electric Drive	Apr. 6, 1943	2,315,490
Apr. 1, 1942	Electric Drive	Apr. 6, 1943	2,315,491
Apr. 17, 1942	Electric Computer	Mar. 11, 1947	2,417,229
Aug. 26, 1942	Speed Control Arrangement for Induction Clutches	Nov. 2, 1943	2,333,458
Nov. 9, 1942	Follow-Up System	Feb. 25, 1947	2,416,562
Jan. 22, 1943	Pulse Echo Apparatus for Spotting Shell Fire	Mar. 1, 1949	2,463,233
Feb. 26, 1943	Electric Ship Propulsion System	Aug. 29, 1944	2,357,087
May 22, 1943	Radio Landing Apparatus (with F. G. Patterson)	Oct. 19, 1948	2,451,793
Apr. 15, 1944	Follow-up Control System (with G. A. Hoyt)	June 14, 1949	2,473,235
Oct. 2, 1944	High-Frequency Wave Transmitting Apparatus	Mar. 30, 1948	2,438,735
Aug. 17, 1945	Radio Detection and Ranging System Employing Multiple Scan (with F. G. Patterson and M. W. Sims)	Oct. 17, 1950	2,526,314
May 13, 1947	Stabilizer for Alternating Current Power Transmission Systems	May 17, 1949	2,470,454
Oct. 18, 1947	Electric Frequency Transformation System (with A. H. Mittag and M. W. Sims)	Oct. 12, 1948	2,451,189
Dec. 30, 1949	Fault Suppressing Circuits (with A. H. Mittag and E. L. Phillipi)	Apr. 10, 1951	2,548,577
Dec. 30, 1949	Locomotive Power System (with B. D. Bedford and A. H. Mittag)	Apr. 17, 1951	2,549,405
Feb. 6, 1951	Electronic Motor and Commutating Means Thereof (with S. P. Nixdorff)	July 7, 1953	2,644,916
Feb. 6, 1951	Electric Motor and Stabilizing Means Therefor (with S. P. Nixdorff)	May 26, 1953	2,640,179
Feb. 28, 1951	Electronic Frequency Changer and Stabilizing Control Means Therefor	Feb. 19, 1952	2,586,498
Feb. 28, 1951	Stabilizer for Alternating Current Power Transmission Systems (with R. W. Kuenning)	July 7, 1953	2,644,898
Apr. 17, 1951	Current Interrupter (with A. H. Mittag and R. W. Kuenning)	Sept. 30, 1952	2,612,629
Feb. 14, 1952	Receiver for Color Television	Feb. 8, 1955	2,701,821
Sept. 17, 1952	Phase Balancing System	Sept. 15, 1953	2,652,529
Apr. 20, 1953	System for Controlling the Flow of Molten Metal	Oct. 30, 1956	2,768,413
July 27, 1953	Magnetic Amplifier Motor Control	June 26, 1956	2,752,549
Sept. 1, 1953	Push-Pull Magnetic Amplifier	July 9, 1957	2,798,904
Mar. 31, 1954	Magnetic Amplifier Motor Control System	July 22, 1958	2,844,779

Date Filed	Title	Date Issued	Patent No.
July 1, 1954	Methods and Systems for Motor Control	Sept. 9, 1958	2,851,647
July 21, 1954	Alternating-Current Motor	Apr. 26, 1955	2,707,257
July 21, 1954	Alternating-Current Motor	June 21, 1955	2,711,502
July 27, 1954	Motor Control System	Mar. 3, 1959	2,876,408
Dec. 9, 1954	Alternating-Current Motor	June 25, 1957	2,797,375
Dec. 24, 1954	Magnetic Computer	May 16, 1961	2,984,414
Sept. 25, 1959	Electric Motor Control Apparatus	Mar. 19, 1963	3,082,367
Apr. 17, 1961	Electric Motor Control Apparatus	Aug. 21, 1962	3,050,672
June 16, 1961	Electric Motor Control System	Jan. 28, 1864	3,119,957
Mar. 19, 1968	Adjustable Speed Motor Control System	May 29, 1973	3,736,481

Bibliographic Note

Books and journal articles, including technical papers by Alexanderson and his colleagues, are cited in endnotes.

Manuscript Sources

Ernst F. W. Alexanderson Papers. Schaffer Library, Union College, Schenectady, New York. This manuscript collection, the most important I used, was augmented in 1975, while my research was in progress and includes an unpublished finding aid.

Samuel P. Nixdorff Papers. Schaffer Library, Union College, Schenectady, New York. A relatively small collection with no finding aid. The Nixdorff notebooks (Nixdorff was an assistant to Alexanderson for many years) were useful in following Alexanderson's activities at certain times; the papers included important information on Alexanderson's department at General Electric.

George H. Clark Radioana Collection. National Museum of American History, Smithsonian Institution, Washington, D.C. Robert S. Harding recently has completed an extensive register that should greatly facilitate use of this collection, which contains much helpful information on the correspondence and records of Reginald A. Fessenden and the National Electric Signaling Company.

William W. Brown Papers. National Museum of American History, Smithsonian Institution, Washington, D.C. Brown worked with Alexanderson and became a leading expert on large, low-frequency radio antennas. After I contacted him, he decided to donate this collection of drawings, correspondence, and notebooks to the Smithsonian. He also permitted the Smithsonian to make copies of his collection of photographs of radio-transmitting and related apparatus.

Lloyd Espenschied Papers. National Museum of American History, Smithsonian Institution, Washington, D.C. This collection arrived at the Smithsonian during my tenure as a Smithsonian Fellow, and I prepared a finding aid in the process of transferring the papers to archival boxes. It contains useful information on the early history of radio and vacuum-tube electronics, including copies of primary source documents.

Irving Langmuir Papers. Manuscript Division, Library of Congress, Washington, D.C. A large collection that includes a register. Langmuir was a close friend and professional colleague of Alexanderson, and his papers provided useful information on work on early radio receivers that were developed for use in the Alexanderson alternator system.

John Hays Hammond Papers. Manuscript Division, Library of Congress, Washington, D.C. This collection includes correspondence and notebooks. Hammond, a friend of Alexanderson and also an engineer-inventor, was an early user of the Alexanderson radio alternator. He played an instrumental role in the beginnings of vacuum-tube research and development at General Electric.

Stanford Caldwell Hooper Papers. Manuscript Division, Library of Congress, Washington, D.C. A large collection containing correspondence, research notes, and speeches. Hooper was a naval officer who influenced government policy on radio communication and played a significant role in the formation of the Radio Corporation of America.

Reginald Aubrey Fessenden Papers. North Carolina Department of Archives and History, Raleigh, N.C. A fairly large collection with an unpublished finding aid. Fessenden was an important outside influence on Alexanderson's early radio work at General Electric.

Bernard A. Behrend Papers. Special Collections, Cooper Library, Clemson University, Clemson, S.C. This collection contains an autobiographical essay, correspondence, laboratory notebooks, pamphlets, and photographs.

Notes

Chapter One: From Uppsala to Schenectady

1. Ernst F. W. Alexanderson (hereafter EA in citations of archived material and patents), typescript (TS) of talk, May 14, 1941, Alexanderson Papers, Schaffer Library, Union College, Schenectady, N.Y. (hereafter cited as AP).

2. Verner Alexanderson, "Family Tree, E. F. W. Alexanderson," September 30, 1977. I am indebted to Verner Alexanderson for a copy. Also see the entry on Alexanderson in *American Biography* (1926), 24:397–405.

3. EA, TS of talk given at an American Society of Mechanical Engineers (ASME) meeting, November 30, 1942, AP.

4. Aubrey D. McFadyen, "Ernst F. W. Alexanderson," *Journal of the Patent Office Society,* 22 (1940): 779–84.

5. EA, "Sailing," TS of short essay, March 3, 1969. This is one of several reminiscences by Alexanderson obtained by the late Philip L. Alger, to whom I am indebted for copies. (Hereafter these essays will be cited as PLA.)

6. EA, TS of talk to "Fortnightly Club," April 5, 1937, AP.

7. EA, "Locomotives—Do They Float?" PLA.

8. EA, interview by Frank E. Hill and C. D. Wagoner, Schenectady, N.Y., February 22, 1951. I have used a TS kindly provided by George Wise. (Hereafter this interview will be cited as Hill-Wagoner interview.)

9. Ibid.

10. Ibid. Also EA to W. H. Rasch, March 6, 1945, AP.

11. Eugene S. Ferguson, "The Mind's Eye: Nonverbal Thought in Technology," *Science,* 197 (1977): 827–36. See also Brook Hindle, *Emulation and Invention* (New York: New York University Press, 1981).

12. Reese V. Jenkins and Keith A. Nier, "A Record for Invention: Thomas Edison and His Papers," *IEEE Transactions on Education,* E-27 (1984): 191–96.

13. Hill-Wagoner interview.

14. Ibid.

15. Ibid.

16. The faculty and students were listed in the school catalog. I am indebted to Dagmar von Perner of the Royal Institute of Technology for providing a copy of several pages of the catalog for the year 1900. Alexanderson's memories of Lindstedt appear in EA, "Theoretical Mechanics," PLA.

17. EA, application to DeLaval Company, March 3, 1902; EA, application for admission to Schenectady Works, July 11, 1902, AP.

18. EA, application to Westinghouse Electric and Manufacturing Company, February 1902, AP.

19. The documents were in folder 3, 1975 acquisition, AP. They may subsequently have been moved.

20. I am indebted to Alexanderson's daughter Amelie Alexanderson Wallace for providing me with a translation of some correspondence from June 1899 and July–August 1900. In these letters, Alexanderson addressed his father as "Lo," while his father addressed him as "Tasse." The letter containing Alexanderson's father's warning was dated July 12, 1900.

21. "Lo" to "Tasse," August 13, 1900, and "Lo" to "Tasse," August 28, 1900, translations provided by Amelie Alexanderson Wallace.

22. Thomas P. Hughes, *Networks of Power* (Baltimore: Johns Hopkins University Press, 1983), chaps. 4 and 5.

23. "Jonas Wenstrom," obituary, *Teknisk Tidskrift*, 24 (1894): 3–4. Also see "The Wenstrom Electric Railway System," *Electrical World*, 16 (1890): 293–94; and "Electric Transmission of Power in Sweden," *Electrical Engineer*, 13 (1894): 424–25.

24. Hill-Wagoner interview.

25. "Ernst Danielson," obituary, *Teknisk Tidskrift*, 37 (1907): 239–41; and obituary in *Electrician*, 60 (1907): 17.

26. "Sweden," *Electrical Engineer*, 25 (1900): 577.

27. "Stockholm Central Station," *Electrical Engineer*, 24 (1899): 178–83.

28. "From the International in 1878 to that in 1900," *Electrical Engineer*, 26 (1900): 198–99.

29. The Adams essay is included in *Changing Attitudes toward American Technology*, ed. Thomas Parke Hughes (New York: Harper & Row, 1975), pp. 168–75.

30. W. E. Dalby, "The Education of Engineers in America, Germany, and Switzerland," *Electrician*, 51 (1903): 80–82.

31. On Slaby see *Electrician*, 71 (1913): 4. See also Friedrich Kurylo and Charles Susskind, *Ferdinand Braun* (Cambridge, Mass.: MIT Press, 1981), pp. 107–8; and Hugh G. J. Aitken, *Syntony and Spark* (New York: John Wiley & Sons, 1976), pp. 217–18.

32. EA, 1900–1901 student notebooks, 1975 acquisition, AP.

33. Hill-Wagoner interview.

34. D. G. Tucker, *Gisbert Kapp* (Birmingham, England: University of Birmingham, 1973). This booklet lists Kapp's publications and British patents. For Kapp's views on differences in consulting engineering practice between Great Britain and Germany, see "Interview with Mr. Gisbert Kapp on the work of the Verband Deutscher Elektrotechniker," *Electrical Engineer*, 18 (1896): 35–36.

35. Hill-Wagoner interview.

36. EA, Berlin notebook, AP. The first dated entry is October 23, 1900.

37. Gisbert Kapp, *Dynamos, Motors, Alternators, and Rotary Converters* (London: Biggs, 1902), p. 13.

38. On Roessler see "Dr. Gustav Roessler," *Elektrotechnische Zeitschrift*, 49 (1928): 633.

39. Hill-Wagoner interview.

40. Charles Proteus Steinmetz, *Theory and Calculation of Alternating Current Phenomena* (New York, 1897).

41. G. Roessler, "The Behavior of Transformers under the Influence of Alternating Currents of Different Wave Forms," *Electrician*, 36 (1895): 124–26. The traditional method of analysis is well illustrated in Roessler's book on electric motors published in 1901, *Elektromotoren für Wechselstrom und Drehstrom* (Berlin: Julius Springer).

42. EA, 1901 student notebook, AP.

43. Cited on EA, application for admission to Schenectady Works, July 11, 1902, AP.

44. Hill-Wagoner interview.

45. EA, "America—A Nation of Inventors," TS of talk, n.d., AP.

46. EA, "The Swedish vs. the American View of Life," TS of talk, n.d., AP.

47. EA, TS of talk for "I'm an American" radio broadcast, February 15, 1942, AP.

48. Franklin D. Scott, *Sweden: The Nation's History* (Minneapolis: University of Minnesota Press, 1977), p. 370. Also see "A Comparison of Norway and Sweden," *National Geographic*, 16 (1905): 429–31.

49. Scott, *Sweden*, pp. 373–75.

50. Hill-Wagoner interiew.

51. Herbert Asbury, "Inventor for Victory," *Popular Science*, July 1942, pp. 89–92, 204–5.

52. EA, entries in 1901 expense notebook, 1975 acquisition, AP.

53. EA, TS of "I'm an American" radio broadcast, February 15, 1942; EA, "Creative Education," TS of talk, n.d., AP.

54. EA to Edison Company, n.d., AP.

55. Hughes, *Networks of Power,* pp. 18–22.

56. Matthew Josephson, *Edison* (New York: McGraw-Hill, 1959), pp. 468–69.

57. On Curtis, see the *National Cyclopedia of American Biography* (hereafter cited as *NCAB*), 42:73. On Crocker, see *NCAB*, 12:424; and *Journal of the American Institute of Electrical Engineers* (hereafter *Journal of the AIEE*), 40 (1921): 707. On Wheeler see *NCAB*, 41:49; and "Francis B. Crocker," *Journal of the AIEE,* 42 (1923): 553. See also "The Factory of the C and C Electric Company," *Electrical World,* 27 (1896): 657–59.

58. Hill-Wagoner interview.

59. EA, address book and 1901 expense book, 1975 acquisition, AP.

60. EA, data sheet with analysis of commutator heating, October 14, 1901, AP. Also see McFadyen, "Ernst F. W. Alexanderson," pp. 779–84.

61. Hill-Wagoner interview; EA, interview with author, Schenectady, N.Y., October 17, 1972. Also see EA to A. L. Rohrer, January 7, 1902, AP.

62. James E. Brittain, "C. P. Steinmetz and Ernst F. W. Alexanderson: Creative Engineering in a Corporate Setting," *Proceedings of the IEEE,* 64 (1976): 1413–17.

63. EA to A. L. Rohrer, January 7, 1902; Rohrer to EA, January 13, 1902, AP.

64. EA to Chief Designer, Westinghouse Company, February 3, 1902; G. Berenteen to EA, February 6, 1902; EA to Berenteen, February 8, 1902, AP.

65. EA to Chief Draftsman, General Electric Company, February 10, 1902; S. L. G. Knox to EA, February 13, 1902; EA to Knox, February 18, 1902, AP.

66. G. Berenteen to EA, February 17, 1902; EA to Berenteen, February 28, 1902, AP.

67. Hill-Wagoner interview. Alexanderson evidently boarded for a time with a Mrs. Noble, who acted as something of a surrogate mother to several young Swedish immigrants. This is indicated in a letter to Alexanderson from his mother dated September 25, 1904, AP.

68. EA to DeLaval Company, March 3, 1902, AP.

69. EA to Chief Draftsman, Westinghouse Company, June 5, 1902; G. Berenteen to EA, June 6, 1902; draft of letter from EA to Berenteen, June 9, 1902, and revision dated June 10, 1902, AP.

70. A sketch initialed by Alexanderson, dated April 9, [1902?], was for a controller for the Allegheny Steel Company. Entries in a notebook for 1902 indicate that he was reading articles in *Electrical World* on railway motors.

71. EA to Chief Designer, Stanley Electric Company, September 20, 1902; R. M. Power to EA, October 9, 1902; EA to Power, October 10, 1902, AP.

72. EA to E. M. Hewlett, January 14, 1903, AP. Also see John Winthrop Hammond, *Men and Volts: The Story of General Electric* (New York: J. B. Lippincott, 1941), p. 268.

73. Description witnessed by Kirster and Burkogt, November 6, 1902, AP. A complete table of Alexanderson's patents is given in Appendix B.

74. The conversation with Steinmetz was mentioned in EA to E. M. Hewlett, December 16, 1902, AP.

75. H. G. Reist to EA, December 10, 1902, AP.

76. EA to E. Lundgren, December 29, 1902, AP.

77. EA to Stanley Electric Company, December 17, 1902; C. C. Chesney to EA, December 23, 1902, AP.

78. EA to W. D. Cooley, January 1, 1903; Cooley to EA, January 6, 1903; Cooley to EA, January 10, 1903, AP.

79. Draft of EA to R. A. Fessenden, January 9, 1903, AP.

80. EA to E. M. Hewlett, January 14, 1903, AP.

81. I am indebted to George Wise for providing a copy of the statement from Whitney's notebook dated January 24, 1903. See also W. R. Whitney to EA, January 24, 1903; and EA to Whitney, January 28, 1903, AP. For background information on the use of electric furnaces in research at the GE Research Laboratory, see George Wise, "Ionists in Industry: Physical Chemistry at General Electric, 1900–1915," *Isis*, 74 (1983): 16. For an informed account of the origins of the Research Laboratory and Whitney's role, see George Wise, "A New Role for Professional Scientists in Industry: Industrial Research at General Electric, 1900–1916," *Technology and Culture*, 21 (1980): 408–29.

82. EA to Wakefield, January 29, 1903, AP.

83. EA to E. Lundgren, February 11, 1903; Lundgren to EA, February 19, 1903; EA to Lundgren, February 20, 1903, AP.

84. Copy of contract, February 23, 1903, signed by E. W. Rice on behalf of GE, AP. Also see EA to J. W. Upp, February 24, 1903, AP.

85. EA, two pages of analysis in 1902 notebook, n.d. [but after dated entry of August 23, 1902], AP.

86. EA to H. G. Reist, May 19, 1903; Reist to EA, May 26, 1903; EA to Reist, June 2, 1903; discussion dated May 18, 1903; and sketch of variable-speed motor, June 24, 1903, AP. The patent, entitled "Motor Winding," was issued on January 15, 1907, as U.S. Patent no. 841,609.

87. EA to A. G. Davis, January 30, 1904, AP.

88. EA, application dated June 23, 1903; EA, transcript of talk, May 14, 1941, AP. For an interesting interpretation of the early history of GE and contrasting managerial styles, see A. L. Rohrer, "The Background of General Electric's Engineering Personnel," *General Electric Review* (hereafter *GE Review*), 42 (1939): 343–45. On Rohrer, see *NCAB*, 42:189; and "Albert Lawrence Rohrer," *Electrical Engineering*, 70 (1951): 1121.

89. Hill-Wagoner interview. Also EA, TS of speech, May 14, 1941, AP.

90. See, for example, G. Kapp to B. A. Behrend, December 7, 1898, Bernard A. Behrend Papers, Special Collections, Cooper Library, Clemson University, Clemson, S.C.

91. Ernst F. W. Alexanderson, "Bau und Beitrieb von Drehumformern in America," *Elektrotechische Zeitschrift*, 24 (1903): 737–39. See also Hughes, *Networks of Power*, pp. 121, 208–9.

92. "The Testing Department of the General Electric Company," *GE Review*, 4 (1904–5): 104–6.

93. Charles P. Steinmetz, "The Individual and Corporate Development of Industry," *GE Review*, 18 (1915): 814; and idem, "Engineering Schools of Electrical Manufacturing Companies," *Bulletin of the National Association of Corporate Schools*, 1 (March 1914): 23–28.

94. David F. Noble, *America by Design* (New York: Alfred A. Knopf, 1977), p. 173.

95. W. E. Ayrton, "The Education of the American Electrical Engineer," *Electrical Age*, 32 (1904): 313–15.

96. "The Testing Department of the General Electric Company," pp. 104–6.

97. Service agreement dated October 9, 1903, AP. At least this wage represented an increase from the 1890s, when those entering the test program had been required to pay an entry fee and received no salary for the first six months. See George Wise, " 'On Test': Postgraduate Training of Engineers at General Electric, 1892–1961," *IEEE Transactions on Education*, E-22 (November 1979): 171–77. Also see service agreement, February 1, 1904, AP.

98. On Potter, see Philip L. Alger, *The Human Side of Engineering* (Schenectady, N.Y.:

Mohawk Development Service, 1972), p. 90. See also "William Bancroft Potter," *Electrical World and Engineer*, 38 (1901): 1014.

99. EA to W. B. Potter, April 14, 1931, AP.

100. M. P. Rice to EA, March 4, 1904, AP.

101. Ernst F. W. Alexanderson, "Method for Measuring the Output of Induction Motors," *Electrical World and Engineer*, 44 (1904): 212–13.

102. For a discussion of the concept of normal and revolutionary activities in technology, see Edward W. Constant, *The Origins of the Turbojet Revolution* (Baltimore: Johns Hopkins University Press, 1980), pp. 10–12.

Chapter Two: Dialectical Engineer

1. See an essay by Thomas P. Hughes, "Inventors: The Problems They Choose, The Ideas They Have, and the Inventions They Make," in *Technological Innovation: A Critical Review of Current Knowledge*, ed. Patrick Kelly and Melvin Kransberg (San Francisco: San Francisco Press, 1978), pp. 168–82.

2. Hughes, *Networks of Power*, pp. 18–20.

3. Ibid., pp. 209–11.

4. For an informed discussion of technological systems and the use of systems as an interpretive theme in the history of technology, see ibid., pp. 5–17.

5. For biographical information on Reist, see *NCAB*, 31:419; and Alger, *Human Side of Engineering*, p. 39.

6. The first patent application, entitled "Alternating-Current Generator," was filed August 19, 1904, and issued as U.S. Patent no. 789,476 on May 9, 1905.

7. For a discussion of the design and construction of commutators, see H. M. Hobart and A. G. Ellis, *Armature Construction* (New York: Macmillan, 1907), pp. 107–28.

8. EA to E. J. Berg, April 25, 1904, AP.

9. EA to A. G. Davis, June 23, 1904; EA to E. F. Collins, June 29, 1904; EA to A. G. Davis, July 1, 1904, AP.

10. EA to W. R. Whitney, July 1, 1904; EA to Whitney, November 10, 1904, AP.

11. For biographical information on Albert G. Davis, see *Who Was Who in America*, 1:299. See also *Electrical Engineering*, 58 (1939): 279; and Wise, "A New Role for Professional Scientists in Industry," p. 414.

12. EA, "Research and Engineering at the G.E. Company," TS of talk presented at Royal Institute of Technology, Stockholm, Sweden, August 1948, AP.

13. Albert G. Davis, "Patents," *GE Review*, 11 (1908): 32–34.

14. EA to E. W. Rice, January 23, 1905, AP.

15. For biographical information on Edwin Rice, Jr., see *Dictionary of American Biography* (hereafter *DAB*), suppl. 1, pp. 627–28; and "E. W. Rice, Jr.," *Electrical World and Engineer*, 38 (1901): 128. Also see George Wise, *Willis R. Whitney, General Electric, and the Origins of U.S. Industrial Research* (New York: Columbia University Press, 1985), pp. 66–67.

16. John T. Broderick, *Forty Years with General Electric* (Albany, N.Y.: Fort Orange Press, 1929), pp. 79–82.

17. Ibid., p. 64.

18. Willis R. Whitney, "A Tribute to Edwin Wilbur Rice, Jr.," *GE Review*, 39 (1936): 3–4.

19. Ibid. An anecdote about Alexanderson and Rice seems to reflect Rice's philosophy. Rice is alleged to have decided that it was more important to observe a test of one of Alexanderson's inventions than to report to GE's board of directors. See Alger, *Human Side of Engineering*, p. 38.

20. Hill-Wagoner interview.

21. Walter G. Vincenti, "The Air-Propeller Tests of W. F. Durand and E. P. Lesley: A Case Study in Technological Methodology," *Technology and Culture*, 20 (1979): 714.

22. Ernst F. W. Alexanderson, "A Self-Exciting Alternator," *Transactions of the American Institute of Electrical Engineers* (hereafter *Transactions of the AIEE*), 25 (1906): 76–77.

23. EA to C. P. Steinmetz, December 30, 1904, AP.

24. EA to A. R. Everest, February 5, 1906; Everest to EA, March 13, 1906; EA to Everest, March 27, 1906, AP.

25. For biographical information on Emmet, see "W. L. R. Emmet," *Electrical World and Engineer,* 38 (1901): 530; and William L. R. Emmet, *The Autobiography of an Engineer* (1931; reprint, New York: American Society of Mechanical Engineers, 1940).

26. EA to H. G. Reist, October 10, 1905, AP.

27. EA, TS of talk, November 7, 1945, AP; and Alger, *Human Side of Engineering,* p. 38.

28. EA to R. W. Pope, December 4, 1905, AP.

29. Alexanderson, "Self-Exciting Alternator," pp. 61–77.

30. Discussion in *Transactions of the AIEE,* 25 (1906): 78–80.

31. W. L. R. Emmet to M. P. Rice, December 17, 1907, AP.

32. William Suddards Franklin and William Esty, *The Elements of Electrical Engineering,* vol. 2, Alternating Currents (New York: Macmillan, 1910), p. 14.

33. Hughes, *Networks of Power,* p. 405.

34. Reginald A. Fessenden, "Electromagnetic Mechanism with Special Reference to Telegraphic Work," *Journal of the Franklin Institute,* 150 (1900): 122.

35. For biographical information on Fessenden, see *Dictionary of Scientific Biography* (*DSB*), 4:601; and Helen M. Fessenden, *Fessenden: Builder of Tomorrows* (New York: Coward-McCann, 1940; reprint, New York: Arno Press, 1974). See also Hugh G. J. Aitken, *The Continuous Wave: Technology and American Radio, 1900–1932* (Princeton, N.J.: Princeton University Press, 1985), pp. 40–60; and Susan J. Douglas, *Inventing American Broadcasting, 1899–1902* (Baltimore, Md.: Johns Hopkins University Press, 1987), pp. 42–45.

36. Reginald A. Fessenden, "Atomic Volume and Tensile Strength," *Electrical World,* 18 (1891): 123–25. See also idem, "Some Recent Work in Molecular Physics," *Journal of the Franklin Institute,* 142 (1896): 187–216.

37. Reginald A. Fessenden, "A Determination of the Electric and Magnetic Quantities and of the Density and Elasticity of the Ether," *Physical Review,* 10 (1900): 1–33, 83–115.

38. Aitken, *Continuous Wave,* p. 49.

39. Ibid., pp. 51–53. As Aitken points out, Edward Constant's concept of a presumptive anomaly appears to apply well to this example. See Edward W. Constant, *The Origins of the Turbojet Revolution* (Baltimore, N.J.: Johns Hopkins University Press, 1980), pp. 15–16; Douglas, *Inventing American Broadcasting,* pp. 45–46.

40. Aitken, *Continuous Wave,* pp. 55–60.

41. Ibid., p. 63. Scott's letter was dated June 6, 1900.

42. R. A. Fessenden to C. P. Steinmetz, June 1, 1900; Fessenden to Steinmetz, July 3, 1900; George H. Clark Radioana Collection, National Museum of American History, Smithsonian Institution, Washington, D.C. (hereafter cited as CRC).

43. C. P. Steinmetz to R. A. Fessenden, January 4, 1901; J. T. Broderick to Manager of Schenectady Works, memorandum, January 23, 1901, CRC.

44. H. G. Reist to C. P. Steinmetz, June 8, 1901, CRC.

45. R. A. Fessenden, "Wireless Telegraphy," U.S. Patent no. 706,737, filed May 29, 1901, issued August 12, 1902.

46. According to a manufacturing order dated February 20, 1901, CRC, the arma-

ture and field windings were to follow standard specifications. Some data on high-frequency alternators, including the Steinmetz machine, were included in W. Duddle, "A High-Frequency Alternator," *Philosophical Magazine and Journal of Science*, 9 (1905): 306–7.

47. R. A. Fessenden to C. P. Steinmetz, July 3, 1901; Steinmetz to Fessenden, July 16, 1901, CRC. Fessenden included wireless speech transmission in a patent application filed September 28, 1901, and issued August 12, 1902, as U.S. Patent no. 706,747.

48. Elihu Thomson to R. A. Fessenden, July 11, 1901, CRC. An item published in 1903 reported that GE had developed a 10-kHz alternator and that units had been delivered to Elihu Thomson and to Fessenden. See *GE Review*, 1 (1903): 22.

49. Aitken, *Continuous Wave*, pp. 70–71. Also see Douglas, *Inventing American Broadcasting*, pp. 80–91.

50. A shipping order dated December 8, 1902, CRC, instructed that the alternator was to be sent to the NESCO station at Old Point Comfort. However, it was not shipped until March 1903. See F. P. Beach to E. B. Raymond, March 19, 1903, CRC.

51. R. A. Fessenden to J. F. Wessel, December 8, 1904, CRC.

52. EA to A. G. Davis, "History of Development of High-Frequency Alternator," memorandum, June 29, 1915, AP.

53. J. F. Wessel to R. A. Fessenden, December 10, 1904; Fessenden to Wessel, December 12, 1904, CRC.

54. J. F. Wessel to R. A. Fessenden, December 13, 1904, CRC.

55. Alger, *Human Side of Engineering*, pp. 34, 37.

56. For biographical information on Berg, see "E. J. Berg," *NCAB*, 35:99; and *Electrical World and Engineer*, 38 (1901): 284.

57. Samuel Sheldon and Hobart Mason, *Alternating-Current Machines* (New York: D. Van Nostrand, 1904), pp. 80–85.

58. EA, "Alternating-Current Generator for 100,000 cycles," memorandum attached to E. J. Berg to R. A. Fessenden, December 21, 1904, CRC.

59. For biographical information on Lamme, see "B. G. Lamme," *Electrical World and Engineer*, 38 (1901): 88. Also see Benjamin Garver Lamme, *An Autobiography* (New York: G. P. Putnam's Sons, 1926).

60. Aitken, *Continuous Wave*, p. 63.

61. Lamme, *Autobiography*, pp. 105, 193. See also B. A. Behrend, *The Induction Motor and Other Alternating Current Motors* (New York: McGraw-Hill, 1921), pp. 187–91.

62. Benjamin G. Lamme, "An Early High Frequency Alternator," *Electric Journal*, 18 (April 1921): 110.

63. Benjamin G. Lamme, "Data and Tests on a 10,000 cycle-per-second Alternator," *Transactions of the AIEE, 23 (1904): 417–28*.

64. Lamme, "Early High Frequency Alternator," p. 110. The Lamme alternator may have influenced the design of the alternator designed by M. C. A. Latour and installed at a transmitting station in France after World War I. See D. G. Little, "Continuous Wave Radio Communication," *Electric Journal*, 18 (April 1921): 124–29.

65. Details of Alexanderson's design are given in his U.S. Patent no. 905,621, filed February 6, 1905, issued December 1, 1908. The examiner cited a German patent, a U.S. patent issued to Hutin in 1894, and a U.S. patent issued to Lindell in 1896. Albert Davis agreed to some minor changes in a few claims and erased claim 17 before Alexanderson's patent was issued.

66. E. J. Berg to R. A. Fessenden, December 21, 1904; Fessenden to Berg, January 12, 1905; EA to Berg, January 26, 1905, AP; Berg to Fessenden, January 27, 1905; Fessenden to Berg, February 2, 1905, CRC; EA to Berg, February 6, 1905, AP.

67. Reginald A. Fessenden, "On the Use of Magnetic Formulae in Electrical Design," *Electrical World,* 26 (1895): 214–15.

68. Fessenden, *Fessenden,* pp. 53–54; R. A. Fessenden, "Magnetic Formulae," *Electrical World,* 23 (1894): 834–35.

69. Fessenden, "Electric and Magnetic Quantities of the Ether," p. 95. See also A. E. Kennelly, "Magnetic Reluctance," *Transactions of the AIEE,* 8 (1891): 485–517.

70. EA to F. G. Vaughan, February 16, 1905, AP; C. D. Haskins to F. J. Wessel, February 23, 1905, CRC.

71. R. A. Fessenden to E. J. Berg, February 24, 1905; EA to Berg, February 27, 1905; Berg to Fessenden, February 27, 1905; Fessenden to Berg, March 31, 1905; D. C. Haskins to Fessenden, March 31, 1905; Berg to Fessenden, March 31, 1905; Berg to Fessenden, April 1, 1905; Fessenden to Berg, April 12, 1905, CRC.

72. NESCO to Faneuil Watch Tool Company, June 15, 1905; NESCO to GE Company, July 3, 1905; R. A. Fessenden to E. J. Berg, July 3, 1905; C. D. Haskins to Fassenden, July 6, 1905, CRC.

73. EA to R. A. Fessenden, August 10, 1905, CRC.

74. R. A. Fessenden to EA, August 16, 1905; EA to Fessenden, August 21, 1905, CRC.

75. EA to R. A. Fessenden, October 17, 1905; Fessenden to EA, October 18, 1905; Fessenden to EA, December 14, 1905, CRC.

76. EA to R. A. Fessenden, December 18, 1905, CRC.

77. EA to R. A. Fessenden, January 15, 1906, CRC.

78. EA to H. Geisenhoner, January 15, 1906, AP; clipping from the *Toronto Star Weekly,* September 19, 1942, AP; note on photograph of alternator by J. W. Lee, CRC.

79. EA, "Strength of Revolving Discs," February 1906, AP.

80. Ibid. EA to E. J. Berg, March 1, 1906; EA to A. G. Davis, March 1, 1906; EA to S. Sheldon, June 27, 1906; F. W. Caldwell to EA, April 19, 1909, AP. Alexanderson's patent, entitled "Method of Manufacturing Rotating Bodies for High-Speed Machinery," was filed August 3, 1907, and issued July 25, 1911, as U.S. Patent no. 998,734.

81. EA to E. Rivett, March 5, 1906, AP.

82. Ernst F. W. Alexanderson, "Alternator for One Hundred Thousand Cycles," *Transactions of the AIEE,* 28 (1909): 403–8.

83. EA to C. D. Haskins, March 26, 1906, AP. Earlier, Fessenden had argued that the DeLaval gear could not be driven in reverse and that he had seen it tried while working for Edison. R. A. Fessenden to EA, December 14, 1905, CRC.

84. EA to R. A. Fessenden, March 27, 1906, AP; Fessenden to C. D. Haskins, April 7, 1906, CRC.

85. EA to C. D. Haskins, April 17, 1906, AP.

86. EA to R. A. Fessenden, June 19, 1906; Fessenden to EA, June 20, 1906; EA to H. G. Reist, June 28, 1906, CRC.

87. EA to R. A. Fessenden, July 5, 1906; H. G. Reist to F. G. Vaughan, August 21, 1906; C. D. Haskins to Fessenden, August 28, 1906; Fessenden to Haskins, September 17, 1906, CRC.

88. R. A. Fessenden to C. Feldman, September 21, 1906, CRC. Clarence Feldman was in Holland and apparently was serving as NESCO intermediary to German companies.

89. R. A. Fessenden to C. D. Haskins, September 20, 1906; Fessenden to A. Dempster, September 28, 1906, CRC.

90. H. G. Reist to R. A. Fessenden, October 1, 1906; Fessenden to C. D. Haskins, October 3, 1906; Fessenden to Reist, November 16, 1906, CRC.

91. R. A. Fessenden to H. G. Reist, October 12, 1906; Reist to Fessenden, October

22, 1906; Fessenden to Reist, November 6, 1906; Fessenden to Reist, December 7, 1906, CRC.

92. NESCO to Bell Telephone Company, December 12, 1906, CRC.

93. Excerpts from Pickard's report, "Wireless Tests at Brant Rock and Plymouth, Massachusetts," were published as an appendix in Ernst Ruhmer, *Wireless Telephony in Theory and Practice*, trans. James Erskine-Murray (New York: D. Van Nostrand, 1908), pp. 205–14.

94. A copy of Pickard's earlier report, "Hertzian Wave Transmission of Speech," dated September 6, 1902, together with photographs of experimental apparatus, is attached to a letter from Pickard to Lloyd Espenschied dated July 22, 1948, in the Lloyd Espenschied Papers, National Museum of American History, Smithsonian Institution, Washington, D.C. The report is also discussed in R. A. Fessenden to H. V. Hayes, December 31, 1906; and Hayes to Fessenden, January 15, 1907, CRC.

95. H. V. Hayes to EA, March 11, 1907, AP.

96. George W. Hilton and John F. Due, *The Electric Interurban Railways in America* (Stanford, Calif.: Stanford University Press, 1960): pp. 3, 226. Also see Albro Martin, *Enterprise Denied: Origins of the Decline of American Railroads, 1897–1917* (New York: Columbia University Press, 1971), p. viii.

97. William S. Murray, *Superpower: Its Genesis and Future* (New York: McGraw-Hill, 1925), pp. 90–94.

98. Ibid., pp. 97–98, 111–15. This battle of the systems may be compared to the well-known earlier battle of electric supply systems between the Edison DC system and the AC system promoted by Westinghouse. This battle was resolved during the 1890s by a synthesis of the systems into a universal polyphase power supply system. See Hughes, *Networks of Power*, chap. 5.

99. See a comment by Charles P. Steinmetz in *Transactions of the AIEE*, 20 (1902): 31.

100. Murray, *Superpower*, pp. 112–13. See also Michael Bezilla, *Electric Traction on the Pennsylvania Railroad, 1895–1968* (University Park: Pennsylvania State University Press, 1980), p. 35.

101. Bezilla, *Electric Traction on the Pennsylvania Railroad*, pp. 36–37.

102. Comment by William S. Murray in *Transactions of the AIEE*, 27 (1908): 1713. Also see Lamme, *Autobiography*, p. 123.

103. Benjamin G. Lamme, "Washington, Baltimore, and Annapolis Single-Phase Railway," *Transactions of the AIEE*, 20 (1902): 15–30. The WB&A included a 31-mile line from Washington, D.C., to Baltimore and a fifteen-mile spur to Annapolis.

104. Comment by Charles P. Steinmetz in *Transactions of the AIEE*, 20 (1902): 31–33. See a discussion of the AC series motor in Franklin and Esty, *Elements of Electrical Engineering*, vol. 2, chap. 14. See also the discussion of series and repulsion AC motors in Ralph R. Lawrence and Henry E. Richards, *Principles of Alternating-Current Machinery* (New York: McGraw-Hill, 1953), pp. 520–41.

105. Comment by Lamme in *Transactions of the AIEE*, 20 (1902): 43. Presumably, his "good reasons" involved proprietary features of the design.

106. Charles P. Steinmetz, "The Alternating-Current Railway Motor," *Transactions of the AIEE*, 23 (1904): 9–25.

107. Walter I. Slichter, "Speed-Torque Characteristics of the Single-Phase Repulsion Motor," *Transactions of the AIEE*, 23 (1904): 1–7. See also the discussion in ibid., 23 (1904): 33.

108. Ernst F. W. Alexanderson, "A Single-Phase Railway Motor," *Transactions of the AIEE*, 27 (1908): 1.

109. Charles P. Steinmetz, "Theory of the Single-Phase Motor," in *Transactions of the International Electrical Congress* (St. Louis, 1904), 3:76–128.

110. Ernst Danielson, "Theory of the Compensated Repulsion Motor," in *Transactions of the International Electrical Congress*, 3:174–85.

111. See the entry on Murray in *NCAB*, vol. E, pp. 402–3; "W. S. Murray," *Electrical World*, 55 (1910): 399; Bezilla, *Electric Traction on the Pennsylvania Railroad*, pp. 95–96. See also Hughes, *Networks of Power*, pp. 296–97.

112. Murray, *Superpower*, p. 91.

113. See Lamme, *Autobiography*, pp. 124–28.

114. On technology transfer, see Hughes, *Networks of Power*, chap. 3 and p. 405.

115. E. D. Priest, "Present Development of Alternating-Current Railway Motors," memorandum, May 6, 1907, AP.

116. EA to W. H. Frost, June 7, 1906; EA to E. D. Priest, June 7, 1906; EA to E. W. Rice, June 21, 1906, AP. In EA to L. J. Magie, June 29, 1906, AP, Alexanderson requested letters of introduction to several German engineers.

117. Hill-Wagoner interview. E. Danielson to EA, August 3, 1906; EA to Danielson, August 15, 1906, AP. Also see "The Swedish Electric Railway Experiments," *Electrical World and Engineer*, 46 (1905): 805–6; Robert Dahlander, "Single-Phase Electric Traction on the Swedish State Railways," *Electrician*, 62 (1908–9): 792–93; Benjamin G. Lamme, "The Use of Alternating Current for Heavy Railway Service," *Electric Journal*, 3 (1906): 97–105.

118. G. H. Hill to E. W. Rice, October 15, 1906, AP. Latour received U.S. Patent no. 841,257 on a compensated series motor. His application was filed November 9, 1905, and the patent was issued January 15, 1907, with assignment to GE. Also see EA to Rice, September 24, 1906, AP.

119. EA to W. B. Potter, November 2, 1906, AP.

120. "Electrification of Steam Railroads," *GE Review*, 8 (1906–7): 118–19.

121. EA to W. B. Potter, May 16, 1907, AP. The railroads listed began operation with single-phase AC but changed to DC by 1915. See Hilton and Due, *Electric Interurban Railways in America*, p. 62.

122. A. S. McAllister, *Alternating Current Motors* (New York: McGraw-Hill, 1907). The book contained a chapter on repulsion motors. Alexanderson ordered copies of an issue of *Scientific American* containing "Opening of Electric Service on the New York Central and the New Haven Railroads," *Scientific American*, 96 (1907): 72–73.

123. The report to A. G. Davis, written by Pritchart and Salmonson, was dated June 5, 1907, AP.

124. EA to F. B. Howell, May 29, 1907, AP. On resistance leads, see Franklin and Esty, *Elements of Electrical Engineering*, 1:311–12.

125. C. P. Steinmetz, Technical Report no. M 373, August 27, 1907, AP. See also the comment by Steinmetz in *Transactions of the AIEE*, 26 (1907): 1399–1402.

126. Frank Sprague, "Some Facts and Problems Bearing on Electric Trunk-Line Operation," *Transactions of the AIEE*, 26 (1907): 691–92, 747–53.

127. Discussion in *Transactions of the AIEE*, 26 (1907): 778–98.

128. W. J. Davis, "High-Voltage Direct-Current and Alternating-Current Systems for Interurban Railways," *Transactions of the AIEE*, 26 (1907): 387–92.

129. Ernst F. W. Alexanderson, "A Single-Phase Railway Motor," *Transactions of the AIEE*, 27 (1908): 1–17.

130. Discussion in *Transactions of the AIEE*, 27 (1908): 19–28.

131. Discussion in *Transactions of the AIEE*, 27 (1908): 28–29, 35–38.

132. William B. Potter, "Economics of Railway Electrification," *GE Review*, 13 (1910): 389–98.

133. Hilton and Due, *Electric Interurban Railways in America*, p. 62. Also see J. J. Doyle,

"The Washington, Baltimore, and Annapolis Electric Railroad from an Operating Standpoint," *GE Review,* 16 (1913): 785–97.

134. William B. Potter and G. H. Hill, "A Review of Electric Railways," *GE Review,* 18 (1915): 444–53.

135. Ernst F. W. Alexanderson, "Single-Phase Motor-Car Equipment for Trunk Line Service," *GE Review,* 16 (1913): 326–32. See also "Alexanderson Single-Phase Motor and Phase Converter for Locomotive Service," *Electric Railway Journal,* 42 (1913): 677–81.

136. EA to C. P. Steinmetz, April 3, 1915, AP.

137. EA to C. P. Steinmetz, December 3, 1915, AP.

138. Ernst F. W. Alexanderson, "Communication," *Electric Railway Journal,* 46 (1915): 1174. The editorial was "The Repulsion-Starting Series Motor," ibid., 46 (1915): 1019–20.

139. "Philadelphia-Paoli Electrification," *Electric Railway Journal,* 46 (1915): 981–89.

140. Bezilla, *Electric Traction on the Pennsylvania Railroad,* pp. 62–65.

141. EA to E. W. Rice, June 18, 1907; Rice to EA, June 24, 1907, AP.

142. EA to S. B. Paine, April 26, 1909, AP. Also see Harold C. Passer, *The Electrical Manufacturers, 1875–1900* (Cambridge, Mass.: Harvard University Press, 1955), pp. 302–5.

143. See the discussion of synchronous speed in Lawrence and Richard, *Principles of Alternating-Current Machinery,* pp. 388–89.

144. Ernst F. W. Alexanderson, "Repulsion Motor with Variable-Speed, Shunt Characteristics," *Transactions of the AIEE,* 28 (1909): 511–15.

145. Discussion in *Transactions of the AIEE,* 28 (1909): 516–24.

146. "Adjustable-Speed Single-Phase Commutator Motor," *Electrical World,* 54 (1909): 1252.

147. Ernst F. W. Alexanderson, comment in *Transactions of the AIEE,* 31 (1912): 2159.

148. EA, "Invention and the Corporation," TS of talk, January 31, 1951, AP.

Chapter Three: High-Frequency Alternators and Wireless Politics

1. J. Stone, report proposing a "dynamo relay," May 26, 1892; Stone, memorandum, July 12, 1892, Espenschied Papers.

2. G. Pickard, "Memorandum on Proposed Telephone Relay," September 23, 1902, Espenschied Papers. For a biographical note on Pickard, see "G. W. Pickard," *Electrical World,* 56 (1910): 1052–53.

3. H. V. Hayes to J. A. Barrett, September 24, 1902, Espenschied Papers.

4. James E. Brittain, "The Introduction of the Loading Coil: George A. Campbell and Michael I. Pupin," *Technology and Culture,* 11 (1970): 36–57. For a discussion of the concept of presumptive anomaly, see Edward W. Constant II, *The Origins of the Turbojet Revolution* (Baltimore: Johns Hopkins University Press, 1980), pp. 15–16.

5. Lillian Hoddeson, "The Emergence of Basic Research in the Bell Telephone System, 1875–1915," *Technology and Culture,* 22 (1981): 512–44. See also Leonard S. Reich, *The Making of American Industrial Research* (Cambridge: Cambridge University Press, 1985), pp. 157–60.

6. Alfred N. Goldsmith, *Radio Telephony* (New York: Wireless Press, 1918), pp. 15–17, 84.

7. H. V. Hayes to EA, March 11, 1907; EA, "Report of Conference with Mr. Gould Regarding Generator for Telephone Relay," March 15, 1907, AP.

8. EA to A. G. Davis, March 20, 1907; Manufacturing Order 3698, March 29, 1907; EA to H. Geisenhoner, April 2, 1907, AP. Alexanderson's patent application, entitled "Telephone Relay," was filed January 25, 1908, and was issued as U.S. Patent no. 902,195, October 27, 1908.

9. R. A. Fessenden to H. G. Reist, February 5, 1907; Fessenden to C. D. Haskins, April 20, 1906; Fessenden to EA, May 8, 1907; Reist to W. F. Howe, May 20, 1907, CRC.

10. EA to R. A. Fessenden, June 13, 1907; Fessenden to EA, June 18, 1907, CRC.

11. R. A. Fessenden to C. D. Haskins, June 26, 1907; Fessenden to EA, July 10, 1907, CRC.

12. C. Robinson to R. C. Miller, August 10, 1907; Robinson to S. S. Forster, August 10, 1907; Robinson to EA, October 2, 1907; Robinson to Miller, October 16, 1907, CRC.

13. EA to C. D. Haskins, September 3, 1907; R. A. Fessenden to Haskins, September 4, 1907, CRC.

14. H. V. Hayes to R. A. Fessenden, September 10, 1907; Fessenden to Hayes, September 25, 1907; Hayes to Fessenden, October 30, 1907, CRC.

15. L. Espenschied to C. F. Elwell, September 23, 1952, Espenschied Papers. Espenschied mentioned a proposal from H. J. Round for an audion investigation in 1906. For an informed analysis of the Bell reorganization and its overall impact, see Reich, *Making of American Industrial Research*, pp. 151–70.

16. R. A. Fessenden to EA, November 6, 1907; Fessenden to C. D. Haskins, December 2, 1907; EA to Fessenden, January 17, 1908; Fessenden to Haskins, January 20, 1908; R. C. Barnes to C. Robinson, January 22, 1908, CRC.

17. C. Robinson to H. G. Reist, January 24, 1908; Robinson to R. C. Barnes, January 24, 1908; Robinson to C. D. Haskins, February 24, 1908, CRC.

18. H. G. Reist to C. D. Haskins, January 27, 1908; Reist to A. G. Davis, January 27, 1908; Davis to C. Robinson, January 28, 1908; E. P. Edwards to F. C. Pratt, February 24, 1908, CRC. Pratt was the son of a founder of the Pratt and Whitney Company. After graduating in mechanical engineering from Yale in 1888, Pratt worked at Pratt and Whitney until 1906, when he joined GE as an assistant to Edwin Rice, Jr. For Pratt's biography, see *NCAB*, 27:233–34. Edmund Edwards worked as an assistant to Caryl D. Haskins and seems to have devoted considerable energy to product diversification. See, for example, Edmund P. Edwards, "The Promise of Electrified Agriculture," *GE Review*, 13 (1910): 546–49.

19. E. P. Edwards to R. A. Fessenden, April 7, 1908; Edwards to F. C. Pratt, April 22, 1908; H. G. Reist to C. Robinson, April 30, 1908; Edwards to E. Clark, April 22, 1908, CRC. Fessenden was billed for $1,775 for the uncompleted Robinson project.

20. See "The Wireless Conference," *Electrical World*, 48 (1906): 629–30; and "Objections to the Wireless Convention," *Electrical World*, 51 (1908): 1332–34. See also Douglas, *Inventing American Broadcasting*, pp. 137–42.

21. R. A. Fessenden to E. Thomson, January 27, 1908; Thomson to Fessenden, January 29, 1908; Thomson to Sen. H. C. Lodge, January 31, 1908, CRC.

22. R. A. Fessenden, "Brief Filed by National Electric Signalling Company in Opposition to the Ratification of the Berlin International Wireless Telegraph Convention," CRC.

23. A copy of Lodge's letter was attached to E. Thomson to R. A. Fessenden, February 4, 1908, CRC.

24. J. H. Hayden to H. Walker, Jr., February 14, 1908, CRC.

25. NESCO, "The Regulation of Wireless Telegraphy by Senate Bill 5949," March 11, 1908; R. A. Fessenden to A. E. Kennelly, June 10, 1908, CRC. See also "Ratification of the Berlin International Wireless Telegraph Convention," *Electrical World*, 59 (1902): 787.

26. R. A. Fessenden, "Wireless Telephony," *Proceedings of the American Institute of Electrical Engineers* (hereafter *Proceedings of the AIEE*), 27 (1908): 1283–1358.

27. Ibid., pp. 1342–52.

28. EA. "Telephone Relay," U.S. Patent no. 902,195, October 27, 1908. Also EA to R. A. Fessenden, March 9, 1908, CRC.

29. Ernst F. W. Alexanderson, "Generating Apparatus for Wireless Telegraphy and Telephony," *GE Review*, 16 (1913): 21–22.

30. Goldsmith, *Radio Telephony*, pp. 192–93.

31. EA to D. Miller, May 13, 1908; EA to R. A. Fessenden, May 5, 1908; EA to Fessenden, July 11, 1908, CRC.

32. EA to R. A. Fessenden, September 11, 1908; E. P. Edwards to Fessenden, September 24, 1908; Fessenden to Edwards, September 25, 1908; EA to Fessenden, December 5, 1908, CRC; "Thrust Bearings with Temperature Expansion," drawing, December 5, 1908, AP.

33. The air-friction formula was contained in EA to R. A. Fessenden, March 7, 1910, CRC. Also see Ernst F. W. Alexanderson, "Magnetic Properties of Iron at Frequencies up to 200,000 Cycles," *Transactions of the AIEE*, 30 (1911): 2434.

34. E. P. Edwards to J. H. Lovejoy, October 22, 1908, CRC.

35. E. P. Edwards to R. A. Fessenden, March 13, 1909; Edwards to H. Walker, Jr., March 23, 1909, CRC.

36. "CQD," *Electrical World*, 53 (1909): 254; and "Wireless and Wreck of the Republic," *Electrical World*, 53 (1909): 265. Also see Douglas, *Inventing American Broadcasting*, pp. 200–202, 219.

37. R. A. Fessenden to EA, June 21, 1909; EA to Fessenden, June 21, 1909; Fessenden to EA, July 21, 1909, CRC.

38. E. P. Edwards to R. A. Fessenden, September 1, 1909; Fessenden to Edwards, October 11, 1909; Edwards to Fessenden, December 7, 1909, CRC.

39. EA to R. A. Fessenden, May 28, 1909; Fessenden to EA, June 3, 1909; Fessenden to EA, June 30, 1909, CRC.

40. Ernst F. W. Alexanderson, "Alternator for One Hundred Thousand Cycles," *Transactions of the AIEE*, 28 (1909): 399–412.

41. Discussion in *Transactions of the AIEE*, 28 (1909): 413–15.

42. Arthur E. Kennelly and Ernst F. W. Alexanderson, "The Physiological Tolerance of Alternating-Current Strengths up to Frequencies of 100,000 Cycles per Second," *Electrical World*, 56 (1910): 154–56.

43. Data on Kennelly's earlier experiments are contained in the first volume of a series of "West Orange Observations, Galvanometer Building," Thomas A. Edison Papers, Edison National Historic Site, West Orange, N.J.

44. EA to R. A. Fessenden, February 11, 1910, CRC.

45. R. A. Fessenden to J. R. Werth, August 13, 1909; Fessenden to E. P. Edwards, August 9, 1909; EA to Fessenden, March 7, 1910; EA to Fessenden, March 10, 1910; EA to J. H. Kelman, June 1, 1910, CRC.

46. R. A. Fessenden to E. P. Edwards, March 5, 1909; Edwards to Fessenden, March 22, 1909; Fessenden to Edwards, March 23, 1909; EA to J. H. Kelman, July 1, 1910, CRC.

47. R. A. Fessenden to T. H. Vail, February 25, 1910, CRC.

48. Lloyd Espenschied, "Concerning the Telephone Company's Early Interest in the Fessenden Radio," memorandum, October 28, 1947, Espenschied Papers.

49. H. Walker to R. A. Fessenden, December 10, 1910, Reginald Aubrey Fessenden Papers, North Carolina Department of Archives and History, Raleigh, N.C. See also Aitken, *Continuous Wave*, pp. 82–85, 142–43, 456; and Douglas, *Inventing American Broadcasting*, pp. 163–67.

50. Hill-Wagoner interview; EA, transcript of banquet speech to radio engineers, January 13, 1942, AP.

51. Alexanderson signed three-year contracts on February 1, 1907, and February 1, 1910. Copies of the contracts are included in AP.

52. From a copy of Alexanderson's certificate of naturalization dated January 27, 1908, AP.

53. Amelie Alexanderson Wallace, interview with author, Garden City, New York, June 19, 1977; Edith Alexanderson Nordlander, interview with author, November 15, 1978; Verner Alexanderson, "Family Tree, E. F. W. Alexanderson," September 30, 1977; Amelie Alexanderson Wallace, personal communication, February 18, 1979.

54. A. Alexanderson to EA, May 18, 1909, AP.

55. A. Alexanderson to E. Alexanderson, July 7, 1909; A. Alexanderson to E. Alexanderson, November 12, 1909, AP.

56. Copy of the Agora's Constitution, n.d.; Agora meeting announcements, September 6, 1907, and October 23, 1911; list of Agora topics for 1907–11, AP.

57. EA, "Women and Labor," notes for talk given October 20, 1911, AP.

Chapter Four: Corporate Consultant

1. Wise, *Willis R. Whitney*, pp. 75–85.

2. James E. Brittain, "C. P. Steinmetz and Ernst F. W. Alexanderson: Creative Engineering in a Corporate Setting," *Proceedings of the IEEE*, 64 (1976): 1413–17.

3. C. P. Steinmetz, "Organization of Consulting Engineering Department and of Laboratories Connected Therewith," memorandum, July 10, 1912, Samuel P. Nixdorff Papers, Schaffer Library, Union College, Schenectady, N.Y. (hereafter cited as NP).

4. Ibid.

5. Ibid.

6. Ibid.

7. See "Notes from the Consulting Engineering Department and the Transformer Department," *GE Review*, 16 (1913): 40.

8. EA, "Report for Consulting Engineering Department," October 9, 1911, AP.

9. EA, memorandum with trip itinerary, October 22, 1910, AP.

10. EA to E. W. Rice, January 13, 1911; EA, "Hunt's Cascade Motor," memorandum, n.d., AP.

11. "Generators of Electric Oscillations for Wireless Telegraphy," *Electrical World*, 57 (1911): 386; Emil E. Mayer, "The Goldschmidt System of Radio Telegraphy," *Proceedings of the IRE*, 2 (1914): 69–91. Also see J. Zenneck, *Wireless Telegraphy*, trans. A. E. Seelig (New York: McGraw-Hill, 1915), pp. 216–19.

12. EA to M. A. Oudin, February 25, 1911; EA to J. C. Close, March 4, 1911; EA to C. D. Haskins, March 29, 1911, AP.

13. Oscar C. Roos, "Concerning American Wireless Engineering," *Electrical World*, 57 (1911): 729–31.

14. EA to C. W. Stone, May 11, 1911, AP.

15. See a biographical item on Squier in "Major G. O. Squier," *Electrical World*, 45 (1905): 928; and National Academy of Sciences, *Biographical Memoirs*, 20:151–59. Also see Paul W. Clark, "Early Impacts of Communications on Military Doctrine," *Proceedings of the IEEE*, 64 (1976): 1407–13.

16. George O. Squier, "Multiplex Telephony and Telegraphy by Means of Electric Waves Guided by Wires," *Transactions of the AIEE*, 30 (1911): 1617–65.

17. Discussion in *Transactions of the AIEE*, 30 (1911): 1666–75. For a discussion of later work on multiplex at AT&T, see Leonard S. Reich, *The Making of American Industrial Research* (Cambridge: Cambridge University Press, 1985), pp. 170–71.

18. EA to A. E. Kennelly, November 1, 1911; Kennelly to EA, November 7, 1911, AP.

19. L. de Forest to EA, June 12, 1911; EA to de Forest, August 16, 1911; E. P. Edwards to H. Walker, August 15, 1911, AP.

20. Ernst F. W. Alexanderson, "Magnetic Properties of Iron at Frequencies Up to 200,000 Cycles," *Transactions of the AIEE*, 30 (1911): 2433–47.

21. Discussion in *Transactions of the AIEE*, 30 (1911): 2450–51.

22. Ibid., 2451–52.

23. "Magnetic Properties of Iron at Abnormal Frequencies," *Electrical World*, 58 (1911): 1276.

24. For a discussion of the role of critical problems as a stimulus to invention, see T. P. Hughes, *Networks of Power*, p. 80.

25. Goldsmith, *Radio Telephony*, p. 183.

26. EA to A. G. Davis, January 19, 1912, AP. Alexanderson filed a patent application on the magnetic amplifier on December 7, 1912; the patent was issued as U.S. Patent no. 1,206,643 on November 28, 1916.

27. John Anderson Miller, *Workshop of Engineers* (Schenectady, N.Y.: General Electric Company, 1953), p. 34.

28. Ernst F. W. Alexanderson and Samuel P. Nixdorff, "Magnetic Amplifier for Radiotelephony," *GE Review*, 19 (1916): 215–22. See also "Magnetic Amplifier in Radio Work," *Electrical World*, 68 (1916): 234–35.

29. EA to C. W. Stone, January 22, 1912; EA to A. A. Buck, February 1, 1912, AP.

30. "The Disaster to the Titanic," *Electrical World*, 59 (1912): 579–80; and "The Control of Wireless," ibid., 59 (1912): 880. See also Douglas, *Inventing American Broadcasting*, pp. 226–33.

31. Verner Alexanderson, "E. F. W. Alexanderson, Family Tree," September 30, 1977; Edith Alexanderson Nordlander, telephone interview with author, November 15, 1978.

32. Based on biographical information in the John Hays Hammond Papers, Manuscript Division, Library of Congress, Washington, D.C. (hereafter cited as HP). Also see Orrin E. Dunlap, *Radio's 100 Men of Science* (New York: Harper & Brothers, 1944), p. 229; and Benjamin F. Miessner, *On the Early History of Radio Guidance* (San Francisco: San Francisco Press, 1964), pp. 1–14.

33. T. S. Bacon to J. H. Hammond, Jr., September 19, 1912; Hammond to EA, September 25, 1912, HP.

34. EA to F. Lowenstein, September 13, 1912, HP.

35. J. H. Hammond, Jr., to L. W. Austin, October 11, 1912, HP. For biographical information on Austin, see Dunlap, *Radio's 100 Men of Science*, p. 143; and Rexmond C. Cochrane, *Measures for Progress: A History of the National Bureau of Standards* (Washington, D.C.: Department of Commerce, 1966), pp. 13, 140. Austin headed a radio laboratory at the National Bureau of Standards from 1908 to 1932.

36. J. H. Hammond, Jr., to EA, October 21, 1912; Hammond to EA, December 9, 1912; EA to Hammond, December 19, 1912, HP. Also see EA to Hammond, January 25, 1913, AP.

37. For a comprehensive discussion of de Forest and the audion, see Aitken, *Continuous Wave*, pp. 194–233.

38. F. Lowenstein to J. H. Hammond, Jr., November 8, 1911; Lowenstein to Hammond, November 13, 1911, Espenschied Papers.

39. EA to J. H. Hammond, Jr., October 21, 1912, HP.

40. J. H. Hammond, Jr., to EA, December 9, 1912, HP; EA to L. A. Hawkins,

February 4, 1913, AP. For biographical information on Pierce, see Dunlap, *Radio's 100 Men of Science,* p. 161.

41. Wise, *Willis R. Whitney,* pp. 149–59. See also Reich, *Making of American Industrial Research,* pp. 82–83.

42. Wise, *Willis R. Whitney,* p. 176.

43. EA to J. H. Hammond, Jr., January 20, 1913; EA to L. A. Hawkins, February 4, 1913, AP.

44. EA to A. G. Davis, February 4, 1913; J. H. Hammond, Jr., to EA, February 13, 1913; EA to L. A. Hawkins, February 14, 1913, AP.

45. EA, with the assistance of E. E. Thomas, "Tuning in Geometrical Progression," memorandum, February 27, 1913, AP. Also see EA to C. P. Steinmetz, March 8, 1913, CRC.

46. For a discussion of the Edison effect and its role in the early history of electronics, see Aitken, *Continuous Wave,* pp. 205–11.

47. I. Langmuir, Notebook no. 457, p. 66, Irving Langmuir Papers, Manuscript Division, Library of Congress, Washington, D.C.

48. Wise, *Willis R. Whitney,* pp. 173–75.

49. I. Langmuir, Notebook no. 457, pp. 124–44, 154, 205, Langmuir Papers.

50. Ibid., pp. 223–31. Alexanderson's patent on a radio circuit using mercury-vapor tubes was issued as U.S. Patent no. 1,042,069 on October 22, 1912.

51. I. Langmuir, Notebook no. 457, pp. 236–41; Langmuir, Notebook no. 458, pp. 9–34, Langmuir Papers.

52. J. H. Hammond, Jr., to EA, April 9, 1913; EA to Hammond, April 11, 1913, AP. Also see I. Langmuir, Notebook no. 458, p. 121, Langmuir Papers.

53. W. C. White, interview by Frank Hill, New York City, December 13, 1950. I am indebted to George Wise for providing me with a transcript of the interview.

54. EA to M. W. Sage, May 14, 1913; EA to A. A. Buck, May 28, 1913; EA to I. Langmuir, June 11, 1913, AP.

55. I. Langmuir, Notebook no. 458, p. 130, Langmuir Papers.

56. EA to J. H. Hammond, Jr., June 12, 1913; I. Langmuir to Hammond, June 14, 1913; Hammond to EA, August 19, 1913; EA to W. R. Whitney, August 21, 1913, AP.

57. Alexanderson filed a patent application on October 29, 1913, and it was issued as U.S. Patent no. 1,173,079 on February 22, 1916.

58. For biographical information on Hull, see the National Academy of Sciences, *Biographical Memoirs,* 41:215–33. Also see Wise, *Willis R. Whitney,* pp. 159–60; and Reich, *Making of American Industrial Research,* pp. 87–88.

59. EA to A. G. Davis, December 29, 1913, AP.

60. James E. Brittain, "Power Electronics at GE, 1900–1941," *Advances in Electronics and Electron Physics,* 50 (1980): 417. For similar reasons, Alexanderson's father, a classical languages scholar, preferred the term *teleopt* to *television.*

61. EA to I. Langmuir, February 18, 1914, AP.

62. Wise, *Willis R. Whitney,* pp. 176–77.

63. Laurens E. Whittemore, "The Institute of Radio Engineers—Fifty Years of Service," *Proceedings of the IRE,* 50 (1962): 534–58. See also A. Michal McMahon, *The Making of a Profession: A Century of Electrical Engineering in America* (New York: The IEEE Press, 1984), pp. 127–32.

64. McMahon, *Making of a Profession,* p. 130. See also James E. Brittain, ed., *Turning Points in American Electrical History* (New York: IEEE Press, 1977), p. 200; and EA to A. N. Goldsmith, October 24, 1913, AP.

65. Charles Proteus Steinmetz, *Theoretical Elements of Electrical Engineering* (New York: Electrical World and Engineer, 1902), pp. 48–57.

66. Charles Proteus Steinmetz, "On the Law of Hysteresis," *Transactions of the AIEE*, 9 (1892): 3–64.

67. EA to H. C. Senior, October 24, 1913; S. P. Nixdorff, "Winding Schemes for Three Speed Quarter Phase Induction Motor," report, February 1, 1912, AP. I also have used a biographical note on Nixdorff provided by Frances Miller, Special Collections, Union College, Schenectady, N.Y.

68. Ernst F. W. Alexanderson, "Dielectric Hysteresis at Radio Frequencies," *Proceedings of the IRE*, 2 (1914): 137–45.

69. EA to E. M. Hewlett, November 17, 1913, AP. Alexanderson applied for a patent on an asbestos protective device on November 26, 1915; and it was issued as U.S. Patent no. 1,266,377 on May 14, 1918.

70. Alexanderson, "Dielectric Hysteresis at Radio Frequencies," p. 145. Also see Edwin T. Layton, "American Ideologies of Science and Engineering," *Technology and Culture*, 17 (1976): 695–99. For a related but somewhat different interpretation, which points to the difficulty in distinguishing between scientific and technological research in industry, see Reich, *Making of American Industrial Research*, pp. 205–6.

71. Alexanderson, "Dielectric Hysteresis at Radio Frequencies," p. 147.

72. EA to C. P. Steinmetz, October 30, 1913, AP.

73. Alexanderson, "Dielectric Hysteresis at Radio Frequencies," p. 147.

74. Ibid., pp. 151–53. Frank W. Peek, "Electrical Characteristics of Solid Insulators," *GE Review*, 18 (1915): 1051. Also see the discussion of Peek's research in Thomas P. Hughes, "The Science-Technology Interaction: The Case of High-Voltage Power Transmission Systems," *Technology and Culture*, 17 (1976): 646–62.

75. Discussion by H. E. Hallborg, *Proceedings of the IRE*, 2 (1914): 157.

76. Ernst F. W. Alexanderson, "Generating Apparatus for Wireless Telegraphy and Telephony," *GE Review*, 16 (1913): 16–22; Michael I. Pupin, "A Discussion of Experimental Tests of the Radiation Law for Radio Oscillators," *Proceedings of the IRE*, 1 (1913): 4.

77. Alexanderson, "Generating Apparatus for Wireless Telegraphy and Telephony," pp. 16–22. Also see Ernst F. W. Alexanderson, "High-Frequency Generator for Wireless Telegraphy and Telephony," *Scientific American Supplement*, 75 (1913): 328–30.

78. EA to E. P. Edwards, February 6, 1913; EA to T. S. Bacon, September 4, 1913; EA to L. Gibson, October 27, 1913, AP. The Marconi Wireless Telegraph Company of America was frequently referred to as American Marconi, while the Marconi Wireless Telegraphy and Signal Company was referred to as British Marconi. My sources often did not distinguish between the two, simply referring to "the Marconi Company."

79. "Practicability of Electrification," *Electrical World*, 62 (1913): 723.

80. Editorial, *Electric Railway Journal*, 42 (1913): 1267.

81. H. M. Hobart, "2400 Volt Railway Electrification," *Transactions of the AIEE*, 32 (1913): 1149–88.

82. Charles P. Kahler, "Trunk Line Electrification," *Transactions of the AIEE*, 32 (1913): 1189–1226; comments by Horace F. Parshall, ibid., 32 (1913): 1245–56.

83. Frank J. Sprague, "The Possibilities of the Electrical Commission," *GE Review*, 17 (1914): 1011–15.

84. "The Paths of Progress," *GE Review*, 17 (1914): 1007–8.

85. John A. Dewhurst, "A Review of American Steam Road Electrification," *GE Review*, 17 (1914): 1144–47.

86. See discussion of the phase converter in Lawrence and Richards, *Principles of Alternating Current Machinery*, pp. 516–19. Alexanderson filed his first two-phase converter patent applications in 1911; they issued as U.S. Patent no. 1,150,652, August 17,

1915, and U.S. Patent no. 1,060,731, May 6, 1913. He filed at least 12 additional patent applications on phase converters during the period 1912–17.

87. Ernst F. W. Alexanderson, "Induction Machines for Heavy Single-Phase Motor Service," *Transactions of the AIEE*, 30 (1911): 1357–60.

88. Ibid., pp. 1360–69; quotation on p. 1360. The phasor diagrams were commonly called vector diagrams by electrical engineers.

89. Ernst F. W. Alexanderson, "The Split-Phase Locomotive," *GE Review*, 16 (1913): 731–33.

90. "Alexanderson Single-Phase Locomotives," *Electrical World*, 62 (1913): 731–33. Also see EA, "Split-Phase Locomotive Electrical Characteristics," report, June 3, 1913, AP.

91. A letter from Alexanderson to the editor requesting three copies of the October 11, 1913, issue of *Electrical World* is dated October 23, 1913, AP. See also "Combination Single-Phase and Polyphase Locomotive for the Norfolk and Western," *Electrical World*, 62 (1913): 367; and F. E. Wynne, "Operation on the Norfolk and Western Railway," *Transactions of the AIEE*, 35 (1916): 147–53.

92. "The Phase Converter Locomotive," *Electrical World*, 62 (1913): 722–23. See also "New Single-Phase Polyphase Motors for the Norfolk and Western," ibid., 62 (1913): 298–99.

93. Ernst F. W. Alexanderson, "Phase Balancer for the Single-Phase Load on Polyphase Systems," *GE Review*, 16 (1913): 962–63.

94. John Liston, "Some Developments in the Electrical Industry during 1915," *GE Review*, 19 (1916): 8. Also see the discussion in *Transactions of the AIEE*, 35 (1916): 1306.

95. Ernst F. W. Alexanderson and George H. Hill, "Single-Phase Power Production," *Transactions of the AIEE*, 35 (1916): 1316–19; "George Henry Hill: In Memoriam," *GE Review*, 20 (1917): 262–63.

96. Alexanderson and Hill, "Single-Phase Power Production," pp. 1316–19.

97. "Award of Contract for Electrification of Melbourne Suburban System," *Electric Railway Journal*, 41 (1913): 255. See also "Melbourne Railway Electrification," ibid., 41 (1913): 1153; "Approval of Melbourne Electrification," ibid., 41 (1913): 71; W. D. Bearce, "Melbourne Suburban Electrified, Australia," *GE Review*, 23 (1920): 662–68; and idem, "Achievements of Ernst F. W. Alexanderson, Chief Engineer of RCA," memorandum, July 14, 1924, AP.

98. Liston, "Some Developments in the Electrical Industry during 1915," p. 8. See also "Some Developments in the Electrical Industry during 1916," *GE Review*, 20 (1917): 11; J. J. Linebaugh, "Regenerative Electric Braking," *GE Review*, 19 (1916): 967–72; and W. D. Bearce, "The Electrification of the Mountain District of the Chicago, Milwaukee, and St. Paul Railway," ibid., 19 (1916): 926–27.

99. See Alexanderson's discussion of a paper by R. E. Hellmund on regenerative braking in *Transactions of the AIEE*, 36 (1917): 60–63.

100. Ernst F. W. Alexanderson, "Critical Speeds of Railway Trucks," *GE Review*, 17 (1914): 1122–24.

101. Ibid., pp. 1124–27.

102. EA to C. P. Steinmetz, July 10, 1914, AP.

103. Albert Rosenfeld, *The Quintessence of Irving Langmuir* (New York: Pergamon, 1966), p. 154.

104. EA to J. H. Hammond, Jr., August 14, 1914, AP; Hammond to EA, September 22, 1914, HP; EA to E. W. Rice, September 11, 1914, AP.

105. EA to C. P. Steinmetz, September 25, 1914, attached to EA, "Development of Apparatus for Radio Communication," memorandum, AP.

106. Ibid.

107. EA to M. W. Sage, August 24, 1914; EA to A. A. Buck, August 25, 1914; EA to I. Langmuir, September 10, 1914, AP.

108. For biographical information on Hogan, see *DAB*, suppl. 6, pp. 298–99.

109. John V. L. Hogan, "Transatlantic Radio Station at Sayville, N.Y.," *Electrical World*, 64 (1914): 615–16. Also see Aitken, *Continuous Wave*, p. 283.

110. "War News by Radio," *Electrical World*, 64 (1914): 597. Also see Fritz van der Wonde and Alfred E. Seelig, "The High Power Telefunken Radio Station at Sayville, Long Island," *Proceedings of the IRE*, 1 (1913): 23–37.

111. John V. L. Hogan, "The Goldschmidt Transatlantic Radio Station," *Electrical World*, 64 (1914): 853–55. See also "Radio Telegraphy to Germany," editorial, ibid., 64 (1914): 843; comment from Alexanderson in *Proceedings of the IRE*, 2 (1914): 95–97; and Aitken, *Continuous Wave*, p. 283.

112. John V. L Hogan, "A New Marconi Transatlantic Service," *Electrical World*, 64 (1914): 425–28. Also see Aitken, *Continuous Wave*, pp. 282–83, 306.

113. F. M. Sammis to GE, December 22, 1914; EA to Sammis, December 24, 1914; E. P. Edwards to J. B. Shelby, December 30, 1914, AP.

114. Verner Alexanderson, "Family Tree, E. F. W. Alexanderson," September 30, 1977; Edith Alexanderson Nordlander, telephone interview with author, November 15, 1978.

Chapter Five: Stentorian Alternator

1. Hughes, *Networks of Power*, pp. 40–45.

2. J. B. Shelby to E. P. Edwards, February 16, 1915; EA to F. C. Pratt, January 6, 1915; EA to H. Farquhar, February 9, 1915; EA to F. M. Sammis, February 20, 1915, AP.

3. EA to F. C. Pratt, April 12, 1915, AP.

4. EA, speech transcript, November 7, 1945, AP.

5. EA to G. Marconi, February 25, 1915; Marconi to EA, April 2, 1915, AP.

6. W. J. Baker, *A History of the Marconi Company* (New York: St. Martin's Press, 1971), p. 171.

7. EA to F. C. Pratt, April 12, 1915, AP.

8. EA, "Notes on Visit to the New Brunswick Station of Marconi Company," May 6, 1915; EA to G. Marconi, May 8, 1915, AP.

9. Gleason L. Archer, *History of Radio to 1926* (New York: American Historical Society, 1938), pp. 129–30.

10. EA, interview with author, Schenectady, N.Y., October 17, 1972.

11. EA to F. C. Pratt, July 7, 1915, AP. Also see Aitken, *Continuous Wave*, pp. 309–11.

12. S. P. Nixdorff, book 1, July 16, 1915, p. 150; and July 30, 1915, p. 195, NP.

13. EA to G. Marconi, September 25, 1915, AP.

14. EA to F. C. Pratt, September 20, 1915, AP.

15. Friedrich Kurylo and Charles Susskind, *Ferdinand Braun* (Cambridge, Mass.: MIT Press, 1981), pp. 203–9.

16. EA to J. Zenneck, June 14, 1915, AP; and S. P. Nixdorff, book 1, June 19, 1915, p. 70, NP.

17. EA to A. G. Davis, April 21, 1915; EA to E. W. Rice, April 22, 1915, AP.

18. Irving Langmuir, "The Pure Electron Discharge and Its Application in Radio Telegraphy and Telephony," *Proceedings of the IRE*, 3 (1915): 261–86; quotation is on p. 283.

19. Discussion in *Proceedings of the IRE*, 3 (1915): 291–92.

20. EA to A. N. Goldsmith, May 10, 1915, AP.

21. Aitken, *Continuous Wave,* pp. 246–48.

22. A. N. Goldsmith to EA, April 14, 1915; EA to Goldsmith, September 24, 1915, AP.

23. EA to M. W. Sage, April 19, 1915 (an attached sketch was entitled "Control of High Frequency Alternator with Tuned Speed Regulator"), AP. Also see EA to M. W. Sage, May 5, 1915, AP.

24. EA to A. G. Davis, June 4, 1915, AP. Also see S. P. Nixdorff, book 1, May 24, 1915, pp. 1–4, NP.

25. EA to A. A. Buck, June 15, 1915, AP. Also see EA, "System of Radio Communication," U.S. Patent no. 1,400,847, issued December 20, 1921.

26. Aitken, *Continuous Wave,* pp. 246–47.

27. "Wireless Speech Heard 2500 Miles," *New York Times,* September 29, 1915; "Washington Talks to Hawaii Again," *New York Times,* October 1, 1915; "Transcontinental Wireless Telephony," *Electrical Review and Western Electrician,* 67 (1915): 678–79; "Transcontinental Wireless Telephony," ibid., 67 (1915): 788.

28. See comment by Lee de Forest in discussion in *Proceedings of the IRE,* 4 (1916): 127. Also see Gerald F. J. Tyne, *Saga of the Vacuum Tube* (Indianapolis, Ind.: Howard W. Sams, 1977), pp. 102–6.

29. Alexanderson mentioned his "keen disappointment" in EA to Edgar H. Felix, December 4, 1926, AP.

30. EA to F. C. Pratt, October 4, 1915; EA to C. P. Steinmetz, October 5, 1915, AP. Also see S. P. Nixdorff, book 2, October 11, 1915, NP.

31. S. P. Nixdorff, book 2, October 25, 1915, pp. 61–73, October 27, 1915, p. 79, and October 30, 1915, pp. 80–86, NP; quotation on p. 86.

32. EA to F. C. Pratt, October 28, 1915, AP.

33. EA to O. J. Ferguson, November 4, 1915, AP.

34. EA to F. C. Pratt, October 29, 1915, AP.

35. M. W. Day to F. C. Pratt, November 3, 1915; EA to A. N. Goldsmith, November 9, 1915, AP.

36. E. P. Edwards to A. G. Davis, December 8, 1915; EA to Edwards, January 8, 1916, AP.

37. S. P. Nixdorff, book 1, July 23–30, 1915, pp. 162, 182, and book 2, December 10–19, 1915, pp. 192–212, 216, NP.

38. Ernst F. W. Alexanderson and Samuel P. Nixdorff, "Magnetic Amplifier for Radiotelephony," *Proceedings of the IRE,* 4 (1916): 101–20.

39. Discussion in *Proceedings of the IRE,* 4 (1916): 122.

40. Ibid., pp. 127–28.

41. "Magnetic Amplifier in Radio Work," *Electrical World,* 68 (1916): 210–11.

42. EA to W. R. Whitney, September 5, 1916, AP.

43. Roy A. Weagant, "Design and Construction of Guy-Supported Towers for Radio Telegraphy," *Proceedings of the IRE,* 3 (1915): 135–53.

44. EA to A. G. Davis, December 15, 1915; EA to Davis, February 5, 1916; EA to Davis, February 11, 1916; EA, "Theory of Multiple-Tuned Antenna," February 24, 1916, AP.

45. A. N. Goldsmith to EA, February 17, 1916; Goldsmith to EA, February 24, 1916, AP. Also see S. P. Nixdorff, book 3, p. 118, NP.

46. William W. Brown, "Major Developments and Accomplishments in Connection with the Alexanderson Alternator," MS, January 1929, p. 2. I used a copy provided by George Wise.

47. "The Multiple Tuned Antenna," RCA Circular no. 425, October 1924. I used a

copy provided by George Wise. Also see EA, U.S. Patent no. 1,360,167, November 23, 1920.

48. EA, "Antenna for Simultaneous Sending and Receiving," attached to EA to A. G. Davis, March 10, 1916; EA to Davis, October 23, 1916, AP. Also see EA, U.S. Patent no. 1,313,042, August 12, 1919.

49. EA to F. C. Pratt, July 21, 1916, AP.

50. S. P. Nixdorff, book 3, pp. 33–40, NP.

51. EA to A. A. Buck, January 30, 1917; EA to Buck, June 5, 1917, AP.

52. Ernst F. W. Alexanderson, "Simultaneous Sending and Receiving," *Proceedings of the IRE,* 7 (1919): 363–71; quotation on p. 363. Portions of the manuscript that were received March 14, 1917, are identified.

53. S. P. Nixdorff, book 5, pp. 194–96, NP; EA to A. N. Goldsmith, March 28, 1917, AP.

54. EA to F. C. Pratt, March 22, 1916, AP.

55. Memorandum on meeting, March 31, 1916, AP.

56. EA, "Organization of Radio Development Work," April 1, 1916, AP.

57. Minutes of Standardizing Committee meeting, April 7, 1916, AP.

58. EA to A. G. Davis, "Radio Telephony without Pliotrons," November 21, 1916, AP.

59. EA to A. G. Davis, November 16, 1916, AP.

60. EA to L. C. Everett, January 23, 1917; Everett to EA, February 9, 1917, AP.

61. S. P. Nixdorff, book 5, February 19, 1917, p. 148, NP; EA to W. W. Brown, February 27, 1917, AP.

62. EA to J. L. R. Hayden, January 10, 1917, AP. Also see "The Reminiscences of William Wilbur Brown," interview by Frank E. Hill, Syracuse, N.Y., February 1951 (hereafter cited as Brown, "Reminiscences"), pp. 1–8. I am indebted to Mr. Brown for a transcript of the interview.

63. F. C. Pratt to E. W. Rice, January 27, 1917, AP.

64. A. N. Goldsmith to EA, April 11, 1917, AP. See also "Control of Wireless," *Electrical World,* 69 (1917): 641.

65. EA to A. G. Davis, April 18, 1917, AP.

66. EA, "Notes on Trip to Washington and Battle Fleet," April 30, 1917, AP.

67. EA to E. W. Rice, May 7, 1917; EA to A. G. Davis, May 24, 1917, AP. Alexanderson was also authorized to continue operating a radio receiver at his home. Authorization form dated May 19, 1917, and description of station dated May 23, 1917, AP.

68. E. W. Rice to Marconi Wireless Telegraph Company, "Cancellation of 'Heads of Agreement,'" May 10, 1917, AP.

69. EA to F. C. Pratt, May 17, 1917, AP.

70. EA to F. C. Pratt, July 2, 1917, AP.

71. EA to E. W. Rice, July 13, 1917, AP.

72. EA, "Basis of Contracts for High-Power Radio Equipment," July 18, 1917; EA to A. G. Davis, July 23, 1917, AP.

73. EA to F. C. Pratt, August 23, 1917; EA, "Developments in Signaling with CW," September 6, 1917; EA to Pratt, September 10, 1917, AP.

74. E. P. Edwards to E. W. Rice, August 1, 1917; Rice to O. D. Young, July 31, 1917, AP.

75. EA to A. G. Davis, August 14, 1917; EA to Davis, July 23, 1917, AP.

76. E. P. Edwards to A. W. Burchard, September 20, 1917, AP.

77. E. P. Edwards to A. W. Burchard, September 29, 1917, AP.

78. H. M. Hobart to EA, October 10, 1917; E. P. Edwards to A. W. Burchard, October 30, 1917, AP.

79. H. M. Hobart to EA, January 30, 1918, AP.

80. EA to A. G. Davis, October 4, 1917; EA to Davis, "Radio Telephony by Phase Variation," September 1, 1916, AP.

81. EA to Lt. Comdr. G. C. Sweet, January 21, 1918; EA to Lt. Comdr. H. P. LeClair, May 8, 1918, AP.

82. EA to A. G. Davis, January 9, 1918; EA to Davis, "Unidirectional Receiver," January 16, 1918, AP. See also EA, U.S. Patent no. 1,465,108, issued August 14, 1923.

83. EA to E. W. Rice, August 23, 1918; EA to Rice, September 10, 1918, AP.

84. Alexanderson, "Simultaneous Sending and Receiving," pp. 364–74; quotations on pp. 364, 374.

85. Ibid., p. 377.

86. See the biographical entry on Hoxie in NCAB, 39:199.

87. EA to A. G. Davis, "Visual Radio Signaling and Secrecy," April 6, 1917; EA to Davis, April 13, 1917, AP. For a discussion of using a vibration galvanometer for secret communication, see Jonathan Zenneck, Wireless Telegraphy (New York: McGraw-Hill, 1915), pp. 330–31.

88. EA to F. C. Pratt, July 2, 1917; EA to A. G. Davis, July 3, 1917, AP.

89. John R. Hewell, "The General Electric Company in the Great World War," GE Review, 22 (1919): 614.

90. EA to F. C. Pratt, October 4, 1917, AP.

91. E. P. Edwards to A. W. Burchard, October 6, 1917, AP.

92. EA to J. S. Conover, November 26, 1917, AP.

93. EA to F. C. Pratt, January 4, 1918, AP; S. P. Nixdorff, book 6, December 1, 1917, p. 243, NP.

94. EA to F. C. Pratt, February 13, 1918; EA to Pratt, February 16, 1918, AP.

95. S. P. Nixdorff, book 7, pp. 88–167, NP. See also Shipping Order, March 1, 1918; EA to W. G. Ely, April 30, 1918, AP.

96. EA to A. H. Taylor, June 15, 1918; EA, "Report on Preliminary Tests of 200 kw Equipment at New Brunswick Station," June 22, 1918, AP.

97. Albert Hoyt Taylor, Radio Reminiscences: A Half Century (Washington, D.C.: Naval Research Laboratory, 1948), pp. 89, 97–98.

98. "Description of the 200 kw Alexanderson Alternator," RCA Circular no. 405, November 1924, pp. 1–7. I used a copy provided by George Wise.

99. Ibid., p. 65.

100. EA to F. C. Pratt, July 9, 1918; EA to E. P. Edwards, July 29, 1918; EA to Pratt, August 12, 1918; EA to E. W. Rice, August 23, 1918; EA to E. W. Rice, September 10, 1918, AP.

101. EA to A. W. Burchard, September 10, 1918, AP.

102. EA to E. W. Rice, September 20, 1918, AP.

103. S. C. Hooper to G. C. Sweet, September 1918, box 2, Stanford Caldwell Hooper Papers, Manuscript Division, Library of Congress, Washington, D.C.

104. EA to M. P. Rice, May 16, 1918, AP.

105. H. R. Webster, "The Call of W Double I," Electrical Experimenter, July 1919, p. 268, clipping in AP.

106. EA to E. W. Rice, October 25, 1918, AP. See also Archer, History of Radio, pp. 145–46.

107. Brown, "Reminiscences," pp. 10–11.

108. EA to L. Gibson, November 30, 1918, AP.

109. S. C. Hooper to G. C. Sweet, November 27, 1918, Hooper Papers.

110. Lt. Paternot to EA, November 18, 1918, AP.

111. EA to F. C. Pratt, April 20, 1918, AP.

112. Ibid.

113. EA to A. W. Burchard, May 15, 1918, AP.

114. EA to H. G. Reist, October 10, 1905, AP. William Le Roy Emmet, *The Autobiography of an Engineer* (Albany, N.Y.: Fort Orange Press, 1931), p. 170.

115. Ibid., pp. 185–87.

116. Ibid., pp. 173–74; Eskil Berg, "Electrical Equipment for the Propulsion of the U.S. Collier Jupiter," *GE Review*, 15 (1912): 490–94.

117. William L. R. Emmet, "Proposed Application of Electric Ship Propulsion," *Transactions of the AIEE*, 30 (1911): 529–44.

118. Discussion in *Transactions of the AIEE*, 30 (1911): 545–49.

119. Editorial, *GE Review*, 15 (1912): 479–80; Berg, "Electrical Equipment for the Propulsion of the U.S. Collier Jupiter," p. 494.

120. William L. R. Emmet, "Electric Propulsion on the U.S.S. Jupiter," *Electrical World*, 17 (1914): 119–24; Emmet, *Autobiography of an Engineer* (1931), pp. 173, 176.

121. Samuel M. Robinson (1882–1972) served on the *New Mexico* as fleet engineer of the Pacific fleet from 1919–22 and was head of the design division of the Naval Bureau of Engineering in 1922–25. He became chief of the U.S. Navy Bureau of Ships in 1940. See *NCAB*, vol. G, pp. 293–94; an obituary in the Washington *Star-News*, November 12, 1972.

122. H. Franklin Harvey and W. E. Thau, "Electric Propulsion of Ships," *Transactions of the AIEE*, 44 (1925): 497–519. Also see Roger E. Bilstein, *Flight Patterns: Trends of Aeronautical Development in the United States, 1918–1929* (Athens, Ga.: University of Georgia Press, 1983), pp. 22–25.

123. William L. R. Emmet, "Advantages and Future of Electric Ship Propulsion," *Electrical World*, 69 (1917): 3–5.

124. "Controversy over Naval Electric Ship Propulsion," *Electrical World*, 69 (1917): 431–32. Also see Emmet, *The Autobiography of an Engineer* (1931), pp. 183–84.

125. Emmett, *Autobiography of an Engineer* (1931), pp. 184–85.

126. Emmett, "Advantages and Future of Electric Ship Propulsion," pp. 3–5.

127. Alexanderson applied for eight patents on ship propulsion during the period 1913–20.

128. EA to A. G. Davis, August 22, 1912, AP. See EA, U.S. Patent no. 1,215,094, February 6, 1917.

129. EA to C. P. Steinmetz, February 18, 1913; EA assisted by J. A. Hepperlin, "Design of Induction Motor with Multiple Squirrel Cage," October 9, 1912, AP. See also EA to E. W. Rice, February 17, 1913, AP.

130. See EA, U.S. Patents no. 1,289,593, issued December 31, 1918; no. 1,304,240, issued May 20, 1919; and no. 1,215,095, issued February 6, 1917.

131. EA to W. L. R. Emmet, May 3, 1918; EA to S. M. Robinson, December 9, 1918, AP.

132. Ernst F. W. Alexanderson, "General Characteristics of Electric Ship Propulsion Equipment," *GE Review*, 22 (1919): 224. Also see Howard I. Becker, interview by A. C. Stevens, March 16, 1977, p. 2. I am indebted to George Wise for providing a transcript of this interview.

133. Alexanderson, "General Characteristics of Electric Ship Propulsion Equipment," p. 224.

134. Ibid., pp. 224–32.

135. A. D. Badgley, "The New Mexico's Motors," *GE Review*, 22 (1919): 255–60. See also C. S. Raymond, "The New Mexico's Generators," ibid., 22 (1919): 244–54.

136. Samuel M. Robinson, "Electric Drive from a Military Point of View," *GE Review*, 22 (1919): 220–22. For a more recent assessment, see "Electrical Propulsion Is Wave

of Future for Next Generation of Navy Warships," *Atlanta Constitution*, September 30, 1988.

137. EA, "Efficiency and Democracy," speech transcript, n.d., AP.

138. Ibid.

139. See EA, U.S. Patent no. 1,400,847, issued December 26, 1921.

140. Hughes, *Networks of Power*, p. 20.

Chapter Six: "Alexanderson the Great"

1. EA to A. G. Davis, December 19, 1918, AP. Also see George O. Squier, "Aeronautics in the United States, 1918," *Transactions of the AIEE*, 38 (1919): 1–62.

2. EA to A. N. Goldsmith, January 7, 1919, AP. The paper was presented on April 2, 1919. See Ernst F. W. Alexanderson, "Simultaneous Sending and Receiving, *Proceedings of the IRE*, 7 (1919): 363–78.

3. Discussion in *Proceedings of the IRE*, 7 (1919): 379.

4. EA to A. N. Goldsmith, February 27, 1919, AP.

5. *Proceedings of the IRE*, 7 (1919): 95–96.

6. EA, transcript of response to the president and members of the institute, AP. See also "Medal for Radio Work of E. F. W. Alexanderson," *Electrical World*, 73 (1919): 1053.

7. EA to L. de Forest, May 20, 1919, AP.

8. EA to A. G. Davis, "Magnetic Control of Electron Discharge," January 14, 1919, AP. See also EA, "Electron-Discharge Device," U.S. Patent no. 1,535,082, filed September 28, 1920.

9. EA to A. W. Hull, February 3, 1919, AP.

10. Albert W. Hull, "The Dynatron: A Vacuum Tube Possessing Negative Resistance," *Proceedings of the IRE*, 6 (1918): 5–35.

11. EA to A. G. Davis, February 5, 1919, AP.

12. EA to A. G. Davis, "Ballistic Electron Valve," April 8, 1919, AP.

13. EA to A. G. Davis, May 14, 1919, AP.

14. The synchronous-resistance detector was described in EA, "Signaling System," U.S. Patent no. 1,549,737, filed September 15, 1919, and issued August 18, 1925. Several years later, Alexanderson explained that he had invented an electronic equivalent of the Goldschmidt tone-wheel synchronous detector after Owen Young had been unable to negotiate for rights to the Fessenden heterodyne patent. EA to A. G. Davis, July 29, 1932, AP. The tone wheel was a toothed-wheel current interrupter that could convert received radio signals to an audio frequency. See a description of the tone wheel in E. E. Bucher, *Practical Wireless Telegraphy* (New York: Wireless Press, 1917), pp. 284–86. In a letter written in 1934, Alexanderson stated that Young would have had to pay $10 million to obtain the Fessenden patent, and that the patent experts believed that his synchronous-resistance detector did not infringe the Fessenden invention. EA to S. M. Kintner, November 5, 1934, AP.

15. EA to A. G. Davis, June 11, 1919, AP.

16. D.C. Prince to A. G. Davis, June 12, 1919, AP. Prince had been with GE since 1912 and had been assigned to the Consulting Engineering Department in 1913. He joined the GE Research Laboratory in 1923 and later did research on high-voltage circuit breakers. In 1941 Prince was elected president of the AIEE, and he became a GE vice-president the following year. He headed the General Engineering Laboratory at GE until his retirement in 1951. "Consulting Engineering Section," memorandum with biographical information on personnel in the Consulting Engineering Section, ca. 1920, AP. Also see Alger, *Human Side of Engineering*, pp. 59–60.

17. EA to W. R. Whitney, September 5, 1919, AP.

18. EA to A. G. Davis, January 8, 1920; EA to L. A. Hawkins, February 6, 1920, AP.

19. Albert W. Hull, "The Magnetron," *Journal of the AIEE*, 40 (1921): 715–23. Also see idem, "The Axially Controlled Magnetron," *Transactions of the AIEE*, 42 (1923): 915–20.

20. EA to W. R. Whitney, December 18, 1918, AP.

21. EA to E. W. Rice, February 24, 1919, AP. Also see "Sea Phone Call Greets Wilson 800 Miles Out," clipping from *New York Sun*, February 23, 1919, AP.

22. EA to E. W. Rice, February 24, 1919, AP.

23. EA to E. P. Edwards, February 25, 1919, AP.

24. EA to E. W. Rice, March 18, 1919; EA to S. Hooper, "Two Way Radio Conversation with the George Washington," n.d., AP. See also "Alexanderson Tells Story of U.S. to Europe by Wireless Phone," clipping from *New York Evening Journal*, March 22, 1919, AP.

25. John H. Payne, "Radiophone Transmitter on the U.S.S. George Washington," *GE Review*, 23 (1920): 804–6.

26. Harold H. Beverage, "Duplex Radiophone Receiving on U.S.S. George Washington," *GE Review*, 23 (1920): 807–12. Also see EA to F. C. Pratt, May 5, 1919, AP.

27. Aitken, *Continuous Wave*, pp. 312–13.

28. S. C. Hooper to D. W. Todd, November 8, 1918, box 2, Hooper Papers. Admiral Bullard served as director of naval communications until his retirement in 1922 and later headed the Federal Radio Commission until his death in 1927. See *DAB*, 3:255. For a biographical sketch on Hooper, see *NCAB*, 47:109–10.

29. Hooper later wrote in an unpublished autobiographical essay that he had not been in a suitable position to make policy and had therefore needed an admiral who could understand the issues. "History of Radio," MS, n.d., Hooper Papers.

30. S. C. Hooper to G. C. Sweet, November 9, 1918; Hooper to Sweet, January 20, 1919, box 2, Hooper Papers. Also see Douglas, *Inventing American Broadcasting*, pp. 279–84.

31. EA to A. W. Burchard, January 15, 1919, AP.

32. EA to E. W. Rice, January 27, 1919, AP.

33. Roy A. Weagant was born in Canada in 1881 and graduated from McGill University in 1905. After working for various electrical companies including Westinghouse and Fessenden's NESCO, Weagant joined American Marconi in 1913 and was named its chief engineer in 1915. See *NCAB*, vol. C, p. 348.

34. EA to A. G. Davis, January 24, 1919, AP.

35. EA to R. A. Weagant, March 24, 1919, box 63, CRC.

36. A. W. Burchard to E. W. Rice, E. P. Edwards, and A. G. Davis, March 26, 1919, box 63, CRC.

37. S. C. Hooper, "The Formation of the Radio Corporation of America," dictated to G. H. Clark, and with preface by E. J. Nally, TS, August 16, 1940, CRC.

38. Ibid., p. 26. Also see Aitken, *Continuous Wave*, pp. 328–30.

39. Hooper, "Formation of the Radio Corporation of America," pp. 31, 42. Hooper referred to the need for a "Monroe Doctrine of Radio" in S. C. Hooper to S. S. Robison, May 28, 1920, box 3, Hooper Papers.

40. Hooper, "Formation of the Radio Corporation of America," pp. 30–42. Also see Aitken, *Continuous Wave*, pp. 331–32.

41. Draft copy of contract for the establishment of a high-power radio service, box 2, Hooper Papers. See also Hooper, "Formation of the Radio Corporation of America," pp. 42–52; Aitken, *Continuous Wave*, pp. 337–44.

42. S. C. Hooper to Chief of the Bureau of Steam Engineering, May 22, 1919, box 2, Hooper Papers.

43. Hooper, "Formation of the Radio Corporation of America," p. 60. A handwritten note on the draft contract states that Hooper told Young to go ahead without Daniels's approval. Also see Aitken, *Continuous Wave*, pp. 344–54, 360–66.

44. B. P. Apkes to W. W. Brown, April 23, 1919, AP. The department was to go through a series of name changes before its eventual merger in 1945 with the General Engineering Laboratory. Between 1922 and 1928 it was the Radio Consulting Department, between 1928 and 1933 the Consulting Engineering Department, and between 1933 and 1945 the Consulting Engineering Laboratory.

45. EA to F. C. Pratt, April 18, 1919, AP.

46. Based on a list of Alexanderson's assistants dated March 17, 1919, and a later, undated list entitled "Radio Engineering Section," AP.

47. Charles M. Ripley, "Hello Europe via Radio," *Electrical Experimenter,* July 1919, copy in CRC.

48. Agreement between American Marconi and GE, July 25, 1919, AP. Also see W. W. Brown to EA, July 25, 1919, AP.

49. A. A. Buck to EA, July 26, 1919, AP. Also see Aitken, *Continuous Wave*, pp. 388–91. EA to Buck, "Negotiations with the British Marconi Company," August 1, 1919, AP. Copies of this memorandum were sent to E. W. Rice, F. C. Pratt, and O. D. Young.

50. EA to A. A. Buck, "Negotiations with the British Marconi Company," August 1, 1919, AP.

51. EA to E. P. Edwards, August 4, 1919, AP.

52. EA to M. P. Rice, August 29, 1919, AP.

53. Ernst F. W. Alexanderson, "Transoceanic Radio Communication," *Transactions of the AIEE*, 38 (1919): 1270–76.

54. Ibid., pp. 1277–85.

55. Archer, *History of Radio,* pp. 173–74; and "Marconi Sale Rumor," clipping from *New York Herald,* September 4, 1919, AP. Also see Aitken, *Continuous Wave*, pp. 391–404.

56. EA to E. W. Rice, October 11, 1919; copy of employment contract, March 1, 1919, AP.

57. A copy of the preliminary agreement is reprinted in the *Report of the FTC on the Radio Industry* (New York: Arno Press, 1974), pp. 116–17. See also "Radio Corporation of America Is Formed," *Electrical World,* 70 (1919): 905; "New Wireless Corporation to Take over Marconi Patents and General Electric Radio Rights," clipping from *New York Herald,* October 23, 1919, AP.

58. O. D. Young to E. J. Nally, October 23, 1919, box 67, CRC.

59. A copy of the "Main Agreement" is in *Report of the FTC on the Radio Industry,* pp. 118–22.

60. Archer, *History of Radio,* pp. 179–80. Also see Aitken, *Continuous Wave*, pp. 410–15.

61. *Report of the FTC on the Radio Industry,* pp. 221–23.

62. EA to J. S. Conover, November 24, 1919, AP.

63. EA to A. G. Davis, December 12, 1919, AP.

64. EA to E. W. Rice, "Engineering Organization of the Radio Corporation," December 2, 1919, AP. An agreement between RCA and Weagant dated January 2, 1920, called for Weagant to serve as consulting engineer at a salary of $12,000 per year through July 1, 1927. He also was to receive $50,000 in payment of rights to his patents. See the *Report of the FTC on the Radio Industry,* pp. 221–23.

65. EA to E. W. Rice, "Engineering Organization of the Radio Corporation," December 2, 1919, AP.

66. EA to A. G. Davis, January 7, 1920, AP; E. J. Nally to A. N. Goldsmith, January 22, 1920, box 65, CRC.

67. A. N. Goldsmith to E. P. Edwards, May 1, 1918; EA to Goldsmith, May 8, 1919, AP. Also see E. J. Nally to Goldsmith, January 22, 1920, box 65, CRC.

68. A. N. Goldsmith to E. J. Nally, December 20, 1919, box 64, CRC.

69. EA to A. N. Goldsmith, April 7, 1920, box 65, CRC.

70. A. N. Goldsmith, "Report of Research Department," May 1920, box 65, CRC.

71. Archer, *History of Radio*, pp. 187–88.

72. EA to E. J. Nally, March 8, 1920, AP.

73. The letter is reprinted in Archer, *History of Radio*, p. 186. It was signed by A. J. Hepburn and dated January 5, 1920.

74. Archer, *History of Radio*, pp. 194–95. The agreement is reprinted in *Report of the FTC on the Radio Industry*, pp. 130–40. See also Reich, *Making of American Industrial Research*, pp. 221–22.

75. S. C. Hooper to O. D. Young, March 2, 1920, box 3, Hooper Papers.

76. O. D. Young to S. C. Hooper, March 29, 1920; Hooper to Young, April 14, 1920; Young to Hooper, May 8, 1920, box 3, Hooper Papers.

77. Archer, *History of Radio*, p. 188.

78. EA to A. G. Davis, April 9, 1920, AP.

79. F. H. Corrigan to O. D. Young, August 3, 1920, AP. See also Thorn L. Mayes, *The Alexanderson 200 Kw High-Frequency Alternator Transmitters*, Society of Wireless Pioneers, Historical Papers, vol. 4 (Santa Rosa, Calif.: Society of Wireless Pioneers, n.d.), p. 37. I am indebted to Thorn Mayes for providing me with a copy.

80. F. H. Corrigan to O. D. Young, August 3, 1920, AP.

81. "A New Day in Communications," *Saturday Evening Post*, February 7, 1920, copy in CRC.

82. Elmer E. Bucher, "The Alexanderson System for Radio Communication," *GE Review*, 23 (1920): 213–39.

83. Ernst F. W. Alexanderson, "Central Stations for Radio Communication," *Proceedings of the IRE*, 9 (1921): 83–84.

84. Ibid., pp. 84–86.

85. Ibid., p. 90. For a discussion of the development of the concepts of load factor and load diversity in the power field, see Thomas P. Hughes, "The Electrification of America: The System Builders," *Technology and Culture*, 20 (1979): 124–61.

86. Alexanderson, "Central Stations for Radio Communication," pp. 92–93.

87. RCA Traffic Chart, January 17, 1921; "Present Capabilities of High Power Stations," RCA internal memorandum, January 19, 1921, AP.

88. EA to F. C. Pratt, January 8, 1921, AP.

89. C. H. Taylor to W. H. Eccles, January 19, 1921; F. C. Pratt to EA, January 29, 1921, AP.

90. EA to A. N. Goldsmith, January 14, 1921, AP.

91. G. H. Clark, transcript of remarks, n.d., CRC.

92. EA, transcript of 1921 presidential address, AP.

93. EA to F. C. Pratt, June 25, 1921, AP.

94. EA to O. D. Young, August 18, 1921; EA to D. Sarnoff, May 3, 1921, AP.

95. EA to A. G. Davis, May 13, 1921; EA to F. C. Pratt, June 18, 1921, AP.

96. S. P. Nixdorff, book 8, January 27, 1921, p. 33, NP.

97. Minutes of Technical Committee meeting, May 24, 1921, box 68, CRC.

98. EA to E. P. Edwards, June 3, 1921, AP. See also P. R. Fortin and Conan A.

Priest, "Twenty Kilowatt Vacuum Tube Radio Telegraph Transmitter," *GE Review*, 27 (1924): 61.

99. EA to E. P. Edwards, June 3, 1921; EA to Edwards, June 17, 1921, AP.

100. EA to C. H. Taylor, October 19, 1921, AP.

101. "Opening the New Long Island Station of the Radio Corporation of America," *GE Review*, 25 (1922): 52–54.

102. Ibid., pp. 55–56.

103. E. D. Sabin, RCA Photo Album, 1921–23, donated by Mr. Sabin to the author.

104. EA, transcript of speech, April 15, 1922, AP.

105. See the essay on Beverage in *NCAB*, vol. E., pp. 431–32. Also see "Wireless Operators Honor Harold Beverage," *IEEE Spectrum*, June 1974, pp. 121–22.

106. H. H. Beverage to EA, August 15, 1919; EA to A. G. Davis, December 11, 1919; EA to Davis, December 19, 1919, AP.

107. Harold H. Beverage, Chester W. Rice, and Edward W. Kellogg, "The Wave Antenna: A New Type of Highly Directive Antenna," *Transactions of the AIEE*, 42 (1923): 215–66. Also see Harold H. Beverage and H. O. Peterson, "Radio Transmission Measurements on Long Wave Lengths," *Proceedings of the IRE*, 11 (1923): 661–73; and John D. Kraus, *Antennas* (New York: McGraw-Hill, 1950), pp. 412–13.

108. William W. Brown, "Radio Frequency Tests on Antenna Insulators," *Proceedings of the IRE*, 11 (1923): 495–522.

109. Discussion in *Proceedings of the IRE*, 11 (1923): 523–24.

110. William W. Brown and J. E. Love, "Designs and Efficiencies of Large Air Core Inductances," *Proceedings of the IRE*, 13 (1925): 755–65.

111. Nils E. Lindenblad and William W. Brown, "Main Considerations in Antenna Design," *Proceedings of the IRE*, 14 (1926): 291–323. Lindenblad, like Alexanderson, was born and educated in Sweden. He came to the United States in 1919 and spent his whole career with RCA. He received more than 300 patents. See his obituary in the *New York Times*, February 20, 1978.

112. C. W. Hansell to D. C. Prince, November 1, 1949, AP.

113. Ibid.; Lindblad and Brown, "Main Considerations." Charles W. Hansell (1898–1967) joined GE after graduating from Purdue in electrical engineering in 1919. He was assigned to work on the Alexanderson radio alternator system while still in the Testing Department and was among the GE engineers who transferred to RCA soon after its formation. He became a prolific inventor and headed the technical staff at the transmitter laboratory at Rocky Point. In an article published in 1947, Hansell was ranked fourth among American inventors in the field of electronics behind Thomas Edison, John Hammond, Jr., and Alexanderson. The latter was credited with being Hansell's mentor. See *NCAB*, 54:213–14; and William P. Vogel, "Inventing Is Vision Plus Work," *Popular Science*, 181 (October 1947): 97–101.

114. EA to A. G. Davis, January 8, 1924, AP.

115. John H. Jensen and Gerhard Rosegger, "Transferring Technology to a Peripheral Economy: The Case of Lower Danube Transport Development, 1856–1928," *Technology and Culture*, 19 (1978): 701.

116. H. V. Bozell, "Poland Building an Electrical Industry," *Electrical World*, 84 (1924): 675–77.

117. W. W. Brown, "High Power Radio Station in Poland," MS, May 1922, AP. William G. Lush, Fred E. Johnston, and J. Leslie Finch, "Transoceanic Radio Station—Warsaw, Poland," *Proceedings of the IRE*, 13 (1925): 571–88.

118. Ibid., quotations on pp. 577, 579.

119. Ibid.

120. W. Wroblewski to EA, January 11, 1924; EA to Wroblewski, January 17, 1924,

AP. See also "GE Engineer Honored by Poland for Radio Aid at Warsaw," clipping from *Albany Press,* February 6, 1924, AP; *Electrical World,* 83 (1924): 305.

121. Director of the Polish Ministry of Posts and Telecommunication to T. Mayes, April 8, 1980. I am indebted to Mr. Mays for a copy of this letter.

122. ? to EA, n.d. [bound with January 1920 letters], AP.

123. E. A. Lof to EA, May 28, 1920, AP.

124. EA to D. Sarnoff, November 23, 1920, AP.

125. EA to G. H. Clark, June 7, 1922, AP.

126. "Sweden to Erect Great Radio Station near Gothenburg," *Electrical World,* 80 (1922): 573. Also see "America Secures Two Million Contract for Swedish Station," clipping from *Springfield Republican,* August 24, 1922; and "Radio Engineer and Family Return Home," clipping from *Brooklyn Standard Union,* August 29, 1922, AP.

127. "New Swedish Transatlantic Radio Station," *Electrical World,* 84 (1924): 1206; H. Kreuger, "Construction of the Tower for the Radio Station at Varberg, Sweden," *Electrical World,* 85 (1925): 88.

128. Hill-Wagoner interview, pp. 27–28.

129. Item in *Electrical World,* 86 (1925): 88.

130. Thorn L. Mayes, "A Live Ghost from the Past," MS, March 24, 1980; EA, "Greetings from America," transcript of radio broadcast, n.d., AP. The remaining alternator at Grimeton was run for two hours on September 29, 1986. Personal communication from Kaye Weedon.

131. Ernst F. W. Alexanderson, Alexander E. Reoch, and Charles H. Taylor, "The Electrical Plant of Transoceanic Radio Telegraphy," *Transactions of the AIEE,* 42 (1923): 707–8.

132. Ibid.

133. Hughes, *Networks of Power,* pp. 296–97.

134. Alexanderson, Reoch, and Taylor, "Electrical Plant of Transoceanic Radio Telegraphy," pp. 710–11.

135. Ibid., pp. 712–15. For further details on the high-speed recorders used by RCA, see Julian Weinberger, "The Recording of High Speed Signals in Radio Telegraphy," *Proceedings of the IRE,* 10 (1922): 176–207.

136. Alexanderson, Reoch, and Taylor, "Electrical Plant of Transoceanic Radio Telegraphy," pp. 716–17.

137. EA to F. C. Pratt, November 13, 1923, AP.

138. EA to J. D. Neel, April 2, 1923, AP.

139. Ernst F. W. Alexanderson, "How Some Problems in Radio Have Been Solved," *GE Review,* 27 (1924): 373–74.

140. Ibid., pp. 378–79.

141. EA to J. L. Bernard, February 10, 1925, AP.

142. Ernst F. W. Alexanderson, "New Fields for Radio Signalling," *GE Review,* 28 (1925): 266.

143. W. J. Butler, "Alexanderson the Great," clipping from *Detroit News,* June 25, 1922, AP.

144. C. W. Hansell to D. C. Prince, November 1, 1949, AP.

145. William W. Brown, interview by F. E. Hill, Syracuse, N.Y., February 1951, pp. 14–15. Mr. Brown provided me with a copy.

146. Howard I. Becker, interview by A. L. Stevens, March 16, 1977, pp. 2–3. I am indebted to George Wise for providing a transcript of the interview.

147. EA to J. S. Connover, December 8, 1922, AP.

148. EA, "Antenna," U.S. Patent no. 1,677,698, issued July 17, 1928.

149. Copy of employment contract, March 31, 1923, AP.

150. See *GE Review*, 25 (1922): 396.

151. *Electrical World*, 80 (1922): 1182. Also see *DAB*, suppl. 4, pp. 359–60.

152. Quoted in Archer, *History of Radio*, p. 247.

153. EA to H. P. Maxim, May 1, 1923, AP; Aitken, *Continuous Wave*, p. 415.

154. Irving Langmuir, "Twenty-Kilowatt Tube, the Most Powerful Ever Made, May Displace Large Alternators Now Used," *Schnectady Works News*, August 18, 1922. A copy of this item was provided to me by George Wise.

155. James E. Brittain, "Power Electronics at General Electric, 1900–1941," *Advances in Electronics and Electron Physics*, 50 (1980): 426.

156. "Vacuum Tubes vs. Alternator," clipping from *New York Evening Mail*, October 21, 1922; "Radio Old in Two Years," clipping from *Brooklyn Citizen*, December 28, 1922, AP.

157. James E. Brittain, "The Evolution of Electrical and Electronics Engineering and the *Proceedings of the IRE*, 1913–1937," *Proceedings of the IEEE*, 77 (1989): 848–49.

158. Nils E. Lindenblad and William W. Brown, "Frequency Multiplication: Principles and Practical Application of Ferro-Magnetic Methods," *Transactions of the AIEE*, 44 (1925): 491–96; Archer A. Isbell, "The RCA World-Wide Radio Network," *Proceedings of the IRE*, 18 (1930): 1732–42. In 1931, Alexanderson's secretary informed a correspondent that GE no longer had printed information on the alternator since alternators were not built anymore and vacuum tubes had replaced them. Letter to F. B. Bloomer, February 5, 1931, AP. See also Harold H. Beverage, Charles W. Hansell, and H. O. Peterson, "Radio Plant of RCA Communications, Inc.," *Transactions of the AIEE*, 52 (1933): 75–82.

159. Mayes, *Alexanderson 200 KW High-Frequency Alternator Transmitters*, p. 38; and F. J. Kishman to T. L. Mayes, July 15, 1976. A copy of this letter was provided me by Mr. Mayes.

160. "RCA Donates Land to N.Y.," clipping from *Pittsburgh Press*, October 2, 1979. I am indebted to Field Curry for a copy of this item.

161. James R. Wait, "Project Sanguine," *Science*, 178 (1972): 272–75. Also see "Project ELF Finally Wins a Vote," *Science*, 221 (1983): 630–31.

162. Ernst F. W. Alexanderson, "New Fields for Radio Signalling," *GE Review*, 28 (1925): 266–70.

163. EA to I. J. Adams, January 5, 1923, AP.

164. EA to I. J. Adams, March 10, 1923; EA to H. E. Dunham, "Television by Radio," July 24, 1924, AP.

165. *Electrical World*, 81 (1923): 1050.

166. Archer, *History of Radio*, p. 301.

167. "Kidnapper Caught by Local Police," clipping from *Fall River Globe*, July 20, 1925, AP. Also see "Suspect Hunted in Alexanderson Kidnapping Caught," clipping from *Albany Press*, October 11, 1927, AP.

168. EA to J. L. Bernard, May 10, 1923, AP; Bernard to J. G. Harbord, May 23, 1923, box 72, CRC.

Chapter Seven: Television, Thyratron Applications, and Engineering Philosophy

1. Brittain, "The Evolution of Electrical and Electronics Engineering and the *Proceedings of the IRE*," p. 848.

2. S. P. Nixdorff, book 11, October 8–23, 1924, NP.

3. EA to H. E. Dunham, January 24, 1925, AP; EA, "Radio Signalling System," U.S. Patent no. 1,790,646, filed May 9, 1925, and issued February 3, 1931. For a discussion

of horizontal and vertical polarization, see Edward C. Jordan, *Electromagnetic Waves and Radiating Systems* (Englewood Cliffs, N.J.: Prentice-Hall, 1950), pp. 139–40.

4. Alexanderson, "New Fields for Radio Signalling," *GE Review*, 28 (1925): 267–68.

5. S. P. Nixdorff, book 11, pp. 145–268, NP.

6. Ibid.

7. EA to A. H. Lovell, November 20, 1925; "Alexanderson Rediscovers Horizontally Polarized Radio Wave," clipping from *New York Herald*, August 23, 1925, AP.

8. Ernst F. W. Alexanderson, "A New System of Short Wave Transmission—Vertical Properties of Wireless Waves," *Wireless World*, 17 (1925): 373–74.

9. EA to S. H. Blake, October 24, 1925, AP.

10. EA, "Engineering Report of Radio Consulting Department for 1925," January 28, 1926, AP.

11. Ernst F. W. Alexanderson, "Discussion on 'Polarization of Radio Waves' by Greenleaf W. Pickard," *Proceedings of the IRE*, 14 (1926): 391–93.

12. Ernst F. W. Alexanderson, "Polarization Changes Caused by Ground Absorption," *GE Review*, 29 (1926): 553.

13. Ibid., p. 554. For a discussion of the Faraday rotation and the Kerr effect, see Simon Ramo, John R. Whinnery, and Theodore Van Duzer, *Fields and Waves in Communication Electronics* (New York: John Wiley & Sons, 1965), pp. 489, 518–19.

14. Orrin E. Dunlap, Jr., "Shooting Radio Concerts into the Sky," *Scientific American*, 134 (1926): 234–35.

15. Orrin E. Dunlap, Jr., "Acres of Radio," *Scientific American*, 135 (1926): 178–79.

16. Ernst F. W. Alexanderson, "Polarization of Radio Waves," *Transactions of the AIEE*, 45 (1926): 698.

17. Robert S. Kruse, "Polarized Transmission: An Interview with Dr. E. F. W. Alexanderson," *QST*, 10 (June 1926): 9–16.

18. EA, two-page TS filed in correspondence for 1926, AP. See also M. L. Prescott, "Tests of Radio Propagation on Short Wavelengths," *GE Review*, 30 (1927): 113–16.

19. "Made Radio Encircle the Globe," clipping from *Brooklyn Eagle*, June 20, 1926, AP.

20. EA to E. P. Edwards, April 12, 1926, AP.

21. EA, memorandum, June 4, 1926, AP.

22. Alexanderson, "New Fields for Radio Signalling," p. 268. See also Ernst F. W. Alexanderson, "Radio Photography and Television," *GE Review*, 30 (1927): 78. Young was reported to have suggested the project during the inauguration banquet for James Harbord as president of RCA. See clipping from *Miami Herald*, December 1, 1924, AP.

23. "Picture Sent 9000 Miles by Radio," clipping from *New York Herald*, November 1, 1923, AP. See also EA and Richard H. Ranger, "Method and Apparatus for Picture Transmission by Wire or Radio," U.S. Patent no. 1,787,851, filed October 20, 1923, and issued January 6, 1931.

24. EA to H. E. Dunham, "Television by Radio," July 24, 1924, AP.

25. EA to F. C. Pratt, August 16, 1924, AP.

26. EA to R. Bown, March 17, 1938, AP.

27. George Shiers, "Early Schemes for Television," *IEEE Spectrum*, 7 (May 1970): 33–34. Also see Joseph H. Udelson, *The Great Television Race: A History of the American Television Industry, 1925–1941* (University, Ala.: University of Alabama Press, 1982), pp. 16–19.

28. EA to E. W. Allen, December 10, 1924, AP.

29. EA to E. W. Allen, December 12, 1924; EA to D. Sarnoff, December 20, 1924, AP.

30. EA to A. D. Lunt, January 17, 1925, AP.

31. S. P. Nixdorff, book 11, January 16, 1925, p. 106, NP.

32. EA to E. W. Allen, October 15, 1925, AP.

33. S. P. Nixdorff, book 11, pp. 268–81, NP.

34. Richard H. Ranger, "Photoradio Developments," *Proceedings of the IRE*, 17 (1929): 966–84.

35. Minutes of conference on facsimile telegraphy, January 25, 1926, AP.

36. EA to A. D. Lunt, March 5, 1926, AP. Also see U.S. Patents no. 1,752,876, no. 1,830,586, and no. 1,694,301.

37. Camp Engineering folder, AP. Also see LRB, "Early History of Association Island," MS, June 20, 1977. I am indebted to George Wise for a copy of this item.

38. EA, "Transmission of Photographs and Moving Pictures," transcript of Association Island talk, July 19, 1926, AP. See also "First Radio Picture Spans the Atlantic in Less than One Hour," clipping from *New York Evening World*, May 1, 1926, AP.

39. EA, "Transmission of Photographs and Moving Pictures," transcript of Association Island talk, July 19, 1926.

40. Ernst F. W. Alexanderson, "Radio Photography and Television," *GE Review*, 30 (1927): 78–79.

41. Ibid., pp. 80–84.

42. Ibid., p. 84.

43. "Radio Movie Shown Here: In Home Soon," clipping from *New York Tribune*, January 11, 1927. Also see "Meeting of the Radio Savants," clipping from *New York World*, January 15, 1927, AP.

44. S. P. Nixdorff, book 12, February 12, 1927, p. 171, NP.

45. "Radio Films and Air Letters Forecast by Television Expert," clipping from *New York Herald*, December 18, 1926; "General Electric Man Invents Device to See across Broad Atlantic," clipping from *Pittsburgh Eagle*, December 18, 1926; "World Television in Decade Is Seen," clipping from *San Antonio Express*, December 19, 1926; "Television Seen as Possibility," clipping from *New Orleans Tribune*, December 16, 1926, AP.

46. "After Radio Telephony, What?" clipping from *New York Times*, January 9, 1927, AP.

47. Clipping from *Tulsa World*, January 2, 1927, AP.

48. "Alexanderson, Pupin, and John L. Baird," clipping from *New York American*, January 2, 1927, AP.

49. "The Turning Point," clipping from *New York Herald*, May 22, 1927, AP.

50. "Electrical Engineers Dip into the Future," *Electrical World*, 89 (1927): 878–79.

51. EA to D. Sarnoff, January 5, 1927, AP.

52. EA, memorandum, January 6, 1927, AP.

53. Ibid.

54. EA, "Radio Research Work of the Associated Companies," memorandum, August 16, 1927, AP.

55. EA to A. D. Lunt, October 28, 1927, AP. Also see EA, "Speed-Control System," U.S. Patent no. 1,736,689, filed March 26, 1928.

56. EA to E. W. Allen, November 2, 1927, AP.

57. C. D. Wagoner to EA, June 26, 1950; EA, "Television Broadcast for the Home and Amateur," TS, 1928, AP.

58. From an item filed with Alexanderson's 1928 correspondence, n.d., AP. Also see Udelson, *Great Television Race*, p. 33.

59. A. P. Peck, "Television Enters the Home," *Scientific American*, 138 (1928): 246–47.

60. EA, "Progress in the Art of Television," transcript of talk, attached to EA to P. L. Alger, May 27, 1928, AP.

61. "General Electric Broadcasts Drama by Television," *Electrical World*, 92 (1928): 580; Robert Hertzberg, "Television Makes the Radio Drama Possible," *Radio News*, 10 (December 1928): 524–27, 587–88. Also see Udelson, *Great Television Race*, p. 33.

62. Hertzberg, "Television Makes Radio Drama Possible," pp. 524–27, 587–88.

63. EA to G. Eastman, August 11, 1928, AP.

64. EA to H. E. Dunham, "Color Television," August 2, 1928, AP.

65. EA to J. Huff, August 6, 1928; EA to G. Day, August 18, 1928, AP.

66. EA to C. E. Tullar, August 16, 1928, AP.

67. EA to J. Huff, October 3, 1928, AP.

68. EA to C. J. Young, October 12, 1928, AP.

69. EA to E. W. Allen, November 5, 1928; EA to J. Liston, November 12, 1928, AP.

70. EA, 1929 Annual Departmental Report, attached to EA to E. W. Allen, January 2, 1930, AP.

71. EA to G. Swope, March 27, 1929, AP.

72. J. G. Harbord to C. H. Taylor, October 31, 1929; minutes of Technical Committee meeting, November 12, 1929, AP.

73. Minutes of Technical Committee meeting, December 3, 1929, AP.

74. Minutes of Technical Committee meeting, January 7, 1930, AP.

75. EA to G. W. Henyon, January 30, 1930, AP.

76. EA to C. A. Priest, February 16, 1939, AP.

77. EA to R. Stearns, September 21, 1927, AP.

78. EA, 1927 Annual Report on Radio Consulting Department; S. P. Nixdorff to H. E. Dunham, February 14, 1928, AP.

79. EA to M. D. Morse, October 31, 1928, AP. Also see EA, U.S. Patents no. 1,969,537 and no. 1,913,148.

80. Ernst F. W. Alexanderson, "Height of Airplane above Ground by Radio Echo," *Science*, 68 (1928): 597–98.

81. "Determination of Altitude by Radio," *Airway Age*, 10 (1929): 55–56.

82. EA, memorandum, December 24, 1928, AP.

83. Ernst F. W. Alexanderson, "Radio Echo Altitude Meter," *Journal of the Aeronautical Sciences*, 3 (1936): 316.

84. Ibid., pp. 316–17. Also see EA, U.S. Patent no. 1,913,148.

85. Peter C. Sandretto, "Absolute Altimeters," *Proceedings of the IRE*, 32 (1944): 167–75.

86. EA to A. G. Davis, September 23, 1921; EA to Davis, October 17, 1921, AP.

87. EA to A. G. Davis, January 13, 1922, AP. See also David C. Prince, "The Inverter," *GE Review*, 28 (1925): 676.

88. S. P. Nixdorff, book 10, January 24, 1922, p. 69, and January 28, 1922, p. 77, NP.

89. EA to A. G. Davis, February 10, 1922; EA to Davis, March 1, 1922, AP.

90. EA to H. E. Dunham, November 13, 1922; EA to A. G. Davis, November 27, 1922; EA, Departmental Report for 1922, December 26, 1922, AP.

91. EA to L. A. Hawkins, memorandum, April 23, 1923, AP.

92. EA to J. S. Conover, November 1, 1923, AP. Also see EA, 1923 Annual Report of Radio Consulting Department, December 26, 1923, AP.

93. Brittain, "Power Electronics at General Electric," p. 429.

94. EA to E. W. Allen, November 14, 1924, AP.

95. EA, 1925 Annual Report of Radio Consulting Department, January 28, 1926, AP.

96. Minutes of Conference on Facsimile Telegraph, January 25, 1926, AP. Also see S. P. Nixdorff, book 11, January 21, 1926, p. 291, NP.

97. John Winthrop Hammond, *Men and Volts: The Story of General Electric* (New York: J. B. Lippincott, 1941), p. 358; Edward M. Hewlett, "The Selsyn System of Position Indicator," *GE Review*, 24 (1921): 210–18. On the theory of the selsyn, see Lawrence and Richards, *Principles of Alternating-Current Machinery*, pp. 471–77.

98. S. P. Nixdorff, book 10, May 26, 1922, p. 155, NP.

99. EA to A. G. Davis, May 31, 1922, AP.

100. EA to E. M. Hewlett, June 2, 1922; EA to F. C. Pratt, October 12, 1922, AP.

101. S. P. Nixdorff, book 11, March 26–June 22, 1925, pp. 209–49, NP.

102. S. P. Nixdorff, "Analysis of Work Done on Account 567," 1933, AP.

103. EA, memorandum, January 6, 1927, AP.

104. EA, 1927 Annual Report of Radio Consulting Department, n.d., AP.

105. Alexanderson mentioned several years later that difficulties with early thyratrons had led to a postponement of efforts to use them in variable speed motors. See EA to H. E. Dunham, "Thyratron Commutation," October 28, 1931, AP.

106. Brittain, "Power Electronics at General Electric," p. 430.

107. Albert W. Hull, "Gas-filled Thermionic Tubes," *Transactions of the AIEE*, 47 (1928): 753–63. Also see David C. Prince, "The Direct-Current Transformer Utilizing Thyratron Tubes," *GE Review*, 31 (1928): 347–50.

108. Albert W. Hull, "Hot-Cathode Thyratrons," *GE Review*, 32 (1929): 213–23.

109. Alexanderson, "New Fields for Radio Signalling," pp. 266–68.

110. EA to C. F. Scott, March 27, 1926, AP.

111. Ibid.

112. EA to E. H. Felix, December 4, 1926, AP.

113. Ibid.

114. EA to E. H. Felix, January 26, 1927, AP.

115. EA, "Research Is Essential," TS of speech, September 1928, AP.

116. EA, "Swedish Engineers in America and Their Contribution to Our Civilization," TS of speech, February 1930, AP; EA, transcript of talk delivered to meeting of radio distributors, Boston, September 3, 1930, AP.

117. EA, "A Review of Electrical Developments from 1920 to 1930," AP.

118. Ibid.

119. Ibid.

120. Ibid.

121. Ibid.

122. EA, "Ignition System," U.S. Patent no. 1,723,908, filed June 14, 1926; EA, "Speed-Indicating System," U.S. Patent no. 1,747,041, filed October 27, 1928.

123. EA to E. Richmond, April 7, 1926, AP.

124. EA, talk prepared for Ericsson Award ceremonies, TS, February 10, 1928, AP. Also see an item in *Electrical World*, 91 (1928): 375.

125. EA to D. H. Sparkman, August 10, 1928; EA to Editor of *Yachting*, April 12, 1928, AP.

126. EA to L. H. Baekeland, September 14, 1928, AP.

127. EA to Taylor Wine Company, November 18, 1924, AP.

128. Edith Alexanderson Nordlander, telephone interview with author, November 15, 1978.

Chapter Eight: Electronic Engineering

1. J. G. Harbord, General Order no. 4, January 17, 1930, AP.

2. Gleason L. Archer, *Big Business and Radio* (New York: American Historical Company, 1939), pp. 339–82. Also see Aitken, *Continuous Wave*, pp. 502–9.

3. Minutes of Technical Committee meeting, April 1, 1930; EA to D. Sarnoff, April 2, 1930, AP.

4. EA, "Television and Its Uses in Peace and War," TS of speech, April 1, 1930, AP.

5. Ibid.

6. D. Sarnoff to EA, April 8, 1930, AP.

7. EA to C. E. Tullar, "Television for War Purposes," May 1, 1930, AP.

8. EA to R. Stearns, May 20, 1930; EA to J. T. Flynn, May 24, 1930; A. C. Reid to EA, May 9, 1930, AP.

9. C. D. Wagoner to M. P. Rice, May 8, 1930, AP. Also see Edgar H. Felix, *Television: Its Methods and Uses* (New York: McGraw-Hill, 1931), p. 225.

10. "Father of Television Sees Future Marvels as Child Increases in Size," clipping from *Providence Tribune*, May 23, 1930, AP. See also "Development of Television Inevitable, Says Scientist," clipping from *Boston Globe*, May 25, 1930, AP.

11. Edgar H. Felix, "Television Advances from Peephole to Screen," *Radio News,* 12 (1930): 228–30, 268–69. See also "Secret of Television Is New Optical Cell," clipping from *Brooklyn Eagle*, May 23, 1930, AP; and Joseph H. Udelson, *The Great Television Race* (University, Ala.: University of Alabama Press, 1982), pp. 33–35.

12. EA to J. T. Flynn, May 24, 1930; EA to J. H. Hammond, June 24, 1930; EA, "Radio Problems Discussed with the Officers of the Saratoga," August 14, 1930, AP.

13. "Famous Engineer Informed by Radio of Father's Death," clipping from *Albany Press*, June 1930, AP.

14. EA to R. Stearns, August 22, 1930, AP.

15. Minutes of Technical Committee meeting, August 5, 1930, AP.

16. Minutes of Technical Committee meeting, November 11, 1930, AP.

17. EA, Budget Request for 1931, October 23, 1930, AP.

18. EA to D. Sarnoff, October 3, 1930; Sarnoff to EA, November 10, 1930; EA to Sarnoff, November 13, 1930; Sarnoff to EA, November 17, 1930, AP.

19. EA, "Requirements of a Commercial Television Picture for the Home," November 24, 1930, AP.

20. EA to L. A. Hawkins, with copies to I. Langmuir, A. Hull, E. M. Hewlett, and C. W. Rice, December 16, 1930, AP.

21. EA, "Television Receiver," U.S. Patent no. 1,889,587; EA, "Television Apparatus," U.S. Patent no. 1,935,427. Also see Ray D. Kell, "Description of Experimental Transmitter Apparatus," *Proceedings of the IRE,* 21 (1933): 1674–91.

22. EA to H. E. Dunham, December 3, 1931, AP.

23. EA, "Colored Television Apparatus," U.S. Patent no. 1,988,931, filed May 2, 1933. Also see EA to C. O. Howard, March 22, 1933, AP.

24. EA, 1930 Annual Report of Consulting Engineering Department, enclosed with letter to C. E. Eveleth, December 1, 1930; EA, "Proposed Budget for 1931," October 23, 1930, AP.

25. EA to C. W. Rice, November 28, 1930, AP.

26. EA to H. E. Dunham, "Synchronous Torque Amplifier," February 16, 1931; EA to Dunham, March 2, 1931, AP.

27. EA to H. E. Dunham, "Synchronous Torque Amplifier for Gun Control," June 30, 1931, AP.

28. EA, "Power Transformation for Vacuum Tubes," TS of speech, Camp Engineering, 1931, AP.

29. EA to H. E. Dunham, "Synchronous Torque Amplifier with Booster," September 24, 1931, AP.

30. Ibid., EA to L. E. Dodds, "Thyratron Follow-up Device Applied to Machine Tools," November 2, 1931, AP.

31. EA, TS of speech, filed with November 1931 correspondence, AP.

32. EA to H. E. Dunham, November 10, 1931; EA to Dunham, "Pantagraph Control of Gun Telescopes," December 3, 1931, AP.

33. EA to Lt. E. Kull, January 28, 1932; EA to E. E. Libman, November 1932, AP.

34. EA, "System for Reproducing Position," U.S. Patent no. 2,550,514.

35. EA, Budget Request for 1933, enclosed with EA to R. C. Muir, October 20, 1932; EA to E. E. Libman, January 18, 1933, AP.

36. EA to C. E. Tullar, May 2, 1933; EA, remarks at meeting of Engineering Council, TS, May 15, 1933, AP. The council members included Edwin Rice, William Emmet, Willis Whitney, William Coolidge, Irving Langmuir, Albert Hull, and William White.

37. EA to C. E. Tullar, May 2, 1933, AP.

38. EA to H. E. Dunham, "Correction Device for Synchronous Torque Amplifier," June 29, 1933, AP.

39. EA to B. L. Delach, August 18, 1933; EA, statement on M. A. Edwards, May 5, 1937, AP. Edwards held B.S. degrees in both electrical and mechanical engineering from Kansas State University.

40. S. P. Nixdorff, "Analysis of Work Done on Account 567," AP.

41. EA, Budget Request for 1934, attached to EA to R. C. Muir, October 23, 1933, AP.

42. EA to I. H. Marshman, "Speed Correction of Torque Amplifier," April 18, 1935, AP; EA, "Follow-up Control System," U.S. Patent no. 2,414,919.

43. EA, "Power Transformation by Vacuum Tubes," TS of talk, Camp Engineering, 1931, AP.

44. Ibid.

45. Ibid.

46. EA to H. E. Dunham, "Variable Speed Induction Motor with Thyratron Commutator," February 19, 1931, AP.

47. "Where is America Going?" clipping from Washington Star, February 8, 1931, AP.

48. EA to H. E. Dunham, "Thyratron Commutation," October 28, 1931, AP.

49. EA to C. E. Eveleth, November 6, 1931, AP.

50. EA, Budget Request for 1932, included with EA to R. C. Muir, October 21, 1931, AP.

51. EA to C. E. Eveleth, December 24, 1931, AP; EA, P. L. Alger, and S. P. Nixdorff, "System of Electric Power Transmission," U.S. Patent no. 1,917,082.

52. Albert W. Hull, "New Vacuum Valves and Their Applications," GE Review, 35 (1932): 622–29.

53. Philip Sporn and G. G. Langdon, "Thyratron Voltage Regulators," Electrical World, 105 (1935): 1182–84. Philip Sporn (1896–1978) graduated from Columbia University in 1917 in electrical engineering, and had been with AG&E since 1920. He later served as president of AG&E.

54. Discussion by Sporn in Electrical Engineering, 54 (1935): 750.

55. EA, "High Power Tubes for Power Transformation," May 15, 1933, AP.

56. EA to R. C. Muir, July 11, 1933; EA, Budget Request for 1934, enclosed with EA to R. C. Muir, October 23, 1933, AP.

57. EA to R. C. Muir, November 3, 1933; EA to Muir, December 7, 1933, AP.

58. Clodiun H. Willis, "Harmonic Commutation for Thyratron Inverters and Rectifiers," GE Review, 35 (1932): 632–38. Also see idem, "A Study of Thyratron Commutator Motor," GE Review, 36 (1933): 76–80.

59. Minutes of Engineering Council meeting, March 5, 1934, AP.

60. EA, Budget Request for 1935, enclosed with EA to R. C. Muir, October 23, 1934, AP.

61. EA to T. A. Worcester, September 12, 1934; EA to M. J. Kelly, September 26, 1934, AP.

62. Ernst F. W. Alexanderson and Albert H. Mittag, "The Thyratron Motor," *Electrical Engineering*, 53 (1934): 1517–23, quotation on p. 1517.

63. Discussion in *Electrical Engineering*, 54 (1935): 750–52.

64. Philip Sporn and G. G. Langdon, "Thyratron Motor Applied," *Electrical World*, 105 (1935): 35–37.

65. EA to R. C. Muir, February 21, 1936, AP; Albert H. Beiler, "The Thyratron Motor at the Logan Plant," *Transactions of the AIEE*, 57 (1938): 19–24.

66. Beiler, "Thyratron Motor at the Logan Plant," pp. 19–24.

67. Discussion in *Transactions of the AIEE*, 57 (1938): 294. Also see Edward L. Owen, Marvin N. Morack, C. Curtis Herskind, and Arthur S. Grimes, "AC Adjustable-Speed Drives with Electric Power Converters: The Early Days," *IEEE Transactions on Industry Applications*, 1A-20 (1984): 298–307.

68. EA to W. D. Coolidge, "Operation of Thyratron," October 17, 1935, AP.

69. EA to C. Maxwell, "Liquidation of Development," January 31, 1936, AP.

70. EA to R. C. Muir, February 21, 1936, AP.

71. EA to C. E. Tullar, May 20, 1935, AP.

72. See an obituary of Robinson in the *Washington Star*, November 12, 1972.

73. S. M. Robinson to EA, December 10, 1938; S. M. Robinson to EA, December 27, 1938, AP.

74. EA, Report of Consulting Engineering Laboratory for 1939, October 9, 1939; EA to C. E. Tullar, April 18, 1940; EA, Report of Consulting Engineering Laboratory for 1940, October 1940, AP.

75. EA to R. C. Muir, September 5, 1933; EA to Muir, October 18, 1933, AP.

76. EA to R. C. Muir, December 1, 1933, AP.

77. EA to H. E. Dunham, "Extinction of Current in Thyratrons," December 6, 1933; EA to Dunham, January 12, 1934, AP.

78. EA to R. C. Muir, "Power Transmission with Direct Current," August 23, 1934, AP.

79. A. E. Knowlton, ed., *Standard Handbook for Electrical Engineers* (New York: McGraw-Hill, 1949), p. 1171. Also see Georg Siemens, *History of the House of Siemens*, trans. A. F. Rodger (Freiburg, Germany: Karl Alber, 1957), 1:116–17; and Alfred Still, "The Thury System of Power Transmission by Continuous Current," *Electrical World*, 60 (1912): 1093–94.

80. H. M. Hobart, "Thury System as a Public Works Project," September 27, 1934; EA to A. R. Stevenson, November 9, 1934, AP.

81. EA, Budget Request for 1935, enclosed with EA to R. C. Muir, October 23, 1934, AP.

82. EA to R. C. Muir, December 4, 1934, AP.

83. Clodiun H. Willis, B. D. Bedford, and F. R. Ellis, "Constant-Current D-C Transmission," *Electrical Engineering*, 54 (1935): 102–8. According to Alexanderson, the monocyclic circuit was a type of inductive-capacitive circuit that Steinmetz had called a "monocyclic square." See Ernst F. W. Alexanderson, "Electronic Engineering," *Electronics*, 9 (June 1936): 25.

84. Discussion in *Electrical Engineering*, 54 (1935): 327–28.

85. EA, Budget Request for 1936, enclosed with EA to R. C. Muir, October 7, 1935, AP.

86. EA to R. C. Muir, October 20, 1936, AP.

87. EA to R. C. Muir, "Power Transmission," February 25, 1936; EA to R. C. Muir, February 21, 1936, AP.

88. Ernst F. W. Alexanderson, "Electronic Engineering," *Electronics*, 9 (June 1936): 25.

89. Ernst F. W. Alexanderson, "Thyratrons and Their Uses," *Electronics*, 11 (February 1938): 10.

90. Alexanderson, "Electronic Engineering," p. 26.

91. EA to R. C. Muir, October 2, 1936; EA to Muir, October 6, 1936, AP.

92. EA to R. C. Muir, "DC Power Transmission," November 9, 1936, AP.

93. EA, Budget Request for 1937, enclosed with EA to R. C. Muir, October 20, 1936, AP.

94. EA, "Power Applications of Vacuum Tubes," TS, April 19, 1937, AP.

95. Ibid.

96. Alexanderson, "Thyratrons and Their Uses," p. 9.

97. *Ibid.*, p. 12.

98. EA to J. D. Ross, March 8, 1938, AP. Also see "Ross Appointed Head of Bonneville Power," *Electrical World*, 108 (1937): 1309; "Bonneville Designed for Problematical Loading," ibid., 108 (1937): 845–51. "Bonneville Dam Power Base Set at 57% of Cost of Commission," ibid., 110 (1938): 1466; and Ross to EA, April 21, 1938, AP.

99. EA to H. E. Dunham, "DC Power Transmission," March 9, 1938, AP. Also see "Research on DC Power Transmission Planned," *Electrical World*, 112 (1939): 529. This item concerned a research project at Bonneville to determine the feasibility of DC transmission over long distances.

100. EA, Engineering Report of Consulting Engineering Laboratory for 1938, November 18, 1938; EA, Report of Consulting Engineering Laboratory for 1939, October 9, 1939; EA, Report of Consulting Engineering Laboratory, October 1940, AP.

101. EA to H. E. Dunham, "The Metadyne Amplifier," September 30, 1937, AP.

102. EA to H. E. Dunham, "Dynamo Electric Machine as an Amplifier," October 20, 1937, AP.

103. Ibid.

104. EA to H. E. Dunham, November 8, 1937; EA, Budget Request for 1938, enclosed with EA to R. C. Muir, October 13, 1937, AP.

105. EA to H. E. Dunham, February 4, 1938; EA to Dunham, "Dynamo Electric Amplifier," November 14, 1938, AP.

106. EA, Engineering Report of Consulting Engineering Laboratory for 1938, November 18, 1938, AP.

107. On the Rosenberg generator, see Alexander S. Langsdorff, *Principles of Direct-Current Machines*, 4th ed. (New York: McGraw-Hill, 1931), pp. 547, 555.

108. EA to C. E. Tullar, 'Dynamo-Electric Amplifier," May 10, 1939, AP.

109. Ibid.

110. Ibid.

111. *Who's Who in Engineering* (1954), p. 1872. Also see "Joseph M. Pestarini," obituary, *Electrical Engineering*, 76 (1957): 910.

112. Joseph Maximus Pestarini, "Théórie élémentaire du fonctionnement statique de la metadyne," *Revue générale de l'electricité*, 27 (1930): 355–65, 395–406. Also see J. M. Pestarini, U.S. Patents no. 1,755,073 and no. 1,749,673, both issued April 15, 1930.

113. Joseph Maximus Pestarini, *Metadyne Statics* (New York: Technology Press of MIT and John Wiley & Sons, 1952).

114. Ibid., p. 250.

115. EA to C. E. Tullar, March 18, 1940, AP.

116. Ernst F. W. Alexanderson, Martin A. Edwards, and K. K. Bowman, "The Amplidyne Generator—A Dynamoelectric Amplifier for Power Control," *GE Review*, 43

(1940): 105. The paper was also published in *Transactions of the AIEE,* 59 (1940): 937–39.

117. Alec Fisher, "The Design Characteristics of Amplidyne Generators," *Transactions of the AIEE,* 59 (1940): 939–44.

118. David R. Shoults, Martin A. Edwards, and Frederick E. Crever, "Industrial Applications of Amplidyne Generators," *Transactions of the AIEE,* 59 (1940): 944–49.

119. Discussion in *Transactions of the AIEE,* 59 (1940): 1136.

120. R. H. Rogers, "From Cobblestone to Amplidyne Generators," *GE Review,* 43 (1940): 103.

121. *GE Review,* 35 (1932): 600; "Sun's Eclipse to be Studied," clipping from *Albany News,* August 17, 1932, AP. Also see A. Brown to EA, August 20, 1932, AP.

122. EA to E. V. Appleton, August 26, 1932, AP.

123. Clipping from *New York Times,* August 31, 1932, AP.

124. Clipping from *Schenectady Gazette,* September 1, 1932, AP. See also "Finds Wide Effect of Eclipse on Radio," clipping from *New York Times,* September 4, 1932, AP.

125. Ernst F. W. Alexanderson, "The Effect of the Sun's Eclipse on Radio Waves," *Radio Engineering,* 12 (October 1932): 19–20. Also see idem, "Eclipse Effect on Radio," *Scientific American,* 147 (1932): 302–3, 307.

126. A. N. Goldsmith to EA, July 30, 1934; EA to Goldsmith, August 7, 1934, AP.

127. Ernst F. W. Alexanderson, "The Eclipse of August, 1932, Observed by Radio Facsimile," *Proceedings of the IRE,* 23 (1935): 454–60.

128. Ibid.

129. EA, "High Frequency Transmission System for Railways," U.S. Patent no. 2,001,514, issued May 14, 1935.

130. EA, Budget Request for 1932, enclosed with EA to R. C. Muir, October 21, 1931; EA, Budget Request for 1933, enclosed with EA to Muir, October 20, 1932, AP.

131. EA, "Indicating System for Aircraft," U.S. Patent no. 1,907,471, issued May 9, 1933; EA and J. H. Hammond, "Airplane Landing Field Using Directional Radio Beams," U.S. Patent no. 2,077,196, issued April 13, 1937.

132. EA to C. E. Eveleth, November 15, 1932, AP.

133. EA to C. E. Tullar, "Printed Book That Reads Aloud," November 16, 1932; EA to H. E. Dunham, "Radio Receiver," November 30, 1932, AP.

134. EA, speech to radio sales managers, TS, February 23, 1933, AP.

135. Ibid.

136. EA, speech to radio dealers meeting, TS, September 19, 1933, AP.

137. Ibid.

138. EA to B. C. Bowe, January 18, 1934, AP.

139. EA, Budget Request for 1935, enclosed with EA to R. C. Muir, October 23, 1934, AP.

140. EA to C. Lang, March 22, 1933, AP.

141. EA to R. C. Muir, "Radio Patent Situation and Comments on Discussion in the Last Radio Advisory Committee," December 21, 1937, AP.

142. Ibid.

143. Ibid.

144. EA to R. C. Muir, February 14, 1938, AP.

145. EA to H. B. Marrin, October 27, 1937, AP.

146. EA to R. C. Muir, August 22, 1938, AP.

147. EA, Engineering Report for Consulting Engineering Laboratory for 1938, November 18, 1938; EA to R. C. Muir, "Broadcast Chains for TV," March 2, 1939, AP.

148. EA to H. E. Dunham, "Negotiation with Respect to Agreement Supplemental to Agreement A-1," June 16, 1939, AP.

149. EA to H. E. Dunham, March 10, 1939; EA to Dunham, March 22, 1939, AP.

150. EA, Report of Consulting Engineering Laboratory for 1939, October 9, 1939, AP.

151. EA to C. E. Tullar, March 6, 1940, AP. See also EA to H. E. Dunham, "Narrow Band FM," February 20, 1940, AP.

152. EA to Radio Advisory Committee, "Radio Altimeter," December 8, 1937, AP.

153. EA to H. E. Dunham, March 3, 1938; EA to Dunham, "Radio Echoes," May 13, 1938, AP. See also EA, "Radio Distance Meter," U.S. Patent no. 2,248,599, filed September 7, 1939, and issued July 8, 1941.

154. EA, 1930 Annual Report of Consulting Engineering Department, included with EA to C. E. Eveleth, December 1, 1930; EA to G. W. Clinger, June 18, 1931, AP.

155. EA to M. M. Boring, October 29, 1934, AP.

156. EA, "Mathematics and Engineering," TS, March 1936, AP.

157. EA, testimony before Federal Communications Commission, TS, May 27, 1937, and June 1, 1937, AP.

158. Irving Langmuir, "Fundamental Research and Its Human Value," *GE Review*, 40 (1937): 569–73.

159. EA to R. C. Muir, February 1, 1939, AP.

160. G. Swope to EA, December 20, 1934; EA to Nobel Committee for Physics, January 14, 1935, AP. Also see an item on Langmuir and the Nobel Prize in *GE Review*, 35 (1932): 600.

161. *Electrical Engineering*, 57 (1938): 325.

162. EA, talk at banquet of Lake George Yacht Club, TS, September 1, 1936, AP.

163. EA to C. E. Bacon, May 13, 1936; EA, "Comments on Sailing," TS, April 5, 1937, AP.

164. Copy of Alexanderson's security clearance form, 1942, AP.

165. EA to H. L. Erlicher, November 15, 1932; EA to E. E. Mayer, November 4, 1937, AP.

166. "Reed and Wilson Named to Head General Electric," *Electrical World*, 112 (1939): 1513.

167. Herbert Asbury, "Inventor for Victory," *Popular Science*, 141 (July 1942): 89–92, 204–6.

Chapter Nine: "Inventor for Victory"

1. C. Norman to EA, December 27, 1939, AP. Norman had been in the United States since 1902 and had taught at Ohio State since 1917. See *Who's Who in Engineering* (1941), pp. 1305–6.

2. Heidenstam to EA, January 30, 1940; EA, talk on behalf of the Finnish Relief Fund, TS, n.d., AP.

3. EA to K. G. Tiselius, August 13, 1940, AP.

4. EA to H. Stockman, April 22, 1940, AP.

5. EA to C. E. Tullar, July 26, 1940; EA to H. E. Dunham, July 26, 1940, AP.

6. EA, Engineering Report of Consulting Engineering Laboratory for 1940, October 1940; EA to W. R. G. Baker, October 8, 1940, AP.

7. EA to R. C. Muir, "Radio Gun Detector," November 12, 1940; EA to C. E. Tullar, November 13, 1940, AP.

8. EA, "Memorandum of Conference at Naval Research Lab, November 28–29," December 6, 1940, AP.

9. EA to C. E. Tullar, "Control of Beam Antenna," January 16, 1941, AP.

10. EA to M. D. Morse, June 5, 1941, AP.

11. EA to R. Stearns, May 15, 1941; EA to Stearns, June 12, 1941; EA, "Report on Gun Director and Computer," June 12, 1941, AP.

12. EA, "Electric Computer for Gun Director," August 14, 1941; EA to H. E. Dunham, "Electric Computer and Gun Director," September 5, 1941, AP.

13. EA to H. E. Dunham, "Control of Motor over a Wide Range of Speeds," September 5, 1941; EA to C. E. Tullar, October 1, 1941, AP.

14. EA to M. D. Morse, "Electric Computer," October 10, 1941, AP. Also see EA, "Electric Computer," U.S. Patent no. 2,417,229, filed April 17, 1942, and issued March 11, 1947.

15. EA to H. E. Dunham, "Selsyn Regulator," October 24, 1941, AP.

16. Ibid.

17. EA to H. E. Dunham, "Aided Tracking for Radar Detectors and Guns," November 14, 1941, AP.

18. Ibid.; EA to H. E. Dunham, "Computer Control of Radar Detectors and Guns," November 21, 1941, AP.

19. EA, "Electric Control and Computer for Radar Detectors and Guns," November 25, 1941, AP.

20. Ibid.

21. Ibid.

22. EA to W. R. G. Baker, January 16, 1942, AP.

23. EA to H. E. Dunham, "Combination of Aided Tracking with Automatic Radio Tracking," February 24, 1942; EA to Dunham, "Gun Laying by Radar," February 26, 1942, AP. Also see EA, "Visit to Radiation Labs of MIT," September 8–9, 1942, AP.

24. Herbert Asbury, "Inventor for Victory," Popular Science, 141 (July 1942): 89–92, 204–6.

25. "An Anniversary and a Noted Radio Inventor," clipping from New York Times, October 25, 1942. Also see W. Holbrook, "Creating New Weapons for Defense: What General Electric's Dr. Alexanderson Is Doing," clipping filed in AP.

26. Roster of Personnel in Consulting Engineering Laboratory, December 10, 1942, AP.

27. "Song of the Consulting Engineering Lab," n.d., AP.

28. Data on Security Clearance, enclosed with EA to H. Mayers, October 30, 1942, AP.

29. EA, Budget Request for 1943, enclosed with EA to H. A. Winne, October 26, 1942, AP.

30. Item from GE News Bureau, n.d. [probably March 1945], 1945 binder, AP.

31. EA, transcript of speech, November 7, 1945, AP.

32. Industrial Engineering Department memorandum, October 26, 1942, AP.

33. Fremont Felix, "The Amplidyne: What It Is and How It Works," GE Review, 46 (1943): 442–45.

34. "Amplidyne Puts Electrical Short Circuits to Useful Work," Scientific American, 170 (January 1944): 29–30.

35. EA to A. N. Goldsmith, June 26, 1943; EA to Goldsmith, November 17, 1943, AP. Whereas the title of the IRE Proceedings had originally used the organization's full name, it was changed to use the acronym after 1939.

36. EA to C. D. Wagoner, January 11, 1944, AP; Ernst F. W. Alexanderson, Martin A. Edwards, and Kenneth K. Bowman, "The Amplidyne System of Control," Proceedings of the IRE, 32 (1944): 517–20.

37. Alexanderson, Edwards, and Bowman, "Amplidyne System of Control," pp. 517–20.

38. W. W. Brown to EA, May 2, 1944, AP.

39. G. J. C. Eshelman to G. H. Clark, July 24, 1942, CRC.

40. EA to H. E. Gieshe, March 22, 1940, AP.

41. EA to R. Klang, February 6, 1941, AP.

42. L. A. Hawkins, TS of remarks, May 14, 1941; EA, TS of talk, May 14, 1941, AP.

43. EA to C. E. Wilson, May 21, 1941, AP.

44. Ibid.; C. E. Wilson to EA, May 29, 1941, AP.

45. EA, TS of speech, January 13, 1942, AP.

46. EA to J. F. Young, "Selection of Students," March 6, 1942, AP.

47. EA to D. J. McGlenn, June 25, 1942, AP.

48. EA, "Intuition, Invention, and Ingenious Faculties in Engineering," TS of talk, November 30, 1942, AP.

49. EA to A. R. Stevenson, January 8, 1943, AP.

50. EA to C. E. Tullar, January 7, 1943. Also see K. K. Paluev to EA, January 11, 1943, AP.

51. J. F. Young to EA, January 1, 1943, AP.

52. EA, "Big Business and Invention," TS of speech, August 6, 1944, AP.

53. Ibid.

54. EA, "Electronics and the Electrical Industry," TS of speech, October 20, 1944, AP.

55. Ibid.

56. EA to C. E. Tullar, November 11, 1944; Tullar to EA, February 16, 1945, AP.

57. "Dr. Alexanderson Gets Edison Medal," *New York Times*, January 25, 1945, p. 13; EA, TS of speech, January 24, 1945, AP.

58. "Medals for Electrical Genius," clipping from *American Swedish Monthly*, March 1945, pp. 11, 25–26, AP.

59. Donald G. Fink, *Television Standards and Practice: Selected Papers from the Proceedings of the National Television System Committee and the Panels* (New York: McGraw-Hill, 1943), pp. 28, 162. Also see idem, "Perspectives on Television: The Role Played by the Two NTSC's in Preparing Television Service for the American Public," *Proceedings of the IEEE*, 64 (1976): 1322–31.

60. EA, "Trails of Progress in Electronics and Radio," TS of talk, January 26, 1943, AP. Also see EA to H. E. Dunham, "Radio Vision with Three Dimensions," May 6, 1943, AP.

61. EA to I. A. Terry, October 22, 1943; EA, "Invention vs Prophetic Imagination," TS of talk, January 19, 1944, AP.

62. EA, TS of talk, June 9, 1944, AP.

63. EA, TS of speech, November 6, 1944, AP.

64. EA, "Television System and Method of Operation," U.S. Patent no. 2,333,969, issued November 9, 1943; Fink, "Perspectives on Television," p. 1328.

65. EA to H. E. Dunham, "Colored TV," September 24, 1940; EA to P. C. Goldmark, October 2, 1940, AP.

66. Ernst F. W. Alexanderson and Earl L. Phillipi, "History and Development of the Electronic Power Converter," *Electrical Engineering*, 63 (1944): 654–57; quotations on pp. 654, 657.

67. EA to H. E. Dunham, February 9, 1944; D. R. MacLeod to EA, January 29, 1945, AP.

68. D. R. MacLeod to EA, February 21, 1945; EA to MacLeod, February 23, 1945, AP.

69. EA, TS of remarks on AC locomotive, 1945, AP.

70. EA, "Power Electronics as a New Industry," October 27, 1945; EA, "War and Innovation," TS of talk, April 12, 1946, AP.

71. EA, memorandum, April 30, 1946; minutes of Engineering Council meeting, May 3, 1946, AP.

72. EA to H. A. Winne, May 6, 1946, AP. For a biographical note on Harry A. Winne, see *Electrical Engineering,* 73 (1954): 1038.

73. EA, "DC Power Transmission: Summary and Recommendations," May 15, 1946, AP.

74. R. C. Muir to E. E. Johnson, May 20, 1946; minutes of meeting on the DC transmission tube, July 31, 1946, AP.

75. H. A. Winne to D. C. Prince, August 29, 1946, AP. Winne cited a report by Selden Crary on the DC power proposal.

76. EA, "Pool Thyratron for the Navy," February 7, 1946, AP.

77. EA to A. K. Bushra, September 24, 1946; EA to D. C. Prince, January 8, 1947; EA to P. S. Mack, "Electronic Ship Propulsion," January 23, 1947, AP.

78. EA to D. C. Prince, January 8, 1947, AP.

79. EA to P. S. Mack, "Wave Transmission," February 18, 1947; EA, "Electronic Stabilizer for Power Transmission," February 26, 1947, AP.

80. EA, "Electronic Stabilizer Tests on 900-Mile Transmission Line," April 11, 1947; EA, "Demonstration of Artificial Synchronizing Force," April 23, 1947, AP.

81. EA, "Electronic Stabilizer for Power Transmission," May 8, 1947, AP; EA, "Stabilizer for Alternating Current Power Transmission Systems," U.S. Patent no. 2,470,454, issued May 17, 1949.

82. D. C. Prince to P. Sporn, March 27, 1947, AP. Prince had served his apprenticeship at GE under Alexanderson many years before and had become an expert on power electronics before becoming vice-president in charge of engineering.

83. EA to A. U. Lamm, May 19, 1947; EA to Lamm, May 26, 1947; EA, note on discussion of transductor, June 5, 1947, AP.

84. Ernst F. W. Alexanderson and David C. Prince, "Electronic Stabilizer for Power Transmission," *Electrical Engineering,* 66 (1947): 1054–57; quotations on pp. 1054, 1057.

85. Discussion in *Transactions of the AIEE,* 66 (1947): 954–55.

86. Ibid., p. 956.

87. A. Uno Lamm, "Some Fundamentals of a Theory of the Transductor or Magnetic Amplifier," *Transactions of the AIEE,* 66 (1947): 1078–85. Also see Katherine Wollard, "Uno Lamm: Inventor and Activist," *IEEE Spectrum,* 25 (March 1988): 42–45.

88. EA to P. S. Mack, "Combined Electronic Condenser and Phase Balancer," August 13, 1947, AP.

89. EA to H. A. Winne, October 6, 1947; EA, TS of remarks at Engineering Council meeting, November 17, 1947, AP.

90. EA to I. A. Terry, October 22, 1943; EA to Terry, November 6, 1943, AP.

91. EA to P. S. Mack, "Magnetic Amplifiers," September 10, 1947; EA to Mack, September 23, 1947, AP.

92. Alexanderson mentioned Greene's paper in a talk given in December 1947. EA, TS of talk, December 11, 1947, AP.

93. Walter E. Greene, "Applications of Magnetic Amplifiers," *Electronics,* 20 (September 1947): 124–28.

94. Ibid.

95. EA to E. S. Lee, October 7, 1947; EA to P. S. Mack, October 14, 1947, AP.

96. EA to P. S. Mack, "Magnetic Amplifier with Positive Feedback," October 23, 1947, AP.

97. EA to P. S. Mack, "Follow-up Control with Magnetic Amplifiers," November 7, 1947, AP.

98. EA to P. S. Mack, "Magnetic Amplifier for Follow-up and Speed Control," November 11, 1947; EA to Mack, "Magnetic Amplifier for Control of Induction Motors," November 14, 1947, AP.

99. EA, "Magnetic Amplifiers," TS of speech, December 11, 1947, AP.

100. Ibid.

101. EA to O. D. Young, December 31, 1947; EA to C. E. Wilson, December 31, 1947, AP.

102. Albert W. Hull, "To Alex." From a copy provided to me by Philip Alger.

103. H. B. Maxwell to EA, January 12, 1948, AP.

104. EA, TS of talk, January 19, 1944; EA, TS of talk, April 12, 1946, AP.

105. Verner Alexanderson, "Family Tree, E. F. W. Alexanderson," September 30, 1977.

106. EA to H. W. Hall, March 18, 1948, AP.

107. Thyra Alexanderson, personal communication, March 15, 1982.

Chapter 10: "Inventors Never Stop!"

1. EA to P. Alger, March 19, 1948, AP.

2. Ibid.

3. EA to A. Hastings, April 14, 1948, AP.

4. EA, "Research and Engineering at the General Electric Company," TS of talk, October 7, 1948, AP.

5. Ibid.

6. Ibid.

7. Ibid.

8. EA, "Invention," TS of talk, April 21, 1949, AP.

9. Ibid.

10. EA to D. C. Prince, September 19, 1949, AP.

11. EA, "The Experimental Approach," TS of talk, February 28, 1950, AP.

12. EA, "Invention and the Corporation," TS of talk, January 31, 1951, AP.

13. EA to D. I. Rogers, September 19, 1951, AP.

14. Ibid.

15. EA to E. E. Johnson, "Pioneer Work of the Laboratory," November 15, 1951, AP.

16. Ibid.

17. Ibid.

18. EA to W. R. Herod, December 6, 1948, AP.

19. EA to W. R. Herod, January 25, 1949, AP.

20. EA to W. Borquist, March 8, 1949, AP.

21. EA, "Magnetic Regulator," memorandum, February 20, 1950, AP.

22. EA to E. Lee, October 4, 1950; EA to F. Dahlgren, September 5, 1951, AP. Also see EA and R. W. Kuenning, "Stabilizer for Alternating Current Power Transmission Systems," U.S. Patent no. 2,644,898, issued July 7, 1955.

23. EA to J. H. Hammond, December 20, 1951; L. A. Hawkins to E. H. Armstrong, February 4, 1952, AP.

24. Alexanderson's folder on miscellaneous topics for the period 1953–55 contains several clippings and sketches on transistors, including the item from the *New York Times*, dated January 13, 1952. The folder also contained a copy of R. L. Wallace, Jr., and W. J. Pietenpol, "Some Circuit Properties and Applications of n-p-n Transistor," *Proceedings of the IRE*, 39 (1951): 753–67.

25. EA to R. Bown, April 9, 1952; Bown to EA, April 21, 1952, AP.

26. EA to A. N. Goldsmith, April 28, 1952; Goldsmith to EA, May 1, 1952, AP.

27. Ernst F. W. Alexanderson, "Control Applications of the Transistor," *Proceedings of the IRE*, 40 (1952): 1509–11.

28. EA to C. B. Jolliffe, June 3, 1953, AP. Also see EA, "Magnetic Amplifier Motor Control," U.S. Patent no. 2,752,549, issued June 26, 1956; and EA, "Push-pull Magnetic Amplifier," U.S. Patent no. 2,798,904, issued July 9, 1957.

29. EA to P. G. Cooper, November 24, 1953, AP.

30. EA to F. Dahlgren, April 16, 1953, AP.

31. H. F. Storm to EA, June 22, 1953, AP.

32. Herbert F. Storm et al., *Magnetic Amplifiers* (New York: John Wiley & Sons, 1955), preface; quotation on p. v.

33. Ibid., pp. 383–88.

34. EA, "Magnetic Amplifier Motor Control System," U.S. Patent no. 2,844,779, issued July 22, 1958.

35. EA, "Alternating-Current Motor," U.S. Patent no. 2,707,257, issued April 26, 1955; EA, "Alternating-Current Motor," U.S. Patent no. 2,797,375, issued June 25, 1957.

36. EA, "Magnetic Computer," U.S. Patent no. 2,984,414, issued May 16, 1961; EA, memorandum on computer, December 14, 1952. Also see EA to C. B. Jolliffe, March 16, 1955, AP.

37. R. W. Babson to EA, October 1, 1954; EA to Babson, October 7, 1954, AP.

38. EA, "Looking Ahead: The Next Fifty Years—Science and Invention," TS of speech, filed with EA to J. H. Hammond, October 18, 1954, Hammond Papers.

39. Ibid.

40. EA to R. W. Babson, February 19, 1955, AP.

41. "Dr. Alexanderson, Video Pioneer, Gets Patent for a Color Receiver," *New York Times*, February 12, 1955, p. 22; EA, "Receiver for Color Television," U.S. Patent no. 2,701,821, issued February 8, 1955.

42. EA to J. Crosby, November 24, 1960, AP.

43. The Alexandersons spent some time with Thyra's brother and his family in Norrkoping and attended a meeting in Stockholm. They also went to Uppsala and to the island of Gotland for official ceremonies inaugurating the DC power cable from the mainland. Thyra Alexanderson, personal communication, March 15, 1982.

44. A. Uno Lamm, "High Voltage D.C. Power Transmission—A Pioneer Project," *ASEA Journal*, 23 (1950): 172–74.

45. A. Uno Lamm, "The First High Voltage D.C. Transmission with Static Converters. Some Notes on the Development," *ASEA Journal*, 27 (1954): 139–40. Also see E. Uhlmann, "The Gotland D.C. Link: The Layout of the Plant," ibid., 27 (1954): 141–54; and Wollard, "Uno Lamm," pp. 42–45.

46 EA to C. A. Jolliffe, July 13, 1955, AP.

47. EA to E. F. McDonald, November 9, 1950; EA to G. E. Gustafson, October 16, 1957, AP.

48. EA to Vice-President of Engineering at Zenith, July 15, 1959; E. B. Passou to EA, July 23, 1959, AP.

49. EA to J. F. Neal, June 22, 1960; EA to W. H. Greenbaer, July 9, 1960, AP.

50. EA to P. J. McGee, May 18, 1963, AP.

51. EA to S. M. Robinson, January 6, 1958, AP.

52. S. M. Robinson to EA, December 26, 1957, AP.

53. EA to S. M. Robinson, October 2, 1960, AP.

54. EA to Mrs. Mittag, n.d., AP.

55. EA to P. S. Mack, August 3, 1960, AP.

56. EA to J. H. Clough, November 17, 1960, AP.

57. EA to J. H. Clough, October 8, 1960, AP. Clough had spent his professional career with GE. See *Who Was Who in America*, 6:82. Also see W. D. Morton, "25 kv A.C. Electric Locomotives for British Railways," *GEC Journal*, 27 (Spring 1960): 71–76.

58. EA, "System of Control," memorandum, December 31, 1960; EA to J. Clough, January 1, 1961, AP.

59. EA to J. Clough, January 14, 1961, AP.

60. EA, "Electric Motor Control Apparatus," U.S. Patent no. 3,050,672, filed April 17, 1961, and issued August 21, 1962.

61. EA, "Electric Motor Control System," U.S. Patent no. 2,119,957, issued January 28, 1964.

62. EA, "Adjustable Speed Motor Control System," U.S. Patent no. 3,736,481, filed March 19, 1968, and issued May 29, 1973.

63. EA to J. H. Hammond, February 11, 1961, AP.

64. EA to J. H. Hammond, June 16, 1961; EA to Hammond, June 23, 1961; Hammond to EA, June 28, 1961, AP.

65. C. Davidson to EA, January 16, 1962; EA to Davidson, January 20, 1962, AP.

66. Alger, *Human Side of Engineering*, p. 132.

67. Ibid., p. 137.

68. Albin Krebs, "Dr. Ernst Alexanderson, Radio Pioneer, Dies at 97," in *New York Times Biography Series*, vol. 6 (May 15, 1975), pp. 547–48. Also see "Ernst Alexanderson: Radio and TV Pioneer at 97," *IEEE Spectrum*, 12 (August 1975): 109.

69. Thyra Alexanderson, personal communication, March 15, 1982.

Epilogue

1. "Centennial Hall of Fame," *IEEE Spectrum*, 21 (April 1984): 64.

2. David A. Hounshell, "The Inventor as Hero in American History," paper prepared for Edison Centennial Symposium, Newark, N.J., October 1979.

3. Wyn Wachhorst, *Thomas Alva Edison: An American Myth* (Cambridge, Mass.: MIT Press, 1981), pp. 6–7.

4. George Wise, "R&D at General Electric, 1878–1985," paper prepared for a conference on R&D Pioneers, Wilmington, Del., October 1985.

5. Ibid.

6. Narian G. Hingorani, "The Reemergence of D.C. in Modern Power Systems," *EPRI Journal*, 3 (June 1978): 7. Also see F. J. Ellert and Narian G. Hingorani, "HVDC for the Long Run," *IEEE Spectrum*, 13 (August 1976): 37–42; "Energy Report," *IEEE Spectrum*, 13 (October 1976): 10; and Glenn Zorpette, "HVDC: Wheeling Lots of Power," *IEEE Spectrum*, 22 (June 1985): 30–36.

7. Gordon D. Friedlander, "Riding Sweden's Slick Rail System," *IEEE Spectrum*, 11 (March 1974): 53–64.

8. Gordon D. Friedlander, "Railroad Revival: On the Right Track," *IEEE Spectrum*, 9 (August 1972): 63–66.

9. D. J. Ben Daniel and E. E. David, Jr., "Semiconductor Alternating-Current Motor Drives and Energy Conservation," *Science*, 206 (1979): 773–76. Also see B. Jayant Baliga, "Switching Lots of Watts at High Speeds," *IEEE Spectrum*, 18 (December 1981): 42–48; Gadi Kaplan, "Thyristors: Future Workhorses in Power Transmission," *IEEE Spectrum*, 19 (December 1982): 40–45; and Arnold P. Fickett, Clark W. Gellings, and Amory B. Lovins, "Efficient Use of Electricity," *Scientific American*, 263 (September 1990): 65–74.

10. Thomas P. Hughes, *American Genesis* (New York: Viking, 1989) pp. 1–4.

Index

Agora (GE club for engineers), 72,
135–36
Air Force, U.S.: use of radio alternator,
178
Alexanderson, Amelie (daughter), 82,
103, 168
Alexanderson, Amelie von Heidenstam
(mother), 71
Alexanderson, Aron (great-great grandfather), 1
Alexanderson, Aron (great-grandfather),
1
Alexanderson, Aron Martin (father), 1–
4, 216
Alexanderson, Edith (daughter), 82, 103,
168, 212
Alexanderson, Edith Lewin (first wife),
71–72
Alexanderson, Ernst F. W., ix, xii; ac-
cepts job at GE, 11; and amplidyne,
237–42; analysis of air-friction loss for
high-speed rotors, 66; analysis of international radio, 151; applies for job
with DeLaval Company, 11; applies for
job with Stanley Electric Company, 11–
12; applies for job with Westinghouse
Company, 10–11; arrives in America,
8; attends Cathedral School in Lund,
3; becomes U.S. citizen, 71; begins
work on gun-control system, 203–5;
boards with Beebe family, 9–10; and
"breathing spell" theory, 206–7, 209;
chief engineer of RCA, 137–38, 153–
54, 159, 175; comparison with Edison,
9; death of, 305; declines job offer
from Edison, 8–9; designs propulsion
motors for the *New Mexico,* 134–35; design style of, 54; develops mechanical-
scan television, 185–98; discusses engineering creativity, 207–9, 267–70,
286–92; discusses radio broadcasting,
163; draftsman with C&C Electric
Company, 9–11; early access to father's
home shop, 2; early years in Sweden,
1–4; examines GE policy options in radio, 131–32; experiments with color
television, 195; first encounter with
Steinmetz's analysis, 7; first patented
invention, 12; as gatekeeper, xi; head
of Radio Engineering Department,
148–49; hearing impairment of, 253–
54, 284, 299–300, 305; home laboratory, 286, 300; on interaction of radio
and power engineering, x; and inventions at boundaries between systems,
20; invents barrage receiver, 122–24;
invents duplex antenna, 114–15; invents phase converter, 95–98; invents
selective tuning system, 82–87; invents
variable-speed AC motor, 12–13, 15;
investigates critical speed of railway
trucks, 99; investigates dielectric hysteresis, 90–93; investigates high-
frequency properties of iron, 77–79;
investigates physiological tolerance of
high-frequency current, 68; investigates shortwave propagation, 181–86;
joins AC Engineering Department, 21;
joins Agora, 72; joins Consulting Engineering Department, 73; joins IRE,
89–90; joins Railway Engineering Department, 43; managerial style of,
174–75; and magnetic amplifier, 79–
81; as master salvager, 81; mastery of
AC principles, 20; and mechanical resonance of rotor, 39–40; and method
for self-excitation of alternators, 22–
27; method of speed regulation, 109;
on military uses of television, 214; and
multiple-tuned antenna, 113–14, 118;
observes radio eclipse, 242–44; offered
job by patent solicitor, 13; opposes
agreements restricting GE radio research, 213, 248–50; opposes excessive
specialization, 271; paper on rotary
converters, 15; participates in GE test
engineering program, 16–19; patent
applications, 311–22; patents on electric traction, 47; president of IRE,
138–39, 159; radio altimeter experiments, 198–201; and radiotelephony
experiments, 101; receives Cedergren
Medal, 272; receives Edison Medal,
272; receives Ericsson prize, 212;

Alexanderson, Ernst F. W. (cont'd)
receives IRE Gold Medal, 139; receives
Order of the North Star, 169; receives
Order of Polonia Restituta, 167; rec-
ommends vacuum-tube transmitter,
161; and regenerative braking systems,
97–99; retirement from GE, 283, 286;
sailing as an avocation, 2; and series-
repulsion traction motor, 44, 46–52;
and shunt-repulsion motor, 52–53; and
split-phase locomotive, 95–97; stepping-stone approach, 185; and stroboscopic method to measure motor
output, 18; studies engineering in
Charlottenburg, Germany, 5–7; study
of high-speed rotary disks, 14; summer work at ASEA, 4–5; supports development of FM radio, 248–51;
technological style of, 28; and technology transfer, 46–47; telephone alternator project, 55–61, 64; tests home
television receiver, 192–93; theory of
commutation in repulsion motors, 49;
on theory of rotating disks, 39; on
types of invention, 290; urges GE to
develop aviation products, 199; use of
analogy by, 99, 123, 151, 238–40; use
of parameter variation, 25; use of semiconductor motor control, 300–303,
309; uses equivalent circuit of pliotron,
89; works on DC traction, 98–99;
works on electric computer, 258–62;
works on electronic power converters,
272–77; works on gun-control system,
219–22; works on radar system, 256–
64; works on thyratron motor, 224–30,
233; works on torque amplifier, 219–
22, 230; works on transistor applications, 293–94
Alexanderson, Fredrik (grandfather), 1
Alexanderson, Gertrude (daughter), 103
Alexanderson, Gertrude Robart (second
wife), 103, 284
Alexanderson, Thyra (Oxehufwud),
284–85, 305
Alexanderson, Verner (son), 103, 178–
79, 247, 284
Alexanderson radio alternator, x–xi, 21,
27–28, 31–43, 63–70, 76–82, 93–94,
100–112, 115–21, 124–30, 136–37,
145, 149–50, 153, 156–58, 160–62,
166, 169–70, 172–73, 176–78, 180,

186–87, 203, 263, 267, 282, 287–88,
295; and beginning of transatlantic
service, 127; cost of, 126; design features of, 67–68, 128; effect of air friction on, 65–66; frequency range of,
128; installed at New Brunswick station, 117–18; used in carrier telephone
experiment, 77; used in dielectric hysteresis study, 91; used in wireless-
control experiments, 82; used in wireless telephony, 41–43; used to send
message to Germany, 130; use of belt
drive on, 41–42; use of middle bearings on, 39–40; voice modulation of,
110
Alexanderson radio system, 103–4, 106,
113, 115–17, 119–22, 125–27, 129–
31, 136–37, 142, 144, 147, 149–51,
176–77; comparison with transatlantic
cable, 120; cost data for, 146
Alger, Philip L., 224, 241, 303
Allen, E. W., 187, 192
Allgemeine Elektrizitäts-Gesellschaft
(AEG), 51
Allmanna Svenska Elektriska Aktiebolaget (ASEA), 4, 15, 278–79, 298
Alternating Current Department (GE),
12, 18, 21–22, 30–32, 40, 42–43, 57,
59–61
Amateur radio operators: contributions
to shortwave research, 185
American Bell Telephone Company, 55–
56
American Electric Company, 24
American Electric Power Company, 278
American Gas and Electric Company
(AG&E), 225, 227–29, 235
American Institute of Electrical Engi-
neers, xi, 16, 22, 24, 26–28, 33–34, 45,
48–49, 51, 63, 67, 76–77, 79, 94, 96–
98, 150, 164, 169, 174, 177, 181–82,
184, 189–90, 206, 225, 227–28, 232,
274, 278
American Marconi Company, 93, 102–7,
111, 117–18, 131, 137–38, 143, 147,
149, 152–54, 157
American Society of Mechanical Engi-
neers, 270
American Society of Swedish Engineers,
211–12
American Telephone and Telegraph
Company (AT&T), 69–70, 77, 108–10,

139, 147, 155, 172; interest in high-frequency alternators, 42–43; supports high-frequency alternator project, 55–61, 64; transoceanic radio experiments, 109–10, 112

Amplidyne, 237–42, 257–59, 261–68, 271, 280–83, 294–95, 300; applications of, 239, 241, 266; regenerative feedback in, 238; use in gun control, 238

Amplification, methods of, 56

Analog computers, xii

Antenna insulators, 164, 170

Antenna system: at New Brunswick station, 106, 113, 118, 120, 128

Antenna tuning coils, 164–65

Apparatus engineering, 268–69

Appleton, Edward V., 242

Appleton layer, 243

Arco, Georg von, 6

Arc transmitter, 102

Arlington, Va.: transmitting station in, 109–10, 112

Armstrong, Edwin H., 140–41, 248–49, 251, 293, 296, 307

Army, U.S., 221, 257–58, 260, 262, 265

Army Ordnance Department, 260

Army Signal Corps, 67, 76–77, 118, 138, 251, 260; sponsors photographic radio project, 124

Artificial ground for antennas, 165, 170, 175

Asbury, Herbert, 262

Association Island, 188, 194, 204, 222, 283

Asynchronous power transmission, 233–34, 236

Atlantic Communication Company, 105

Audion, 60, 78, 83, 89, 110, 112, 176

Austin, Louis W., 83

Automatic machine tools, 282

Babson, Roger W., 296–97

Babson Institute, 296–97

Baird, John L., 187

Baker, Walter R. G., 188, 257

Ballistic valve. *See* Magnetron

Bar Harbor, Maine: receiving station at, 163

Barrage receiver, 122–24

Barretter wireless detector, 30

Battle of electric traction systems, 44–52, 94–95, 274–75

Becker, Howard I., 175

Bedford, B. D., 232–33, 263

Beiler, Albert H., 228

Bell, Alexander G., 307

Bell Telephone Laboratories, 260, 286, 291–94

Belmar, N.J.: receiving station at, 128, 156

Benford, F. A., 188

Benford crystal, 187

Berg, Ernst J., 10, 32, 34, 36–37

Berg, Eskil, 32

Berlin Convention on international wireless, 62–63

Beverage, Harold H., 144, 163–64, 303

Beverage antenna, 163–64, 167

Bloomington, Pontiac, and Joliet Railroad: electrification of, 47

Bolinas, Calif.: transmitting station at, 156

Bonneville hydroelectric plant, 236–37

Bonneville Power Administration, 279

Boulder Dam: power transmission from, 230–32

Bowman, Kenneth K., 237, 239, 241, 266

Bown, Ralph, 293

Brant Rock, Mass.: wireless station at, 41–43, 55, 57, 60, 69

Braun, Ferdinand, 107

Breit, Gregory, 181

British Marconi Company, 93, 119, 121–22, 137, 145–46, 149, 152, 159–60, 168

British Thomson-Houston Company, 75

Brown, H. D., 202

Brown, William, 113, 117, 130, 135, 144, 164–66, 174, 177, 267

Bucher, Elmer E., 157

Bullard, William H. G., 145, 147–48, 155, 176, 349 n. 28

Burchard, Anson W., 121–22, 125, 129, 132, 145–46

Bureau of Engineering (U.S. Navy), 229–30

Bureau of Ships (U.S. Navy), 230

Butler, William J., 174

C&C Electric Company, 9–11

Camp Engineering, 188, 194, 204, 222, 224

Carnarvon, Wales: transmitting station at, 145
Carnegie Institution, 181
Carnegie Steel Company, 265
Carter, P. S., 165
Chatham, Mass.: receiving station at, 156
Chesney, Cummings C., 13
Chicago, Milwaukee, and St. Paul Railroad: electric traction on, 98–99
Clark, George H., 159
Clough, John H., 301
Coffin, Charles A., 175
Cohen, Louis, 112
Color television, 195, 217–18
Comet valve. *See* Magnetron
Compound alternator. *See* Self-exciting alternator
Computer, 258, 262, 296; applications of, 260
Consulting Engineering Department (GE), x, 55, 73–75, 84, 90–92, 118, 132, 192, 218, 222, 233, 244, 256, 265
Consulting Engineering Laboratory (GE), 237, 239, 249, 251–52, 263–65, 289
Coolidge, William D., 226, 228–29, 245, 253–54, 291
Crandall, S. G., 179
Crever, Frederick E., 241
Crocker, Francis B., 9
Crocker-Wheeler Company, 134
Curtis, Charles G., 9, 133
Curtis turbine: used in ship propulsion, 133
Curtis turboelectric generator, 21, 26

Dahlander, Gustav R., 4
Daniels, Josephus, 137, 142–43, 147–48
Danielson, Ernst, 4–5, 15; and the repulsion motor, 45–47
Davidson, Carter, 303
Davis, Albert G., 23–24, 79, 117, 122–24, 138, 140–41, 145, 147–49, 152–53, 156, 160, 163, 166, 201–2, 265, 288
Day, Maxwell, 111
de Forest, Lee, 60, 83, 89, 108, 110–12, 139–41, 307; interest in radio alternator, 77–78; patents of, 131
DeLaval speed-reduction gear, 40
DeLaval steam turbine, 33
DeLaval Steam Turbine Company, 11
Dempster, Alex, 41

Direct current power transmission, 203, 265–66, 275–79, 305, 308
Direct current traction system, 48–50
Drafting Department (GE), 11–13
Dunham, H. E., 251
Dunlap, Orrin E., Jr., 184
Duplex antenna, 114–15, 118, 138–39
Dushman, Saul, 88–89, 110
Dynamo amplifier. *See* Amplidyne
Dynatron, 140, 142

Eastman, George, 195
Eccles, W. H., 159
Edison, Theodore, 9
Edison, Thomas A., xii, 8–9, 29, 272, 303, 307–8; and dialectical exchange between electrical power and wireless communication systems, 20
Edison effect, 86
Edison Electric Light Company, 1, 68
Edison General Electric Company, 18, 22
Edwards, Bill, 105
Edwards, Edmund P., 61, 66–67, 69, 78, 111, 121–22, 125, 143, 147, 161
Edwards, Martin A., 221–22, 237, 239, 241, 257, 263–66, 360 n. 39
Eickemeyer, Rudolf, 50
Eickemeyer electric motor, 45
Eilvese, Germany: wireless station in, 102
Elder, F. R., 232–33
Electronic power converter, 141, 202, 210, 223–24, 229–37, 273–77, 298, 305, 308
Electronic tubes: metal envelopes for, 226–27
Electronic voltage regulator: for power plant, 205, 210
Elektrotechnische Zeitschrift, 6, 15
Emmet, William L. R., 26–27, 150, 265, 289, 303, 308; works on electric ship propulsion, 132–34
Engineering and science, interaction between, 287–88, 291–92
Engineering Council (GE), 221, 225–26, 275–76, 280, 289, 360 n. 36
Engineering creativity, 251–52
Eshelman, G.J.C., 267

Fairbanks, Harry C., 179
Faneuil Watch Tool Company, 37
Faraday effect, 184
Farnsworth, Philo T., 307

Federal Communications Commission, 252
Federal Telegraph Company, 111, 131, 172
Felix, Edgar H., 207–8, 215
Felix, Fremont, 266
Ferguson, Eugene S.: on role of nonverbal thought in engineering, 3
Fessenden, Reginald A., 13, 21, 82, 100, 131, 139, 141, 191, 296, 303; and continuous-wave system, 29–31, 41–43; critic of U.S. Navy policy on radio, 63; electrostatic doublet theory, 29, 35; and high-frequency properties of iron, 78–79; opposes Berlin Convention, 62–63; and radio alternator, 30–32, 34–43, 55, 57, 59–61, 65–70; on technological style, 28
Finch, J. Leslie, 166
Finnish Relief Fund, 256
Fisher, Alec, 241
French Thomson-Houston Company, 240
Frequency converters, 224, 227, 233, 236, 265
Frequency-modulation radio (FM), 248–51
Frequency multiplier: used with radio alternator, 177

Garr, Donald, 263–64
Geisenhoner, Henry, 57
General Electric Company, ix, xi, xii, 1, 104, 106–35, 213–19; and the amplidyne, 237–42; consulting engineering tradition at, x, 265; corporate style of, xi; develops all-wave radio receiver, 246; electric ship propulsion system of, 132–35; and electric traction, 95; facsimile system of, 197; and military secrecy, 220–21; patent strategy of, 24; and professors as consultants, 253; and radio alternator, 31–32; radio broadcasting station of (WGY), 179; radio research facilities of, 184; and railroad electrification, 43–51; refrigerators of, 247; role in creation of RCA, 144–49; support of radio-television research, 197–98; thyratrons produced by, 224, 228, 231
General Electric Research Laboratory, x, 13–14, 23–25, 73, 84–89, 100, 110, 112–13, 116–17, 140–43, 176, 188, 195, 200, 202, 205–6, 217, 221, 224, 226, 228, 233, 253, 268, 275, 283, 287–88, 291–92
General Engineering and Consulting Laboratory (GE), 265, 282
General Engineering Laboratory (GE), 265, 291–92, 295
Geometry as indicator of engineering creativity, 3
George Washington, the: radio communication with, 142–44, 146
Germany: electric railroads in, 76
Gesellschaft für drahtlose Telegraphie. See Telefunken Company
Gibbs, George, 52
Given, Thomas H., 31
Goldmark, Peter C., 273
Goldschmidt, Rudolph, 76
Goldschmidt radio alternator, 76, 82, 101–2
Goldsmith, Alfred N., 90, 108, 111, 113–16, 118, 138–39, 159, 161, 243, 266, 294–95; director of RCA Research Department, 154; laboratory of, 113–16, 118
Gould, W. M., 57
Greene, Walter E., 281–82
Grimeton, Sweden: radio alternator station in, 169
Grodzisk, Poland: receiving station at, 167
Gun-control system, 203, 219–22, 230, 235, 256–63

Hammond, John H., Jr., 82–88, 94, 100, 244, 293, 296, 303
Hansell, Charles W., 165–66, 174, 352 n. 113
Harbord, James, 176, 179, 196, 213
Harding, Warren G., 162
Harper, J. W., 220
Hartly, L. J., 243
Haskins, Caryl D., 59
Hauffman, Bertil, 9
Hawkins, Laurence A., 85, 188, 217, 268, 285, 293
Hayden, James H., 62
Hayden, J. LeRoy, 74
Hayes, Hammond V., 43, 56–57, 60
Heaviside layer, 181, 194
Heidenstam, Amelie von (mother), 1

Heidenstam, Karl Verner von (grand-
father), 1–2
Hertz, Heinrich, 181
Hertzberg, Robert, 194
Heterodyne receiver, 30, 101–2, 131, 141
Hewlett, Edward M., 12–13, 91, 204
Hill, George H., 97–98
Hobart, H. M., 94, 122, 231
Hogan, John V. L., 102
Hooper, Stanford, 115, 129–30, 137,
143–48, 155, 214
Hounshell, David A., on inventor as
hero, 307
Hoxie, Charles A., 124–25, 187
Hoyt, Gerald A., 263–64
Hughes, Thomas P., on technological
style, 28
Hull, Albert W., 88, 117, 140–42, 188,
205–6, 224, 226–27, 229–30, 233,
236, 283–84, 286
Hulmann locomotive, 133
Hunt-Sandycroft motor, 75–76
Hysteresis insulator (asbestos), 91

Incandescent detector (audion), 85–86
Independent inventors and corporate in-
ventors, 20–21
Inductor alternator, 32–34
Industrial Engineering Department (GE),
220, 266
Institute of Radio Engineers, xi, 28, 89–
93, 108, 110–12, 114, 123, 138–40,
145–46, 150, 154–55, 157, 159, 164–
65, 183, 243, 262, 266, 269, 271–72,
293–94
Intensity coupler, 123
International Electrical Congress, 45
Inverted rectification, 201–2
Ion controller (audion), 83–85, 87
Ion-sheath theory, 203

J. G. White Company, 127
Jenkins, Charles F., 187
Jewett, Frank B., 77
Jolliffe, Charles B., 294, 298
Jupiter, the: electric propulsion system of,
132–33, 135

Kahler, Charles P., 94
Kahuku, Hawaii: transmitting station at,
156
Kapp, Gisbert, 6–7, 15, 326 n. 34

Karolus, August, 187, 217
Karolus cell, 188, 194, 215
Kell, Ray D., 192, 216, 218
Kellogg, Edward W., 164
Kennelly, Arthur E., 27, 77; and mag-
netic circuits, 35; study of physiological
tolerance of high-frequency current,
68
Kennelly-Heaviside layer, 242–43
Kenotron (diode), 89
Kerr effect, 184, 188–89, 191, 195
Koenigliche Technische Hochschule
(Charlottenburg, Germany), 5–7
Kruse, Robert S., 184
Kuenning, R. W., 293
Kungsbacka, Sweden: receiving station
in, 169

Lake George, N.Y., 103, 212, 286, 297;
competitive yachting at, 2
Lake George Yacht Club, 254
Lamm, Uno, 278–82, 298
Lamme, Benjamin G., 33; designs series
traction motor, 44–46, 48–50; and
high-frequency inductor alternator,
33–34
Lamme high-frequency alternator, 331 n.
64
Langdon, G. G., 228
Langmuir, Irving, 101, 103, 108–10,
116, 176, 195, 200, 202–3, 205, 218,
222, 226, 230–31, 236, 242, 245, 253–
54, 268, 285, 291, 308; begins research
on vacuum tubes, 84–89
Larmor, Joseph, 181
Latour, Marius, 47
Latour radio alternator, 150
Layton, Edwin: on difference between
engineering and science, 91–92
Lee, Everett S., 265, 282
Leblanc, Maurice, 33–34
Lewin, Creighton E., 71
Lewin, Edith B. (Alexanderson), 71
Lewin, Minnie Porter, 71
Lexington, the: electric propulsion of, 134
Lindenblad, Nils E., 165–66, 177
Lindstedt, Anders, 4
Lodge, Henry C., 62
Loftin, E. H., 147–48
Logan, W. Va.: thyratron motor used in,
228

Long Island, N.Y.: RCA station on, 156, 158
Love, J. E., 165
Lowenstein, Fritz, 82–84
Lush, William G., 166–67

McAllister, A. S., 47
Machine tools, automatic control of, 220–21
MacLeod, Donald R., 274–75
McMeen, Samuel G., 77
Magnetic amplifier, 65, 79–81, 83–84, 101, 103, 105–6, 108, 110–12, 116, 119, 126–27, 278–83, 288–89, 294–96, 301–2; feedback in, 280–83; German applications of, 281–83; use as power stabilizer, 293
Magnetic computer, 296
Magnetic hysteresis, 35–36
Magnetron, 117, 139–42, 203
Marconi, Guglielmo, 105–7, 139, 156, 162, 262, 272, 303, 307; visit to Schenectady, 106–7
Marconi spark system, 29
Marconi wheel, 106
Marconi Wireless Telegraph and Signal Company. See British Marconi Company
Marconi Wireless Telegraph Company of America. See American Marconi Company
Marion, Mass.: transmitting station in, 129, 156, 168–69, 177–78
Mechanicville, N.Y.: DC power transmission from, 233, 235, 237, 275
Melbourne, Australia: electric traction in, 98
Mercury-arc tubes, 229, 232–34, 236–37
Mercury-vapor rectifier, 86–87
Metadyne, 237, 239–41
Miessner, Benjamin F., 82
Military enterprise and technological change, xii
Mitchell, Charles, 159
Mittag, Albert H., 135, 202, 205, 227, 263–64, 286, 300
Mohawk Development Services, 293
Monocyclic circuit, 361 n. 83
Monroe, N.C.: proposed alternator station in, 129–30
Moore, D. McFarlan, 193
Morse, S.F.B., 307

Muir, Roy C., 226, 230–34, 248–49, 257, 269, 276
Multiple-tuned antenna, 113–14, 118, 122, 126–27, 151, 166
Murray, William S., 44, 46
Mutator, 234

Nally, Edward J., 131, 148–49, 152–55, 162, 166, 176
National Academy of Sciences, 198
National Broadcasting Company, 191
National Bureau of Standards, 83
National Defense Committee, 257
National Electrical Signaling Company (NESCO), 31–32, 36–37, 41–43, 55, 57, 59–62, 66–67, 69–70, 76, 78, 100, 106, 108
National Research Council, Radio Division of, 183
National Television System Committee, 272–73
Nauen, Germany: transmitting station in, 102, 130, 145
Naval Research Laboratory, 229, 243, 257, 262
Navy, U.S., 104, 108–9, 111, 118, 120, 123, 137, 143–48, 152, 178, 180, 214–16, 219, 221, 229, 238, 240, 257–58, 260–62, 265, 277, 282; control of radio transmitters, 118; electric ship propulsion in, 132–35; Radio Division of, 129, 137, 144, 146–47; role in creation of RCA, 144–48, 155, 176
New Brunswick, N.J.: transmitting station in, 102, 104, 106–7, 111, 113, 117–20, 123–31, 137–38, 142–44, 146, 149, 151–53, 155–56, 169
New Haven Railroad: electrification of, 44, 46–47, 49–51
New Mexico, the: electric propulsion of, 133–35, 300
New York Central Railroad: electrification of, 44, 46–48, 276
Nipkow, Paul, 187
Nixdorff, Samuel P., 90–91, 106, 110–12, 114, 126, 135, 160, 181, 187–88, 190, 198, 201–2, 204–5, 222, 224, 263–64
Nordic, the, 212
Norfolk and Western Railroad: use of split-phase locomotive on, 97
Norman, Carl, 256

Office of Naval Research, 281–82
Optical scanner, 245
Oxehufwud, Anders, 285

Paine, Sidney B., 52
Pan American Communication Company, 131
Parson, Charles, 133
Patent Department (GE), 12, 14–15, 23–24, 32, 57, 79, 85, 87–88, 108–9, 111, 134, 138, 140, 178, 181–82, 187–88, 192, 195, 198, 201–2, 204, 218–21, 223–24, 227, 229–31, 237–40, 245, 251, 256, 258–59, 265, 269, 273–74, 277, 281–82
Patent Department (RCA), 295
Paternot, Lt., 122–23, 130–31
Patterson, F. G., 199–200, 257, 263–64
Payne, John H., 143–44
Peek, A. P., 193
Peek, Frank W., 92
Pennsylvania Railroad: electrification of, 52, 274
Pennsylvania, the: turbine drive system of, 135
Penny, F. H., 220
Pestarini, Joseph M., 239–41
Phase converter, 95–98
Phase rotator, 123
Philadelphia Electric Company: uses phase converter, 97
Philadelphia-Paoli line: electrification of, 51–52
Phillipi, Earl L., 243, 263, 274, 303
Photographic radio receiver, 124–25
Pickard, Greenleaf W., 43, 183
Pierce, George W., 84, 88
Pliotron (triode), 89, 101, 103, 108, 114, 116–17, 121, 140, 142, 144, 175–76, 203–4
Poland: transmitting station in, 162, 166–68, 170
Polarization of radio waves, 181–85
Portland, the: gun-control system on, 221
Potter, William B., 18; and traction motors, 47–50
Poulsen-arc transmitter, 108, 112, 121, 129–30, 145–46, 172
Power converter. *See* Electronic power converter
Power engineering and radio engineering, interaction of, 157

Power engineering and wireless engineering, as distinctive cultures, 28, 35–36, 242
Power stabilizer, 277–80, 292
Pratt, Francis C., 61, 104–10, 114–15, 118–20, 126–27, 129, 131, 149, 159–60, 172, 176, 186, 336 n. 18
Prince, David C., 141, 269, 277–80, 289, 348 n. 16
Proctor's Theater, 213, 215–16
Production Department (GE), 57, 65, 68, 143
Publicity Department (GE), 246
Pupin, Michael I., 93

Radar, 256–64
Radiation Laboratory (MIT), 262
Radio altimeter, 180, 198–201, 217, 251
Radio broadcasting, 163
Radio Central (Long Island, N.Y.): transmitting station at, xi, 161–65, 169–70, 172, 175, 177
Radio Consulting Department (GE), 180, 183, 191–92, 203, 265
Radio Corporation of America, ix, xi, 61, 122, 136–38, 144, 148–49, 152–63, 166–79, 180, 182, 187, 191, 196–97, 200, 211, 213, 215–18, 248–50, 294, 297–98; cost data for radio traffic, 156, 158; transoceanic system, 172–74
Radio Development Committee (GE), 186
Radio eclipse observations, 242–44
Radio Engineering Department (GE), 148–50, 153, 161, 265
Radio facsimile, 124, 180, 183, 185–86, 188–91, 196–97, 203, 217–18, 242–45
Radio Relay League, 179
Railroad communication system, 244
Railroad electrification, 21, 43–52, 94–95, 200, 203, 210–11, 219, 223, 229, 233, 271, 274–75, 277, 287, 290, 292, 301, 305, 309
Railway Engineering Department (GE), 18, 43, 47, 57, 73, 94, 97
Ranger, Richard H., 186
Rectifying commutator, 23, 25
Reed, Philip D., 255
Regenerative braking systems, 97–99
Regulatory policy: impact on traction and communication systems, 64
Reist, Henry G., 21–22, 30, 40–42, 61

Reoch, Alexander E., 169–72
Republic disaster, 66–67
Repulsion motor: armature control of, 52–53; field control of, 53
Repulsion traction motor, 44
Research Department (RCA), 154
Rice, Chester W., 116, 121, 164, 181–82, 219, 226, 230
Rice, Edwin W., Jr., 24–25, 46–48, 52, 83, 100, 104, 108, 110, 116, 118–21, 123, 129–30, 142–43, 145, 147, 152–53, 176, 202, 265, 268, 288–89, 303; and "development jobs" system, 25
Richardson, Owen: thermionic emission theory of, 86
Ripley, Charles M., 149
Riverhead (Long Island, N.Y.): receiving station at, 156, 164, 169–71, 177–78
Rivett, Edward, 37
Robinson, Conway, 59–61
Robinson, Samuel M., 133, 135, 229–30, 257, 300, 347 n. 121
Rockefeller Foundation, 258
Rocky Point (Long Island, N.Y.). *See* Radio Central
Roessler, Gustav, 7
Rogers, R. H., 242
Rohrer, Albert L., 10, 15
Roosevelt, Franklin D., 144, 147–48
Roosevelt, Theodore, 62
Ross, James D., 236–37
Royal Institute of Technology (Sweden), 3–4, 256, 287; graduates employed at GE, 8
Royal Technical Institute (Sweden), 272
Rural electrification, 210–11

Sabin, Edward D., xiv, 162–63
Sandycroft Foundry Company, 75
Saratoga, the: electric propulsion of, 134
Sarnoff, David, 153, 160–62, 168, 187, 191, 213–14, 217, 249
Sayville (Long Island, N.Y.): transmitting station at, 100, 102, 146–47
Science and engineering: distinctions and interaction, x, 73, 246
Scott, Charles F., 27, 30, 207
Self-exciting alternator, 22–27
Self-synchronizer. *See* Selsyn
Selsyn, 203–4, 219, 222, 259–60
Series traction motor, 44
Shaw, George Bernard, 189–90

Ship propulsion: electric, 26, 132–35, 277; use of thyratron motor for, 225, 229–30
Shortwave propagation, 180–86
Shortwave radio relay system, 249–50, 273
Shoults, David R., 241
Silicon-controlled rectifier, 300, 302–3, 309
Silicon diodes, 298
Slaby, Adolf, 5–6
Slaby-Arco wireless system, 6
Slichter, Walter I., 45
Slot-cutting machine, 69
Society of Wireless Telegraph Engineers, 90
Spark transmitters, 102, 145
Speed regulation: of radio alternator, 108–9, 128, 130, 151
Sperry, Elmer, 20
Sperry Company, 260
Split-phase locomotive, 95–97
Split-phase motor, 123
Sporn, Philip, 225, 227–28, 233, 278, 360 n. 53
Sprague, Frank, 48, 95, 97, 134
Squier, George O., 76–77, 115, 118, 138
Standardizing Laboratory (GE), 16, 116, 265
Stanley Electric Manufacturing Company, 11–12
Steadman, Sidney, 121
Stearns, Ray, 216
Steinmetz, Charles P., x, 5, 10, 12, 24, 26, 32, 55, 86, 100–101, 110, 132, 134, 192, 212, 232, 240, 265, 281, 286–89, 303–4, 307–8; complex-number method of, 7; designs high-frequency alternator, 30–31; and formation of Consulting Engineering Department, 73–75; on GE test program, 16; and high-frequency alternator, 33–34; investigation of dielectric hysteresis by, 90, 92; investigation of properties of iron by, 78; and magnetic hysteresis, 35–36; and traction motors, 44–45, 47–51
Stentor, 112
Stillwell, Lewis B., 48–49
Stone, Charles W., 147, 226
Stone, John S., 55–56
Storm, Herbert F., 295

Sweden: DC power transmission in, 279, 298, 305; electric railroads in, 47, 76, 160, 275; emigration in early twentieth century, 8; hydroelectric power in, 5; radio alternator station in, 168–70, 178
Swedish Academy of Engineering Science, 182
Swedish Academy of Sciences, 253
Swedish Telegraph Administration, 169
Sweet, George C., 130, 145, 147
Switch Board Department (GE), 143
Switching Department (GE), 91
Swope, Gerard, 176, 196, 255
Synchronous-resistance detector, 141, 348 n. 14
Systems engineering, 268–69

Taylor, Albert H., 128, 229, 243–44, 257, 262
Taylor, Charles H., 159–60, 169–72, 197
Taylor Wine Company, 212
Technical Committee (RCA), 196–97, 213–14, 217
Technological style, 28, 54
Technology transfer, 166–67
Teknisk Tidskrift, 4
Telefunken Company, 6, 76, 79, 103, 105, 107, 168, 188
Telephone amplifier: alternator as, 79, 81
Television, 124, 138, 178, 180, 183, 185–98, 272–73, 297–98; cathode-ray tube for, 195, 218; large-screen system for, 213–16; light-beam carrier in, 218; mechanical-scan system, 213–16, 218; military applications of, 214–15; photo-electric cell for, 187–88, 194; rotary scanning disk for, 186–87, 192–94
Television Broadcasters Association, 273
Television projector, 194, 196
Tennessee Valley Authority, 231
Tesla, Nikola, 4, 82
Test Department (GE), 11, 13–19, 75, 90, 118, 252
Thomson, Elihu, 24; interest in high-frequency alternator, 31; opposes Berlin Convention, 62; and repulsion motor, 45, 54
Thomson-Houston Company, 18, 21, 25
Thury power system, 231
Thyratron, 180, 200, 202–3, 205–6, 219–20, 224, 229, 300; applications of, 206; with thermionic cathode, 205–6

Thyratron commutation, 223–24, 226
Thyratron motor, 203, 224–30, 233, 235–36
Thyratron stabilizer: for power systems, 223–24
Thyratron voltage regulator, 225
Thyristor. See Silicon-controlled rectifier
Tirrell regulator, 27
Titanic disaster: and regulation of wireless, 81–82
Todd, David W., 145
Toledo and Chicago Railroad: electrification of, 47
Torque amplifier, xii, 261, 300; applications of, 219–22, 226–27, 235–38
Transductor. See Magnetic amplifier
Transfer of technology, 75
Transistor, 286, 293, 296; power applications of, 293
Triode amplifier, 87
Tuckerton, N.J.: transmitting station in, 101–2, 156, 267
Tullar, C. E., 239, 259, 270
Tuning by geometric progression, 82–87, 108–9
Turbine Department (GE), 105

Ultrashort wave research, 216–18
Union College (New York), 163, 211, 303
United Fruit Company, 161
United Wireless Telegraph Company, 66
University of Lund (Sweden), 2
University of Uppsala (Sweden), 254
Uppsala, Sweden, 1

Vacuum tube, 141–42, 157, 159–61, 168, 172, 176–77, 295, 300
Vacuum-tube radio transmitter, 161, 168, 172, 176–77, 180
Vail, Theodore, 69
Verband Deutscher Electrotechniker, 6
Vincenti, Walter: on parameter variation, 25
Vreeland, Frederick, 141

Wachhorst, Wyn: on Edison mythology, 308
Walker, Hay, Jr., 31, 66, 70
Wallace, Amelie (Alexanderson), 325 n. 20
Walling, Karl, 4
War Department, U.S., 214

Warren and Jamestown Railroad: electri-
fication of, 47
Washington, Baltimore, and Annapolis
Railroad (WB&A): electrification of,
44, 50
Waterbury gear, 222
Wave transmission: on power lines, 218–
19, 232
Weagant, Roy, 145–46, 153–54, 160, 349
n. 33
Weather Bureau, U.S., 29, 31
Weaver, Warren, 258
Webster, H. R.: poem on New Brunswick
Station, 130
Weintraub, Ezekiel, 121
Wenstrom, Jonas, 4–5
Werth, J. R., 69
Western Electric Company, 155, 168, 254
Westinghouse Electric and Manufactur-
ing Company, 10–11, 30, 33–34, 191,
196–97, 207, 213, 264; and railway
electrification, 43–46, 48–52, 95, 97;
turbine ship propulsion system of, 132

Wheeler, Schuyler S., 9, 134
White, William C., 87–88, 100, 226,
228–29
Whitney, Willis R., 13, 25, 85, 88–89,
110, 112–13, 116, 141–42, 308
Willis, Clodiun H., 226, 232–33
Wilson, Charles E., 255, 268–69, 283–84
Wilson, Woodrow, 130, 142–44, 146–47,
149
Winne, Harry, 276–77
Wireless Institute, 90
Wise, George: on change in research
strategy at GE, 89; on Edison and
Thomson styles at GE, 308

Young, J. F., 269
Young, Owen D., 147–49, 152–53, 155,
160, 162, 175, 185–86, 255, 283

Zenith Radio Corporation, 299
Zenneck, Jonathan, 107–8
Zworykin, Vladimir K., 307

Books in the Series

The Mechanical Engineer in America, 1830–1910: Professional Cultures in Conflict, by Monte Calvert

American Locomotives: An Engineering History, 1830–1880, by John H. White, Jr.

Elmer Sperry: Inventor and Engineer, by Thomas Parke Hughes (Dexter Prize, 1972)

Philadelphia's Philosopher Mechanics: A History of the Franklin Institute, 1824–1865, by Bruce Sinclair (Dexter Prize, 1978)

Images and Enterprise: Technology and the American Photographic Industry, 1839–1925, by Reese V. Jenkins

The Various and Ingenious Machines of Captain Agostino Ramelli, edited by Eugene S. Ferguson, translated by Martha T. Gnudi

The American Railroad Passenger Car, New Series, no. 1, by John H. White, Jr.

Neptune's Gift: A History of Common Salt, New Series, no. 2, by Robert P. Multhauf

Electricity before Nationalisation: A Study of the Development of the Electricity Supply Industry in Britain to 1948, New Series, no. 3, by Leslie Hannah

Alexander Holley and the Makers of Steel, New Series, no. 4, by Jeanne McHugh

The Origins of the Turbojet Revolution, New Series, no. 5, by Edward W. Constant II (Dexter Prize, 1982)

Engineers, Managers, and Politicians: The First Fifteen Years of Nationalised Electricity Supply in Britain, New Series, no. 6, by Leslie Hannah

Stronger Than a Hundred Men: A History of the Vertical Water Wheel, New Series, no. 7, by Terry S. Reynolds

Authority, Liberty, and Automatic Machinery in Early Modern Europe, New Series, no. 8, by Otto Mayr

Inventing American Broadcasting, 1899–1922, New Series, no. 9, by Susan J. Douglas

Edison and the Business of Innovation, New Series, no. 10, by Andre Millard

What Engineers Know and How They Know It: Analytical Studies from Aeronautical History, New Series, no. 11, by Walter G. Vincenti

Alexanderson: Pioneer in American Electrical Engineering, New Series, no. 12, by James E. Brittain

Steinmetz: Engineer and Socialist, New Series, no. 13, by Ronald R. Kline

Designed by Sue Bishop
Composed by Brevis Press in Baskerville
Printed by The Maple Press Company on 50-lb. M. V. Eggshell Cream,
and bound in Holliston Roxite